ONSHORE STRUCTURAL DESIGN CALCULATIONS

ONSHORE STRUCTURAL DESIGN CALCULATIONS

Power Plant and Energy Processing Facilities

MOHAMED A. EL-REEDY, PhD

Consultant Structural Engineer and International Lecturer

ELSEVIER

AMSTERDAM • BOSTON • HEIDELBERG • LONDON
NEW YORK • OXFORD • PARIS • SAN DIEGO
SAN FRANCISCO • SINGAPORE • SYDNEY • TOKYO

Butterworth-Heinemann is an imprint of Elsevier

B H

Butterworth-Heinemann is an imprint of Elsevier
The Boulevard, Langford Lane, Kidlington, Oxford OX5 1GB, United Kingdom
50 Hampshire Street, 5th Floor, Cambridge, MA 02139, United States

British Library Cataloguing-in-Publication Data
A catalogue record for this book is available from the British Library

Library of Congress Cataloging-in-Publication Data
A catalog record for this book is available from the Library of Congress

ISBN: 978-0-08-101944-3

For Information on all Butterworth-Heinemann publications
visit our website at https://www.elsevier.com

Publisher: Joe Hayton
Acquisition Editor: Ken McCombs
Editorial Project Manager: Peter Jardim
Production Project Manager: Kiruthika Govindaraju
Designer: Victoria Pearson Esser

Typeset by MPS Limited, Chennai, India

DEDICATION

This book is dedicated to my wife and my children, Maey, Hisham and Mayar and to the soul of my mother and my father.

. . . It is also dedicated to everyone who likes to gain knowledge and teach others.

CONTENTS

ABOUT THE AUTHOR

Mohamed A. El-Reedy's background is in structural engineering. His main area of research is the reliability of concrete and steel structures. He has provided consulting services to engineering companies and oil and gas industries in Egypt and to international companies such as the International Egyptian Oil Company (IEOC) and British Petroleum (BP). Moreover, he provides different concrete and steel structure design packages for residential buildings, warehouses, and telecommunication towers and electrical projects with WorleyParsons. He has participated in Liquefied Natural Gas (LNG) and Natural Gas Liquid (NGL) projects with international engineering firms. Currently, Dr. El-Reedy is responsible for reliability, inspection, and maintenance strategy for onshore concrete structures and offshore steel structure platforms. He has performed these tasks for hundreds of structures in the Gulf of Suez and the Red Sea.

Dr. El-Reedy has consulted with and trained executives at many organizations including the Arabian American Oil Company (ARAMCO), BP, Apachi, Abu Dhabi Marine Operating Company (ADMA), the Abu Dhabi National Oil Company and King Saudi's Interior Ministry, Qatar Telecom, the Egyptian General Petroleum Corporation, Saudi Arabia Basic Industries Corporation (SABIC), the Kuwait Petroleum Corporation, Petroleum Development Oman (PDO), PETRONAS, and Qatar Petrochemical Company (QAPCO). He has taught many technical courses worldwide especially in the Middle and Far East about assessment and repair for concrete structures, design, construction and maintenance for concrete and steel structures onshore and offshore structures also for the oil and gas and energy sectors.

Dr. El-Reedy has written numerous publications and presented many papers at local and international conferences sponsored by the American Society of Civil Engineers, the American Society of Mechanical Engineers, the American Concrete Institute, the American Society for Testing and Materials, and the American Petroleum Institute. He has been on the OMAE Conference Technical Committee for 10 years. He has published many research papers in *International Technical Journals* and has authored seven books about total quality management, economic management for engineering projects, the assessment and repair of reinforced concrete structures, advanced concrete industry, reliability analysis, offshore structure design, and construction, maintenance, and project management for oil and gas projects. He received his bachelor's degree from Cairo University in 1990, his master's degree in 1995, and his PhD in 2000.

PREFACE

The development of any country depends on its rate of industrial growth. Currently, there is a worldwide race in industrial projects. The development of the industry depends on the development of energy reserves by investing in petroleum projects with new developments in oil and gas on petroleum projects. At the same time, there should be parallel projects to deliver electricity from electrical power or nuclear plants.

In this book, the industrial structure terms refer to all the reinforced concrete and steel structures from a small factory to a nuclear plant. This book will be an overview of the management of project engineering, presenting design calculations with examples of construction, and providing ways of structure assessment and a maintenance plan. In addition, most industrial projects are huge, with billions of dollars spent on a single project; so the client, engineering firm, and contractor will be in the same boat until they finally succeed in the project through a strong management system. Therefore, this book examines each of the items related to all three of these partners.

On the other hand, in these types of projects, all the engineering discipline are working together, but unfortunately the structure or civil engineer is usually the last one who obtain the accurate data from the other disciplines, and the civil work is the first activity start on site, so there will be a challenge to the structure engineers in these type of projects to work fast and accurately.

This book focuses on the structural engineering on all of these projects. The aim of this book is to provide the current methodologies and guidelines to design, construct, and maintain reinforced concrete and steel structures in industrial projects.

This book is intended to serve as a guidebook to junior and senior engineers working in the design, construction, repair, and maintenance of reinforced concrete and steel structures, and to assist them in all the stages for an industrial projects.

The other challenge that faces the structure engineer is that most of the undergraduate courses he studied in university were focused on real estate projects and housing. However, the characteristics of industrial projects are different, so this book provides a guide for the project, engineering, and construction managers to lead the project to success by achieving the owner requirements, which match with current industrial technical practice. Therefore, this book covers the topics related to design calculations for industry such as blast resistant buildings, foundations under vibrating equipment, and foundations under most of the facilities in oil and gas plants.

From a technical point of view, this book describes from the first principal the codes and standards that are most used in industrial projects with the industrial practice and the most applicable method in designing the steel and reinforced concrete structures which serve the static equipment, tanks, tower, and vibrating equipment.

The soil investigation is very essential in the industrial projects, and so this book provides the main key to selecting the reasonable type of test, the way to prepare the scope of work for geotechnical activities and define the required number and depth of bore holes based on oil and gas industrial technical practices, and the main features for designing the pile foundation.

This book provides a practical guide to designing reinforced concrete and steel structures and foundations in industrial projects. The maintenance plan and assessment of the existing structure are vital to maintaining safe operations, so Chapter 9, Assessment for Structures in Industrial Plants, discusses the up-to-date methodology to obtain a maintenance plan through a whole integrity management system.

Mohamed A. El-Reedy, Ph.D
Cairo, Egypt
Email:elreedyma@gmail.com
www.elreedyman.tk

CHAPTER 1
Engineering Management

1.1 INTRODUCTION

Oil, gas, and power generation projects have special economic and technical characteristics. Managing these types of projects requires both special skills and to an overview of all project phases. Economically, facilities are dependent on oil and gas production, which is directly correlated to global investment, which is in turn affected by the price of oil. In 2008, e.g., oil prices increased worldwide; as a result, many projects were started during that time period. During the writing of this book, the price has dramatically decreased, which has caused many problems for investments in many projects worldwide.

Only a limited number of faculties of engineering focus on concrete and steel structure in industrial projects, as it needs special consideration, especially compared to the engineering faculties focused on residential building, which is based on more traditional activities in the market. The engineering design for the industrial facilities in general depends on special codes and standards specific for these types of activities.

All the major multinational companies that work in oil, gas, or energy in general are interested in the industrial facilities. These companies provide continuous support for research and development that will enhance the ability of their engineering firms and construction contractors to support their business needs.

1.2 OVERVIEW OF FIELD DEVELOPMENT

Estimated global oil reserves based on geological and geophysical studies and oil and gas discoveries as of January 1996 indicate that about 53% of current reserves are in the Middle East. Overall, 60% of reserves are controlled by the Organization of Petroleum Exporting Countries (OPEC). Obviously, OPEC and the Middle East are vital to the world's current energy needs.

During this phase, due to the absence of detailed information about reservoir characteristics, future market conditions, and field-development

Onshore Structural Design Calculations.
DOI: http://dx.doi.org/10.1016/B978-0-08-101944-3.00001-2

alternatives, experts make judgments based on their past experience, on cost, and based on data available from previous history. The success of oil and gas companies depends on this expertise, so most companies keep experts on hand and compete with each other to recruit them. When the experiential data is not enough, decisions are made through joint brainstorming sessions between experts and management, which are greatly affected by a company's culture and past experiences.

1.2.1 Project Cost Versus Life Cycle

Field development for a new project or for extending existing facilities is a multistep process. The first step is gathering input data, such as the reservoir and environmental data; the selection and design of major system components, such as the production drilling, wells, facilities, and off-take system; and the decision criteria, such as the economics. The next step is evaluating the different field-development options that satisfy the input requirements and establishing their relative merits with respect to the decision criteria. In this design process loop, alternatives for both each system and for field-development systems need to be taken into account.

For the next stage, a conceptual design for the selected system is required. In this phase, the selection activity focuses on the system components and detail elements. Design iterations are generated until each member of the engineering and operations team are satisfied from a technical point of view. All the system components and construction activities must be well defined.

Once the design is completed, few changes to the system and its components can be made without suffering delays and cost overruns.

All project activities that precede the start of the detailed design phase are called the front-end engineering design (FEED) phase. FEED is the most important phase of a project life cycle.

An ideal field-development schedule should allow sufficient lead time for performance of all FEED work before detailed design starts.

Experience shows that the FEED phase will identify viable options; develop, evaluate, and select concepts; and provide a conceptual design. The FEED phase usually consumes only about 2–3% of the total project cost (TPC) of field development (El-Reedy, 2011) and has the highest impact on cost, schedule, quality, and success. It is common to observe major cost overruns when a full FEED phase is not performed.

Reanalysis of projects that did not have a satisfactory FEED phase because of political factors indicates that a 50% TPC reduction could have been achieved if a satisfactory FEED phase had been performed.

The FEED phase shows the variation in the accuracy of the TPC estimates in different phases of a project. In general, our understanding of the economics and other features of a field-development system improve as we move along the project life cycle. At the start of the FEED phase, a number of options are available and identifying the right field-development concept will profoundly influence a project's success. Selection and definition of the system components and subcomponents also have a significant impact on costs and/or the schedule. At this phase, the accuracy of the TPC estimates is approximately $\pm$15% to $\pm$25%.

The ability to influence cost and savings decreases with progress along the project life cycle. At the concept-development stage, the selection and development of the right concepts have a major impact on the TPC. Savings in the detailed design and construction phases generally stem from good project control and execution, which result in TPC optimization. The final design phase provides detailed engineering analysis and design, approved-for-construction (AFC) drawings. Efficient project management in the execution of the plan; cost, schedule, and quality control; verification; safety assurance and quality; purchasing and documentation have some effect on the TPC, but not as profoundly as the FEED phase. Past experience indicates that at the start of the construction phase, the accuracy of the TPC estimates is approximately $\pm$5% to $\pm$10%.

1.2.2 Concept and Screening Selection

At the concept stage, external factors such as country requirements and characteristics, technology transfer, and environmental pollution potential, as well as the culture, politics, economics, and infrastructure of the country who own the land and the operating oil company and its partners, may have major influences on the concept of selection. Not easily comparable criteria, such as the economics, design completeness, and external factors, have to be weighed against each other and used for ranking and concept selection. In a screening process, the goal of the exercise is defined first, after which the viable field-development options are identified. Next, a multitude of selection criteria are grouped and prioritized by experts, who determine the importance of each criterion by comparing them.

The comparisons are then passed through an analytical process to obtain rankings for each comparison and the alternatives.

The basic design defines the equipment plot plan with drawings for the main structures system in enough detail to allow the detailed design to start.

Basic design results enable reliable cost and schedule estimates, and the ordering of long-lead major item and structural components.

The conceptual drawings show major components of the plant with layouts of main buildings and facilities, with preliminary calculation and preliminary sizes for the facility's equipment.

A concept cost estimate in average (±40%) as an order of magnitude. A conceptual design package that includes this information is prepared by the owner or specialized engineering team (this is called the appraise phase) and is given to the design contractor as input to the preliminary design phase to begin the FEED stage.

1.3 FEED ENGINEERING PHASE

In the feed engineering phase (FEED), after the facilities required for the plant are defined it will be preliminarily locate in the plot plan. A preliminary process calculation will be run by process software to define the equipment size. The structure system of the building will be defined in this stage if it will be a steel or concrete structure. For a steel structure, the members will be designed and the connection will be designed in execution phase with an Engineering, Procurement and Construction (EPC) contractor. In the management of a new project, this is called the select phase.

The FEED process ends with the completion of the preliminary engineering design. At this point, the FEED phase has provided the following deliverables:

- Basic design drawings for all major facilities in the plant. These should contain enough detail to enable reliable field-development cost and schedule estimates. This information is particularly important if owners wish to solicit bids and to enter into a lump-sum EPC agreement with a contractor.
- A basis of detailed design (BOD) document for the detailed design phase. The BOD defines the detailed design requirements, including the environmental parameters (e.g., metocean, seismic, ice, etc.).
- Site-specific information.
- Main process data.

- Design life for the structure should be defined considering the operation requirement and fatigue effect along its lifetime.
- Definition of accidental loads (e.g., collisions with boats, dropped objects, fire, and explosions).
- Preferred material classes.
- Design regulations, codes and recommended practices.
- Corrosion protection requirements (e.g., sacrificial anode or impressed current, and types of anodes).
- A narrative of the construction methods and procedures that will affect the construction in case there is or is not a pile foundation.
- Any other owner requirements that will affect the detailed design.
- This document, which contains most of the information just listed, is issued to all qualified bidders. It is very important to note that most contractors will not be interested in bidding a lump-sum EPC contract if a basic design package is not available.

  The following tasks should also be completed in this phase:
- Preparation of detailed and final process flow diagrams (PFDs). If a detailed design is attempted without final PFDs, process changes during the final design phase may cause significant rework, schedule, and cost overruns.
- Preparation of Process and Instrument diagrams (P&IDs).
- Preparation of equipment and material lists, data sheets, and specifications.
- Preparation of detailed engineering, precommissioning, and commissioning scopes of work.
- Preparation of detailed design schedules and cost estimates.

1.4 DETAIL ENGINEERING

At the end of this phase, the engineering office will deliver the full construction drawings and specifications for the whole project, which contain all the details that enable the contractor to execute his function. In this phase, there will be a huge number of engineering hours, so there must be good coordination between the different disciplines. This will happen if there is good organization in the engineering office and if the client provides a free mode of communication between the different parties through a system channel with continuous coordination.

The complexity of this phase is such that it needs a quality system.

A system of quality assurance is important at this stage because it organizes the work. The target of the project and each team member's responsibility is clear. The concept of quality is defined by supporting documents.

The documents are regarded as the executive arm of the quality assurance process. For example, any amendment or correction in the drawings should be made through the agreed procedure and system. Moreover, the drawings should be sent in a specified time to the client for review and discussion, with an official transmittal letter to control the process time. If there are any comments or inquiries, they should be done through agreement between the two technical parties; then the modification should be done by the engineering firm and re-sent to the client through the same communication procedure.

The development of a system to avoid older copies of the drawings becoming confused with the current copies can prevent human error. The most current set of drawings may be ensured by the establishment of a system for continuous amendment of the date and number of the drawings and engineering reviews until the final stage of the project. The approval of the final set of drawings shall be sealed with a stamp ("Approved for Construction") indicating that these are the final drawings approved for the construction.

After the completion of the detailed engineering phase, the specifications and drawings are ready to be used in the execution phase. In some projects the documents may reach hundreds of volumes, especially the specifications, operation manual, and volumes of maintenance and repairs.

The main challenge for this phase is for the civil engineers, as the last input data will be sent to them from different disciplines and they will be the first discipline to deliver the document to start on site, as the civil activity is the start for any project.

1.5 ENGINEERING DESIGN MANAGEMENT

The goal of engineering management for the engineering firm is to control the design stage to provide high quality with a lower cost to increase the revenue.

The design input comprises all technical information which is necessary for the design process. The basis of this information is the owner of the Statement of Requirement (SOR) document, so the engineering firm should review this document carefully; if there is any confusion or misunderstanding, it should be clarified in the document and through meetings. The main item in the SOR is the scope of work, so you need to describe it in detail and provide the boundaries for this scope of work, taking into consideration the interfaces between the engineering construction and installation.

- Instructions for control of the design, whether the client controls the whole process or requests some specific actions (e.g., a representative from the audit during the design phase) are often provided in the contract.
- The designer must take into account the available materials in the local market in relation to the project and its location and match these with the capabilities of the owner. The designers must have a realistic view as well as full and up-to-date knowledge of the best equipment, machinery, and available materials.
- The design must be matched to the project specifications; The permissible deviation and tolerance should be in accordance with the specifications and requirements of the owner.
- Health, safety, and environment are the most critical objectives, so every design should conform to health, safety, and environmental regulations.
- The computer is one of the basic tools used now in the design process as well as in the recording and storage of information. It also offers the ability to change the design of the work easily. It is now easy to modify the drawings by using computer-aided design (CAD) software in order to obtain more precise information with access to various tables and diagrams.
- The design output must be compatible with all design requirements, and the design should be reviewed through an internal audit. The design should be compared with old designs that have been approved before for similar projects. This is a simple procedure for checking designs. Any engineering firm should have a procedural checklist for reviewing designs.

The audits of the design review are intended to be conducted on a regular basis at important stages in the design. The audit must require complete documentation and can take analytical forms such as the analysis of collapse with an assessment of the risk of failure.

In addition, the owner responsibility within the engineering firm is to collect all the expectations and the requirements in detail from all stakeholders, and specifically from the operations because they are the main client to you as the project or engineering project managers. All their requirements should be discussed with them and defined in the engineering phase. It is important also to have an operation representative to review the drawings specific to processes, piping, and mechanical parts, so they can agree that it matches with their expectations and requirements, as they are the end user for your project.

The main challenge of the engineering firm by their engineering manager or the project manager in the design phase is to control the different disciplines to let them match with the project objective and target. The project manager has to do two very important meetings. The first meeting is after obtaining the bidding document form the client to put a price on the project. This is a meeting with all the discipline departments or leads to clearly present to them the scope. The project manager must be sure that all the disciplines will attend the meeting, as it will be a discussion forum so that each discipline will know his scope in the project so the error of estimating their man hours is less. The second meeting is after winning to have a kick-off meeting with the disciplines to start to take any risk during implementation.

1.5.1 Engineering Stage Time and Cost Control

There are two methods of controlling the engineering phase for lump-sum contracts, and it is usual for small projects to have a time period of 3 months maximum. Most engineering contracts depend on cost, time, and resources (CTR) sheets, which list the cost, time, and resources to be delivered in the offer from the engineering firm to the client to review and approve. This sheet is vital even for small projects as it defines the deliverables that will be received and the cost and time for each deliverable. The sample of a CTR form is shown in Table 1.1.

Table 1.1 CTR form sample

<table>
<tr><td>Date</td><td colspan="2">Project name</td><td>CTR</td><td>Project number</td><td>Discipline</td></tr>
<tr><td colspan="6">Deliverables:</td></tr>
<tr><td colspan="6">Input data:</td></tr>
<tr><td colspan="6">Comments:</td></tr>
<tr><td>Item</td><td>Activity</td><td>Discipline</td><td>Hours of labor</td><td>Cost/hr</td><td>Cost</td></tr>
<tr><td></td><td></td><td></td><td></td><td></td><td></td></tr>
<tr><td colspan="5">Total cost</td><td></td></tr>
</table>

1.5.1.1 Time Schedule Control

Table 1.2 is a guideline for actual projects for the percentage of hours of labor.

This depends on the engineering firms approach in their proposal, as some engineering firms reduce the number of hours of labor and increase the cost per hour, or vice versa, according to their strategy and the market share at the time of bidding.

The progress of engineering activity is required to be identified, so you should decide on some guidelines to control when the engineering firm will do the work; the progress measurement should be identified and agreed upon by both the client and the engineering firm before starting the project. For example, in case study reports the following is an example for the progress measured as shown in Table 1.3.

For different engineering discipline activities, the measured progress should be defined. Just as in oil and gas projects the preparation of P&ID is very critical, so the most practical progress measurement will be as follows in Table 1.4.

For procurement services, Table 1.5 presents the traditional progress-measuring percentage.

Table 1.2 Percentage of hours of labor in engineering phase

Discipline/activity	Percentage of hours of labor, %
Project management	16
Process	12
Mechanical	14
Piping and layout	25
Electrical	8
Control and instrument	3
Civil engineering	10
Pipeline	5
HSSE	7

Table 1.3 Engineering report progress measurement guideline

Activity	Progress percentage
Issue draft to the client	30
Completed document review by the client	50
Incorporate the client comments	80
Final report approval	100

Table 1.4 Engineering preparation for P&ID

Activity	Progress percentage
Study and prepare draft or sketches	20
Completed document review by the client	60
Issue for design	75
Issued for construction finally without hold	100

Table 1.5 Procurement progress measuring percentages

Activity	Progress percentage
Issue request for quotation (RFQ) to bidders	20
Bid evaluation and selection	30
Issue purchase order to vendor	40
Approval of vendor prints	50
Materials successfully pass tests/inspection	85
Materials ready for FOB	90
Materials received on site	100

Table 1.6 Manufacturing delivery progress measuring guideline

Manufacturing and delivery milestone	Equipment	Bulk materials
All vendor drawings and vendor data approved	10%	
All subordered materials received by vendor	25%	10%
Materials successfully inspected, tested, and accepted	50%	75%
All materials delivered on site	15%	15%

The requisition of bulk materials like steel structure or piping or electrical cables are based on the material's take offs MTO and in most cases, it will be issued in three MTOs. The following weights are to be applied for bulk MTO:

First MTO requisition	65%
Second MTO requisition	25%
Third and final MTO requisitions	10%

Table 1.6 presents a guideline for the equipment delivery milestone progress; it defines the percentage progress after finishing the milestones.

Table 1.7 Overall project progress measuring guideline

Activity	Percentage
Detailed engineering	10
Procurement services	5
Delivery of equipment materials	25
Construction	25
Installation	25
Professional services	10

Table 1.8 Breakdown for engineering phase

Activity	Percentage of the engineering cost, %
Design and drafting	57
Proposal	1
Project management	19
Procurement	12
Civil/structure engineering	4
Project control, estimating, and planning	7

For overall project progress, the following measurements in Table 1.7 can be considered as a guideline.

1.5.1.2 Engineering Cost Control

For most projects, the engineering phase is not more than 10% of the cost. Most of the cost is in design and drafting, which is around 58% of the total engineering cost. Table 1.8 presents the breakdown of the engineering phase cost. In addition, the engineering contractor's fee or the administration overhead and profit costs usually average 5%. It is shown in Table 1.8 that design and drafting are the most significant items of technical labor costs, so in most cases, it is wise to calculate the required design and drafting cost and then estimate the other items as a percentage.

1.5.2 Engineering Interfaces

The highest potential risk in any project is the interfaces, and there are many strategies to achieve the interface management. The main goals of the project management is to reduce the effect of the interfaces and to manage it in appropriate manner. In case of engineering procurement and construction contract EPC strategy, it will reduce some risks of interfaces and reduce the risk to the owner. In most cases the EPC will come after the FEED engineer, so it is very important to have an interface between the EPC contractor and the engineering firm in case of FEED. It is required to start this phase after you establish an execution strategy plan (noting, of course, that this document is dynamic during the project life).

The interface here is allocated to the the contractor project manager; he should take care of the interfaces between the different departments in his company. In case you have different contractors for detailed engineering, construction, and installation, the owner should precisely define the interfaces in the tender package; this should be done through an assignment matrix after discussion with the potential construction and installation, but the detailed engineering in most cases will be performed by the same people who do the FEED to terminate the interfaces between the two phases on engineering. If you decided that the detailed engineering will be done by another company rather than the company who develop the FEED engineering, you should decide it as early as possible. The best solution is to let the company who will develop the detailed engineering receive the FEED document and detail. You may pay some money for their review of the required hours of labor, but you will gain all the FEED stages and will avoid arguments, redoing work, and change order hassles.

REFERENCE

El-Reedy, M.A., 2011. Construction Management for Industrial Projects. Wiley-Scrivener.

CHAPTER 2

Loads on the Industrial Structures

2.1 INTRODUCTION

The philosophy for designing and constructing any structure is that it must have the capability to carry all the expected loads which will affect the structure along its lifetime. The structure's probability of safety depends on the load fluctuations the and probability of increasing its value over the design value.

The types of loads and their values vary according to the geographic location. In European countries, e.g., ice loads are the most critical load that affect the design; this is much different than in Middle Eastern countries, whose hot summer climates mean that the thermal load due to increase in temperature is the critical factor in design.

To define the loads we must stop and answer a very important question. What is the building design lifetime?

The answer for this question is easy in case of residential building; but for commercial, industrial, and special building it is not quite so easy. It depends on the project economy as a whole. This will be discussed in the last chapter.

Knowing the structure lifetime is very important, especially in the case of loads that correlate the probability distribution with time; these include loads such as wind and earthquake loads, and in the case of offshore structures, the wave load.

2.2 LOADS

The main factors that are included in the structure design are to define the load on the structure and calculate the structure's resistance to these loads. Defining the load values is the first step in the design process. The load values are defined according to the codes and specifications that the designer will consider in his designing. There are different types of loads, as we will indicate in the following sections.

Onshore Structural Design Calculations.
DOI: http://dx.doi.org/10.1016/B978-0-08-101944-3.00002-4

2.2.1 Dead Load

The dead load is considered the structure self-weight, which can be defined by knowing the structure element dimensions and the density of the materials. The structural members will be made from concrete, steel, wood, or other materials. For the first step, we need to define the structure's preliminary dimensions and define the material from which the structure will be constructed.

In addition to that, the dead load also includes the weight of finishing such as plastering, tiles, or walls constructed from bricks or other finishing materials that are not considered the main part.

It is very important to know that the dead load has a very high value from the total load during design; therefore, the weight of the building members must be calculated precisely, especially in the case of a reinforced concrete structure.

The density for different materials usually used in building is shown in the Table 2.1. These densities are to be referred to when calculating the dead load.

2.2.1.1 General Design Loads

The main concept for designing the new facilities, buildings, and other structures, including floor slabs and foundations, is to design them to resist the loads along their lifetime.

The minimum live load should be mentioned clearly on the client document as SOR or in the Basis of design that will initiate by the engineering firm and in some cases it called design criteria. In this document should mention the project standard and code or it can be refer to the local building code for the traditional building such as administration building. In general, the more traditional code for minimum load in industrial building is ASCE 7, as it is focused on industrial structures.

In addition to the loads in this chapter, other loads shall be considered as appropriate. These loads shall include, but are not limited to: snow, ice, rain, hydrostatic, dynamic, upset conditions, earth pressure, vehicles, buoyancy, and erection.

The rain and snow load should be defined according to the environmental condition of the location of the project.

It is worth mentioning that predicting future loads should be considered in the design when specified by the owner in according to their experience in operations or when highlighted by the engineering companies based on their past experience.

The dead loads are the actual weight of materials forming the building, structure, foundation, and all permanently attached appurtenances which are calculated from the dimensions and the material density as laid out in Table 2.1.

The dead loads are the weights of fixed process equipment and machinery, piping, valves, electrical cable trays, and their contents.

Throughout this book, the dead load's nomenclature will be as follows:

$$D_s,\ D_r,\ D_e,\ D_o,\ \text{and } D_t$$

where

D_s = Structure dead load is the weight of the structure materials (not the empty weight of process equipment, vessels, tanks, piping, or cable trays), foundation, soil above the foundation resisting uplift, and all permanently attached appurtenances. This includes, e.g., lighting; instrumentation; heating, ventilation, and air conditioning (HVAC); sprinkler and deluge systems; fireproofing; and insulation.

D_r = Erection dead load is the fabricated weight of process equipment or vessels.

D_e = Empty dead load is the empty weight of process equipment, vessels, tanks, piping, and cable trays.

D_o = Operating dead load is the weight of process equipment, vessels, tanks, piping, and cable trays full of fluid under normal operating conditions.

D_t = Test dead load is the weight of process equipment, vessels, tanks, and/or piping when full of water (or on in some cases nitrogen) for test procedures.

To consider all the cases during the structure's lifetime, the structure engineer should have all the available data by cooperation with static equipment or pressure vessel engineers and the operations engineers, as the above parameters mostly depend on the mode of operation.

2.2.1.2 Pipe Rack

The more widely known structure in any industrial project is the pipe rack, which carries the piping and the cable tray. In the beginning stages of design, the actual piping and cable tray load will not be defined precisely. Unfortunately, the structure engineer will not have the precise actual load data until the end of the project, and the first activity onsite is the civil and structure work. Below is an estimate load for the piping and

Table 2.1 Self-weight of various materials

Material type	kg/m^3
Plain concrete	2200
Reinforced concrete	2500
Lightweight concrete	1000–2000
Air-entrained concrete	600–900
Heavy concrete	2500–5500
Cement (loose)	1100–1200
Clinker	1500–1800
Big aggregate	1700
Sand	1500
Foamed aerated slag	1700
Granulated	1200
Expanded clay	200–900
Pumice stone	350–650
Fly ash	600–1100
Water	1000
Liquid or powder	1000–1200
Granite	2800
Basalt	3000
Basalt lava	2400
Trechzte	2600
B-Sedimentary rocks	
Lime stone	2700
Marbles	2800
Sand stone	2700
C-Transformed rocks	
Slate	2800
Gneiss	3000
Serpentine	2700
Marble	2700
Masonry bricks	
Red brick	1600–1800
Solid brick	1850
Hollow brick	1400
Light brick	700–800
Refractory brick for general purposes	
Fire clay	1850
Silica	1800

(*Continued*)

Table 2.1 (Continued)

Magnesite	2800
Chrome-magnesite	3000
Corundum	2600
Brick antacid	1900
Glass brick	870
Masonry blocks	
Concrete blocks	1400–1900
Hollow concrete blocks	1150
Leca concrete blocks	600–800
Gypsum blocks	950
Lime	
Limestone powder	1300
Calcined in lumps	850–1300
Calcined	600–1300
Calcined slaked	110
Gypsum	800–1000
Mortar	
Cement mortar	2100
Lime mortar	1800
Lime cement mortar	750–1800
Gypsum mortar	1400–1800
Bitumen mortar with sand	1700
Wood types	
A-Hardwood	
Beech	680
Oak	790
B-Softwood	
Pitch pine	570
White wood	400
C-Fiberboard	
Hard	900–1100
Medium-hard	600–850
Porous insulating	250–400
Plywood	750–850
Core board	450–650

(*Continued*)

Table 2.1 (Continued)

Other building materials	
Dry earth	1700
Wet earth	2000
Rubber floor	1800
Asphalt	3200
Bitumen	1000–1400
Tar	1100–1400
Cement tile	2400
Mosaic tile	2200
Epoxy resin	
Without fills	1150
With mineral materials	2000
With fiberglass	1800
Plastic tile	1100
Polyester resin	1350
Polyethylene	930
PVC hard board	1400
PVC flooring board	1600
PVC flooring tile	1700
Fiberglass	160–180
Glass wool	100–110
Slag wool	200–300
Cork	60
Plaster	1100–1500
Glass in sheets	2500
Wired glass	2600
Acrylic glass	1200
Baled linen	600
Leather in piles	900–1000
Paper	
In stocks	1200
In rolls	1100
Metal materials	
Steel	7850
Wrought iron	7850
Cast iron	7250
Iron ore	3000
Aluminum	2700
Aluminum alloy	2800

(*Continued*)

Table 2.1 (Continued)

Lead	
White lead (powder)	9000
Red lead (powder)	8000
Fuels	
Mineral coal	900–1200
Coke	450–650
Charcoal	250
Cool dust	700
Oils	
Diesel oil	800–1000
Crude oil	950
Petrol (gasoline)	800–850
Petroleum	800
Liquid gas	
Propane	500
Butane	580
Wood	
Hardwood, chopped	400–600
Hardwood logs	500
Softwood, chopped	250
Softwood logs	300
Firewood	400
	1000–1100

cable tray load in order to start work on the design; these estimates are nearly identical with the actual load, and just need to be checked against the actual load in the final phase. The following is the guideline to estimate the piping and cable tray load:

Piping Loads

1. Dead loads for piping on pipe racks shall be estimated as follows, unless actual load information is available and requires otherwise:
   a. Operating dead load (D_o): A uniformly distributed load of 40 psf (1.9 kN/m^2) for piping, product, and insulation. This value is equivalent to 8-in. (203-mm) diameter, Schedule 40 pipes, full of water, at 15-in. (381-mm) spacing.
   b. Empty dead load (D_e): For checking uplift and components controlled by minimum loading, 60% of the estimated

piping operating loads shall be used if combined with wind or earthquake, unless the actual conditions require a different percentage.

c. Test dead load (D_t): This is the empty weight of the pipe plus the weight of the test medium contained in a set of simultaneously tested piping systems. The test medium shall be as specified in the contract documents or as specified by the owner.

2. Unless otherwise specified, a minimum specific gravity of 1.0 shall be used for the test medium.
3. For any pipe larger than 12-in. (304-mm) nominal diameter, a concentrated load, including the weight of piping, product, valves, fittings, and insulation shall be used in lieu of the 40 psf (1.9 kN/m^2). This load shall be uniformly distributed over the pipe's associated area.
4. Pipe racks and their foundations shall be designed to support loads associated with full utilization of the available rack space and any specified future expansion.

You should know that structural engineers have a harder time working in industrial projects than any other discipline team.

As they will receive the actual final load data from the vendor which is the last stage of procurement cycle and also the final piping load after performing the final piping stress analysis simulation results.

Therefore, the below guidelines of load values is from the experience of engineering offices and should be verified by engineering judgment when the below data is referenced by process industrial practice and other owners' specifications.

Pipe Rack Cable Tray Loads

Dead loads for cable trays on pipe racks shall be estimated as follows, unless actual load information is available and requires otherwise:

1. Operating dead load (D_o): A uniformly distributed dead load of 20 psf (1.0 kPa) for a single level of cable trays and 40 psf (1.9 kPa) for a double level of cable trays.

   Note that these values estimate the full (maximum) level of cables in the trays.
2. Empty dead load (D_e): For checking uplift and components controlled by minimum loading, a reduced level of cable tray load (i.e., the actual configuration) should be considered as the empty dead load. Engineering judgment shall be exercised in defining the dead load for uplift conditions.

2.2.1.3 Ground-Supported Storage Tank Loads

Dead loads for ground-supported storage tanks are indicated the same nomenclature as dead loads in the previous sections. The individual load components making up the dead loads may have to be separated for actual use in design, discussed as follows:

1. Operating dead load (D_o): Operating dead load for a ground-supported storage tank is made up of the metal load from the tank shell and roof, vertically applied through the wall of the tank, and the fluid load from the stored product. The fluid load acts through the bottom of the tank and does not act vertically through the wall of the tank. Therefore, the metal dead load and the fluid load must be used separately in design.
2. Empty dead load (D_e): This is for checking uplift and components controlled by minimum loading; the corroded metal weight (when a corrosion allowance is specified) should be considered as the empty dead load.
3. Test dead load (D_t): Test dead load for a ground-supported storage tank is made up of the metal load from the tank shell and roof, vertically applied through the wall of the tank, and the fluid load from the test medium. The fluid load acts through the bottom of the tank and does not act vertically through the wall of the tank. Therefore, the metal dead load and the fluid load must be used separately in design. The test medium shall be as specified in the contract documents or as specified by the owner. Unless otherwise specified, a minimum specific gravity of 1.0 shall be used for the test medium.

The storage tank load and design principal with examples is illustrated in Chapter 4, Steel and Concrete Storage Tank.

2.2.2 Live Loads

The live loads (L) are gravity loads produced by the use and occupancy of the building or structure. These include the weight of all movable loads, such as personnel, tools, miscellaneous equipment, movable partitions, wheel loads, parts of dismantled equipment, stored material, and others.

Areas specified for maintenance (e.g., heat exchanger tube bundle servicing) shall be designed to support the live loads.

The minimum live loads shall be in accordance with *SEI/ASCE 7*, applicable codes and standards, and, unless otherwise specified, in Table 2.2.

The uniform and concentrated live loads listed in Table 2.2 shall not be applied simultaneously.

Table 2.2 Minimum live loads

Structure type	Uniform[a]	Concentrated[a]
Stairs and exits	100 psf (4.8 kN/m^2)	1000 lb (4.5 kN)
Operating, access platforms, and walkways	75 psf (3.6 kN/m^2)	1000 lb (4.5 kN)
Control, HVAC room floors	100 psf (4.8 kN/m^2)	1000 lb (4.5 kN)
Manufacturing floors and storage areas:		
1—Light	125 psf (6.0 kN/m^2)	2000 lb (9.0 kN)
2—Heavy	250 psf (12.0 kN/m^2)[b]	3000 lb (13.5 kN)
Ground-supported storage tank roof	25 psf (1.2 kN/m^2)	NA

[a]The loads provided in this table are to be used unless noted otherwise on the owner's data sheet.
[b]This 250 psf (12.0 kN/m^2) live load includes small equipment.

According to *SEI/ASCE 7*, concentrated loads equal to or greater than 1000 lb (4.5 kN) may be assumed to be uniformly distributed over an area of 2.5 ft (750 mm) by 2.5 ft (750 mm) and shall be located to produce the maximum load effects in the structural members. Stair treads shall be designed according to OSHA regulations or the building code, as applicable.

Live load reductions shall be in accordance with *SEI/ASCE 7*. For manufacturing floor areas not used for storage, the live load reduction specified by *SEI/ASCE 7* for lower live loads may be used.

The loadings on handrails and guardrails for process equipment structures shall be in accordance with OSHA 1910.

The loadings on handrails and guardrails for buildings and structures under the jurisdiction of a building code shall be in accordance with the building code.

2.2.3 Wind Loads

The wind load (W) is the most critical influence in some projects due to its location. In industrial structures, the wind load shall be computed and applied in accordance with *SEI/ASCE 7*. The recommended guidelines for open frame structures, pressure vessels, and pipe racks based on ASCE guidelines for wind loads and anchor-bolt design for petrochemical facilities, shall be followed unless otherwise specified by the client in his requirement.

The owner shall be consulted for the determination of the classification category, which will be based on the location of the project.

2.2.3.1 Basic Wind Load Formula

By using computer software in structure analysis, you can calculate the self weight of the structure member after you have defined its materials and dimensions. However, the finishing materials and the dead load that make up the building should be entered manually into the program.

Wind loads shall be computed and applied in accordance with ASCE 7, and the recommended guidelines for pipe racks, and Open Frame Structure in ASCE's "Wind Loads and Anchor Bolt Design for Petrochemical Facilities."

Wind load calculations shall be based on basic wind speed, V, of 3-s gust speed at 33 ft (10 m) above the ground in Exposure (C) and associated with an annual probability 0.02 of being equaled or exceeded (50-year mean recurrence interval). The basic wind speed V for each site should be obtained from the owner or the country code from the country in which these facilities will be built.

Based on experience, the importance factor (I) shall be category IV.

In most cases for industrial projects such as oil and gas plant are always away from the urban and there is no building surrounding it, so it usually consider exposure category (C) shall be used, in some cases the facilities will be near the shore line so it will be exposure category (D), as defined in ASCE, the Exposure D is defined as Flat, unobstructed areas exposed to wind flowing over open water for a distance of at least 1 mi (1.61 km).

The exposure categories in ASCE7-02 are as follows:

Exposure B: Urban and suburban areas, wooded areas or other terrain with numerous closely spaced obstructions having the size of single-family dwellings or larger.

Exposure C: Open terrain with scattered obstructions having heights generally less than 30 ft (9.1 m). This category includes flat open country, grasslands, and all water surfaces in hurricane-prone regions.

Exposure D: Flat, unobstructed areas and water surfaces outside hurricane-prone regions.

This category includes smooth mud flats, salt flats, and unbroken ice.

The applied wind force F shall be determined by the basic equation from ASCE-7:

$$F = q_z \, G \, C_f \, A_e \tag{2.1}$$

where

q_z = Velocity pressure at height z above ground
G = Gust effect factor
C_f = Net force coefficient

A_e = Projected area normal to wind

The velocity pressure q_z is determined in accordance with the provisions of Section 6.5.10 of ASCE 7.

$$q_z = 0.613\, K_z\, K_{zt}\, K_d V^2\, I\, (\mathrm{N/m^2})$$

where:

K_z is the velocity pressure exposure coefficient as per ASCE7-05 as shown in Table 2.3.

K_{zt} is the topographic factor as defined in ASCE 7. K_{zt} is equal to 1.0 for pipe racks and open frame structures in cases of normal topography.

Table 2.3 Velocity pressure exposure coefficients, K_h and K_z

Height above ground level (z),m	**Exposure (Note I)**			
	B		**C**	**D**
	Case 1	**Case 2**	**All**	**All**
0–4.6	0.70	0.57	0.85	1.03
6.1	0.70	0.62	0.90	1.08
7.6	0.70	0.66	0.94	1.12
9.1	0.70	0.70	0.98	1.16
12.2	0.76	0.76	1.04	1.22
15.2	0.81	0.81	1.09	1.27
18	0.85	0.85	1.13	1.31
21.3	0.89	0.89	1.17	1.34
24.4	0.93	0.93	1.21	1.38
27.4	0.96	0.96	1.24	1.40
30.5	0.99	0.99	1.26	1.43
36.6	1.04	1.04	1.31	1.48
42.7	1.09	1.09	1.36	1.52
48.8	1.13	1.13	1.39	1.55
54.9	1.17	1.17	1.43	1.58
61.0	1.20	1.20	1.46	1.61
76.2	1.28	1.28	1.53	1.68
91.4	1.35	1.35	1.59	1.73
106.7	1.41	1.41	1.64	1.78
121.9	1.47	1.47	1.69	1.82
137.2	1.52	1.52	1.73	1.86
152.4	1.56	1.56	1.77	1.89

Where
Case 1: a. All components and cladding.
b. Main wind force resisting system in low-rise buildings.
Case 2: a. All main wind force resisting systems in buildings except those in low-rise buildings designed.
b. All main wind force resisting systems in other structures.

K_d is the wind directionality factor as per Table 2.4, which is adopted from ASCE 7-02. When used with load combinations, K_d is equal to 0.85 for Pipe Racks and Open Frame Structures.

V is basic wind speed of 3-s gust speed at 11 m (33 ft) above the ground.

I is the Importance factor depending on the occupation category of the building as shown in Table 2.5, concurrent with Table 2.6, for the important factor from ASCE 7-05 and the I equal 1.15 for Category IV structures. In most cases the pipe racks and open frame structures are considered Category IV structures.

Table 2.5 indicates that for process industry facilities, *SEI/ASCE* 7 Category III is the most likely classification because of the presence of hazardous materials. Category II may be used if the owner can demonstrate that release of the hazardous material does not pose a threat to the public.

- *Gust Effect Factors, G*

  Gust effect factor G is determined in accordance with ASCE 7. This contains the equation to calculate G. For rigid structures, the gust effect factor shall be taken as $G = 0.85$ as illustrated in ASCE 7. This value should be used for all Pipe Racks and Open Frame Structures.
- *Force Coefficient,* C_f

  The force coefficient C_f for the various structures shall be as listed in the following sections based on the shape of the equipment guideline.
- *Projected Area,* A_e

  The projected area, A_e, which is normal to wind direction for the various structures in question, shall be as defined in this guideline.

Table 2.4 Wind directionality factor, K_d

Structure type	Directionality factor, K_d
Buildings	
Main wind force resisting system	0.85
Components and cladding	0.85
Arched roofs	0.85
Chimneys, tanks, and similar structures	
Square	0.90
Hexagonal	0.95
Round	0.95
Solid signs	0.85
Open signs and lattice framework	0.85
Trussed towers	
Triangular, square, rectangular	0.85
All other cross sections	0.95

Table 2.5 Occupancy category of buildings and other structures for flood, wind, snow, earthquake, and ice loads

Occupancy category	Description of buildings categories
I	Buildings and other structures that represent a low hazard to human life in the event of failure, including, but not limited to: • Agricultural facilities • Certain temporary facilities • Minor storage facilities
II	All buildings and other structures except those listed in occupancy Categories I, III, and IV.

III	Buildings and other structures that represent a substantial hazard to human life in the event of failure, including but not limited to:
	• Buildings and other structures where more than 300 people congregate in one area.
	• Buildings and other structures with daycare facilities with a capacity greater than 150.
	• Buildings and other structures with elementary school or secondary school facilities with a capacity greater than 250.
	• Buildings and other structures with a capacity greater than 500 for colleges or adult education facilities; healthcare facilities with a capacity of 50 or more resident patients, but not providing surgery or emergency treatment facilities; jails and detention facilities.
	• Buildings and other structures, not included in occupancy Category IV, with the potential to cause a substantial economic impact and/or mass disruption of day-to-day civilian life in the event of failure, including but not limited to:
	  ◦ Power-generating stations.
	  ◦ Water treatment facilities.
	  ◦ Sewage treatment facilities.
	  ◦ Telecommunication centers.
	• Buildings and other structures not included in occupancy Category IV (including but not limited to: facilities that manufacture, process, handle, store, use, or dispose of such substances as hazardous fuels, hazardous chemicals, hazardous waste, or explosives), containing sufficient quantities of toxic or explosive substances such as would be dangerous to the public if released.
	• Buildings and other structures containing toxic or explosive substances.

(Continued)

Table 2.5 (Continued)

Occupancy category	Description of buildings categories
IV	Buildings and other structures designated as essential facilities, including but not limited to: • Hospitals and other health care facilities having surgery or emergency treatment facilities. • Fire, rescue, ambulance, and police emergency vehicle garages. • Designated earthquake, hurricane, or other emergency shelters. • Designated emergency preparedness, communication, and operation centers and other facilities required for emergency response. • Power generating stations and other public utility facilities required in an emergency. • Ancillary structures (including, but not limited to, communication towers, fuel storage tanks, cooling towers, electrical substation structures, fire water storage tanks or other structures housing or supporting water, or other fire-suppression material or equipment) required for operation of Occupancy Category IV structures during an emergency. • Aviation control towers, air traffic control centers, and emergency aircraft hangars. • Water storage facilities and pump structures required to maintain water pressure for fire suppression. • Buildings and other structures having critical national defense functions. Buildings and other structures (including but not limited to facilities that manufacture, process, handle, store, use, or dispose of such substances as hazardous fuels, hazardous chemicals, or hazardous waste) containing highly toxic substances where the quantity of the material exceeds a threshold quantity established by the authority having jurisdiction. Buildings and other structures containing highly toxic substances shall be eligible for classification as Occupancy Category II structures.

Table 2.6 Importance factor, I (wind loads)

Category	Nonhurricane-prone regions and hurricane-prone regions with $V = 85-100$ mph (134–161 km/h) and Alaska	Hurricane-prone regions with $V > 100$ mph (161 km/h)
I	0.87	0.77
II	1.00	1.00
III	1.15	1.15
IV	1.15	1.15

2.2.3.2 Wind Loads on Pipe Racks and Open Frame Structures

2.2.3.2.1 Pipe Racks

Wind on the pipe rack structure itself should be calculated based on lack of shielding.

For all structural members, $C_f = 1.8$ shall be used, except for columns, for which $C_f = 2.0$ shall be used.

Tributary Area for Piping The tributary area for piping should be based on the diameter of the largest pipe plus 10% of the width of the pipe rack. This sum is multiplied by the length of the pipes (bent spacing) to determine the tributary area.

Tributary Area for Cable Trays The tributary area for cable trays should be based on the height of the largest tray plus 10% of the width of the pipe rack. This sum is multiplied by the length of the pipes (bent spacing) to determine the tributary area.

Force Coefficients for Pipes The force coefficient $\boldsymbol{C_f} = 0.7$ should be used as a minimum. The force coefficient C_f is taken from ASCE 7, as shown in Table 2.7 for a round shape, with $h/D = 25$, $D\sqrt{q_z} > 2.5$ and a moderately smooth surface; that is $C_f = 0.7$. If the largest pipe is insulated, then consider using a C_f for a rough pipe dependent on the roughness coefficient of the insulation (D'/d).

Force Coefficients for Cable Trays For cable trays, the force coefficient $\boldsymbol{C_f = 2.0}$ shall be used. The force coefficient C_f for cable trays is taken from ASCE 7, Table 2.3 for a square shape with the face normal to the wind and with $h/D = 25$; that is $C_f = 2.0$.

2.2.3.2.2 Open-Frame Structures

Wind loads should be calculated in accordance with the general procedures and provisions of ASCE 7 for wind loads on "Other Structures."

The force coefficients for components for wind loads for the design of individual components, cladding, and appurtenances (excluding vessels, piping, and cable trays) should be calculated according to the provisions of ASCE 7.

Table 2.8 presents the force coefficients for several items as a guideline.

Table 2.7 Force coefficient, C_f, factor for chimneys tanks, rooftop, and similar structures

Cross section	Type of surface	h/D		
		1	7	25
Square (wind normal to face)	All	1.3	1.4	2.0
Square (wind along diagonal)	All	1.0	1.1	1.5
Hexagonal or octagonal	All	1.0	1.2	1.4
Round ($\grave{D}\sqrt{q_z} > 2.5$)	Moderately smooth	0.5	0.6	0.7
($\grave{D}\sqrt{q_z} > 5.3$, D in	Rough ($D/D = 0.02$)	0.7	0.8	0.9
m, q_z in N/m^2)	Very rough ($D/D = 0.08$)	0.8	1.0	1.2
Round ($\grave{D}\sqrt{q_z} > 2.5$) ($\grave{D}\sqrt{q_z} \leq 5.3$, D in m, q_z in N/m^2)	All	0.7	0.8	1.2

Notes:

1. The design wind force shall be calculated based on the area of the structure projected on a plane normal to the wind direction. The force shall be assumed to act parallel to the wind direction.
2. Linear interpolation is permitted for h/D values other than shown.
3. Notation:

D: diameter of circular cross section and least horizontal dimension of square, hexagonal, or octagonal cross sections at elevation under consideration, in feet (m); and

D: depth of protruding elements such as ribs and spoilers, in feet (m); and

h: height of structure, in feet (m); and

q_z: velocity pressure evaluated at height z above ground, in pounds per square foot (N/m^2)

Table 2.8 Force coefficient, C_f, for miscellaneous structures.

Item	C_f	Projected area
Handrail	2.0	0.24 m^2/m
Ladder without cage	2.0	0.15 m^2/m
Ladder with cage	2.0	0.23 m^2/m
Solid rectangles and flat plates	2.0	
Stair with handrail	2.0	Handrail area plus channel depth 50% gross area
Side elevation		
End elevation	2.0	

Source: Adapted from ASCE—Wind Load and Anchor Bolt Design for Petrochemical Facilities.

Open-Frame Load Design wind forces for the main wind force resisting system for open frame structures should be determined by the Eq. (2.1).

The structure is idealized as two sets of orthogonal frames.

The maximum wind force on each set of frames is calculated independently. C_f accounts for the entire structure in the direction of the wind.

Design wind forces are calculated for the structure as a whole.

The method is described for structures which are rectangular in plan and elevation.

Solidity ratio, ε, for pipe rocks:

$$\varepsilon = A_s / A_g \tag{2.2}$$

where

A_g is the gross area of the windward frame and

A_s is the effective solid area of the windward frame defined by the following:

a. The solid area of a frame is defined as the solid area of each element in the plane of the frame projected normal to the nominal wind direction. Elements considered as part of the solid area of a frame include beams, columns, bracing, cladding, stairs, ladders, handrails, etc. Items such as vessels, tanks, piping, and cable trays are not included in calculations of the solid area of the frame; wind load on these items are calculated separately.

b. The presence of flooring or decking does not cause an increase in the solid area beyond the inclusion of the thickness of the deck.

c. For structures with frames of equal solidity, the effective solid area as should be taken as the solid area of the windward frame.

d. For structures where the solid area of the windward frame exceeds the solid area of the other frames, the effective solid area should be taken as the solid area of the windward frame.

e. For structures where the solid area of the windward frame is less than the solid area of the other frames, the effective solid area should be taken as the average of all the frames.

The calculation of force coefficient C_f depends on the spacing between a frame and its width and the ratio between the area of the steel frame and the total applied wind load.

The force coefficient for a set of frames shall be calculated by

$$C_f = C_{Dg}/\varepsilon \qquad (2.3)$$

where

C_{Dg} is the force coefficient for the set of frames given in Fig. 2.1, and from Figs. 2.2–2.5 and ε is the solidity ratio calculated in accordance with Eq. (2.2).

Force coefficients are defined for wind forces acting normally towards the wind frames, irrespective of the actual wind direction.

The force coefficient can be calculated as shown in Fig. 2.1 for plan frame configuration, taking the following notes into consideration:

1. Frame spacing (*SF*) is measured from centerline to centerline.
2. Frame width (*B*) is measured from outside edge to outside edge.
3. Frame spacing ratio is defined as *SF*/*B*.
4. Number of frames (*N*) is the number of framing lines normal to the nominal wind direction ($N = 5$ as shown in Fig. 2.1).

The area of application of force (A_e) shall be calculated in the same manner as the effective solid area, except that it is for the portion of the structure height consistent with the velocity pressure q_z.

Design Load Cases The total wind force acting on the structure in a given direction, F_T, is equal to the sum of the wind load acting on the

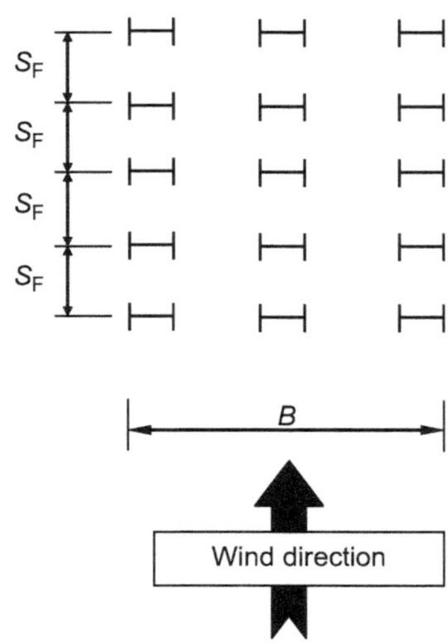

Figure 2.1 Force coefficient, C_{Dg}, for an open-frame structure. *Adapted from ASCE.*

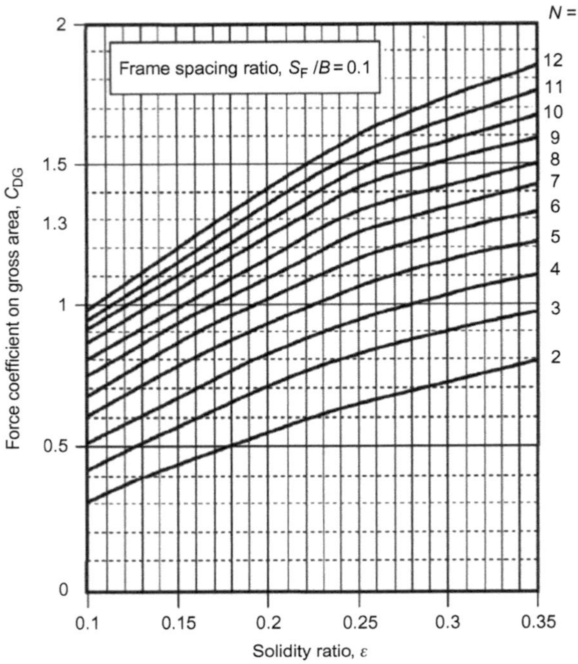

Figure 2.2 Force coefficients, C_{Dg}, for open frame structure with ($S_F/B = 0.1$). *This figure is adapted from ASCE, Wind loads and anchor bolt design for petrochemical facilities.*

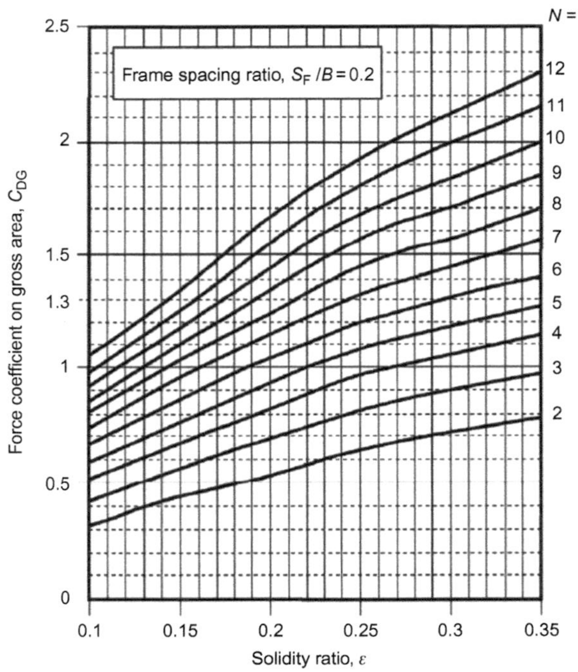

Figure 2.3 Force coefficients, C_{Dg}, for open frame structure with ($S_F/B = 0.2$). *This figure is adapted from ASCE, Wind loads and anchor bolt design for petrochemical facilities.*

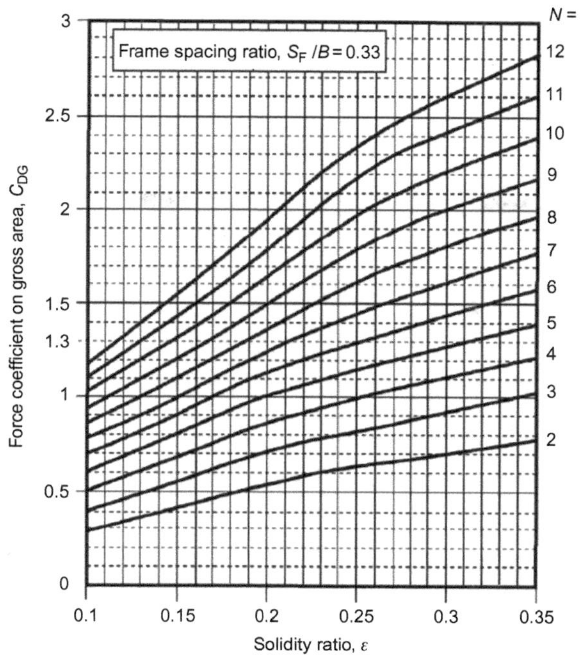

Figure 2.4 Force coefficients, C_{Dg}, for open frame structure with ($S_F/B = 0.33$). *This figure is adapted from ASCE, Wind loads and anchor bolt design for petrochemical facilities.*

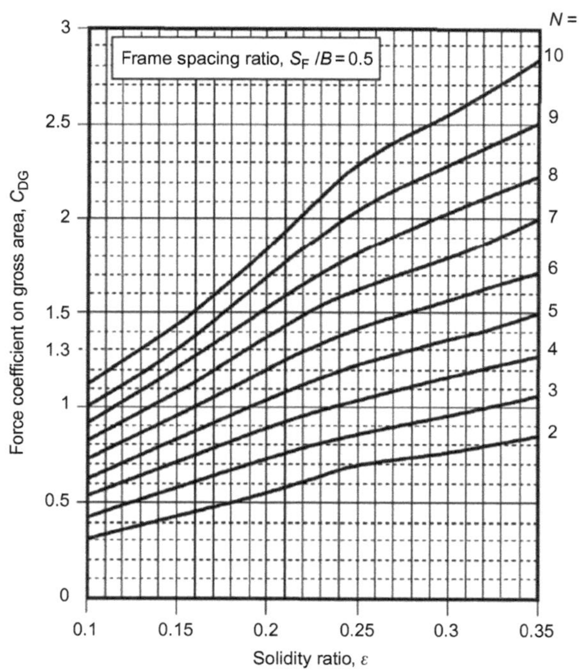

Figure 2.5 Force coefficients, C_{Dg}, for open frame structure with ($S_F/B = 0.5$). *This figure is adapted from ASCE, Wind loads and anchor bolt design for petrochemical facilities.*

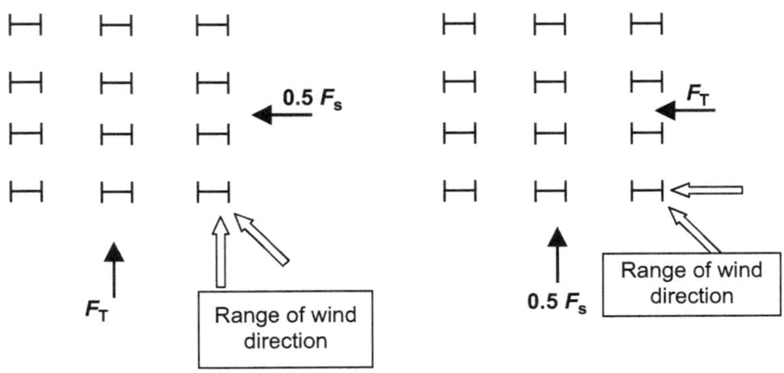

F_S denotes the wind force on the structural frame and appurtenances in the indicated direction and excludes wind load on equipment, piping, and cable tray

F_T denotes the total wind force on the structure in the direction indicated, which is the sum of forces on the structural frame and appurtenances, equipment, and piping.

Figure 2.6 Design load cases from wind load according to ASCE.

structure and appurtenances (F_S), plus the wind load on the equipment, vessels and piping. See Fig. 2.6 for complete definitions of F_T and F_S.

- If piping arrangements are not known, the engineer may assume the piping area to be 10% of the gross area of the face of the structure for each principal axis. A force coefficient of 0.7 should be used for this piping area.

  The following two load cases must be considered as a minimum:
- Frame load + equipment load + piping load (F_T) for one axis, acting simultaneously with 50% of the frame load (F_S) along the other axis, for each direction. These two combinations are indicated in Fig. 2.6.

Partial Wind Load A solid width of 1.5 ft (350 mm) shall be assumed when calculating the wind load on ladder cages.

Partial wind load (WP) shall be based on the requirements of *SEI/ASCE 37-02*, Section 6.2.1, for the specified test or erection duration. The design wind speed shall be (109 kph) (which is 0.75 × [145 kph]) according to *SEI/ASCE 37* for test or erection periods of less than 6 weeks).

For test or erection periods of 6 weeks or more or if the test or erection is in a hurricane-prone area and is planned during the peak hurricane season (from August 1 to October 31 in the United States), refer to *SEI/ASCE 37-02*.

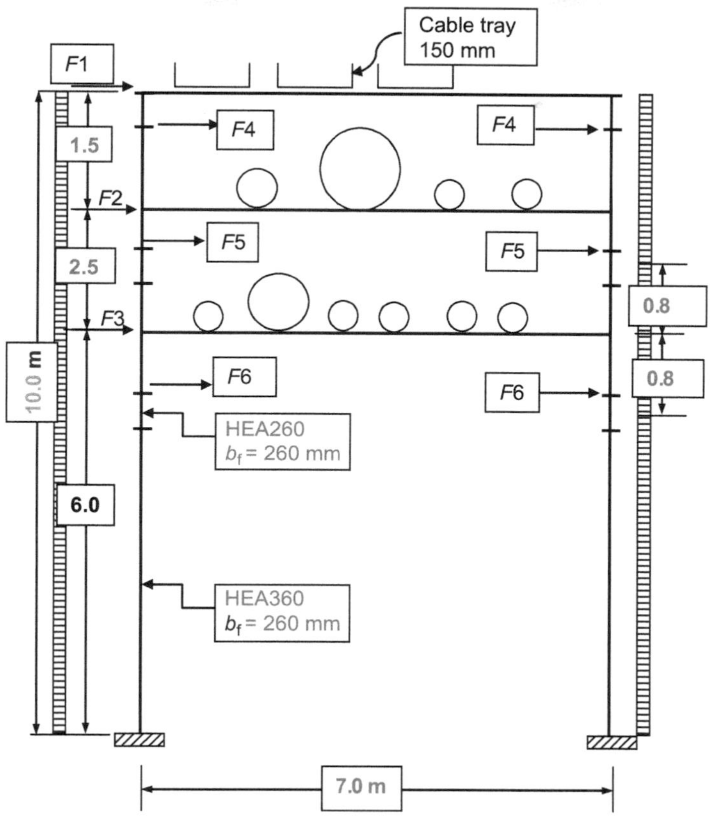

Figure 2.7 Sketch for Example 1 configuration.

Example 1: Wind Load on Pipe Rack

Design a pipe rack in an oil plant. The pipe rack case shall be as shown in Fig. 2.7 with (Bent spacing = 5.5 m), with a 3-s gust wind speed of 155 km/h.

Solution:

Design wind forces are determined by the following equation where F is the force per unit-length of the piping or cable tray:

$$F = q_z \, G \, C_f \, A_e \tag{2.4}$$

Design wind pressure, for 10-m elevation:

$$q_z = 1.4 \text{ kN}/m^2 \tag{2.5}$$

Gust effect factor, $G = 0.85$

- *Force Coefficients*

For structural members	$C_f = 1.8$
For columns	$C_f = 2.0$
For pipes	$C_f = 0.7$
For cable trays	$C_f = 2.0$

- *Projected Area*

Projected Area per meter of pipe rack, A_e = Largest pipe diameter or cable tray height + 10% of pipe rack width.

Part A—Piping and Cable Tray

The guidelines require the consideration of the piping or cable trays separately from the structural members. The following calculations are only for piping and cable trays without the structural support members:

Force calculation force (kN)

Cable tray 150 mm deep (level 10 m)

$C_f = 2.0$

$A_e = 0.15 + (10\% \times 7.0) = 0.85\ m^2/m$

$F1 = q_z\ G\ C_f\ A_e$

$F1 = [(1.4)(0.85)\ (2.0)\ (0.85)] \times 5.5$ m frame spacing **$F1 = 11.13$ kN**

F2 Pipe level 8.5 m—24″ Max. O.D.

$C_f = 0.7$

$A_e = 0.61 + (10\% \times 7.0) = 1.31\ m^2$

$F2 = q_z\ G\ C_f\ A_e$

$F2 = [(1.4\ kN/m^2)\ (0.85)\ (0.7)\ (1.31)] \times 5.5$ m bent spacing $F2 =$ ***6.0* kN**

Pipe level 7 m—18″ Max. O.D.

$C_f = 0.7$

$A_e = 0.45 + (10\% \times 7.0) = 1.15\ m^2$

$F3 = q_z\ G\ C_f\ A_e$

$F3 = [(1.4)(0.85)\ (0.7)\ (1.15)] \times 5.5$ m bent spacing $F3 =$ **5.27 kN**.

Part B—Structural Members

For structural members, assume 7.0 m wide rack with bent spacing of 5.5 m centers, all stringers not shielded.

Stringers at elevations 10 m, 6.8 m, and 5.20 m (as shown in Fig. 2.8).

Assume $q_z = 1.4\ kN/m^2$ for all three levels of stringers (conservative).

$C_f = 1.8$

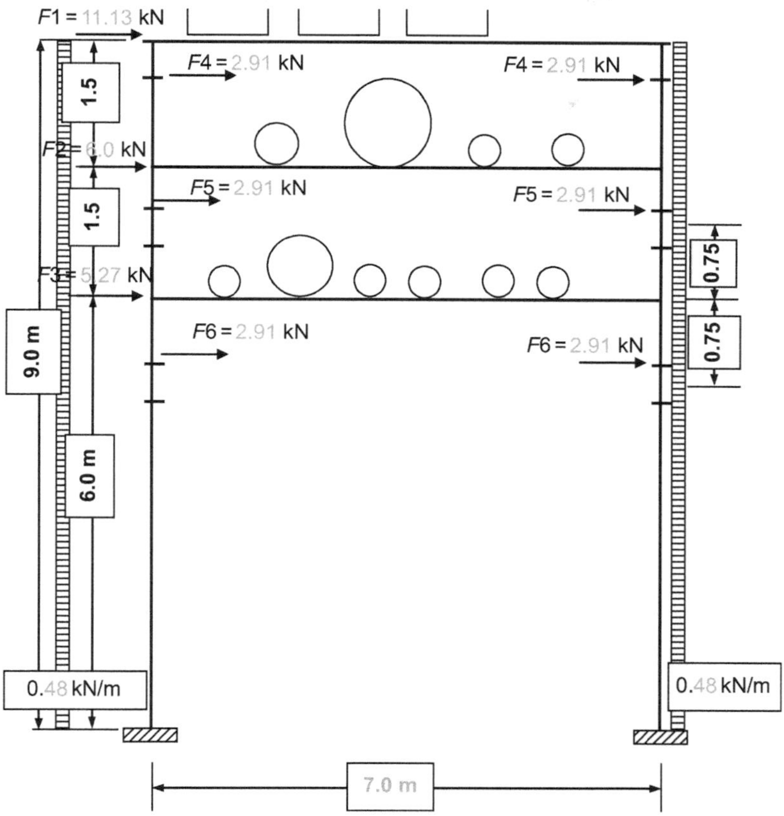

Figure 2.8 Sketch for loads on the pipe rack for Example 1.

$A_e = 0.247$ (beam depth) * 5.5 m (beam length) = 1.36 m^2

$F = q_z\ G\ C_f\ A_e$

$F4 = F5 = F6 = (1.4) \times 0.85 \times 1.8 \times 1.36 = 2.91$ kN

Columns:

$q_z = 1.4$ kN/m^2 at elev. 10 m

$q_z = 1.35$ kN/m^2 at elev. 8.5 m

$q_z = 1.27$ kN/m^2 at elev. 6 m

Use $q_z = 1.35$ kN/m^2 for the whole column (conservative)

$C_f = 2.0$

$A_e = 0.2$ m^2 (column width) $\times$ 1 m = 0.2 m^2/m

$\boldsymbol{F = q_z\ G\ C_f\ A_e}$

Force per column $\boldsymbol{F} = (1.4) \times 0.85 \times 2.0 \times 0.2 =$ ***0.48 kN/m²***

A wind directionality factor $K_d = 0.85$ shall be applied to calculate q_z values above, or otherwise multiply the above-calculated loads $F1$ through $F6$ (including column wind loads) by a factor of 0.85 when used with service and factored load combinations.

K_d is equal to ***0.85*** for pipe racks and open-frame structures.

For pipe rack wind load values on the bent, refer to Fig. 2.8.

Example 2: Wind Load on Pipe Rack

The pipe rack case shall be as shown in Fig. 2.9 (frame spacing = 5.5 m), with a 3-s gust wind speed of 50 m/s.

Solution:

Design wind forces are determined by the following equation (repeated here) where F is the force per unit length of the piping or cable tray:

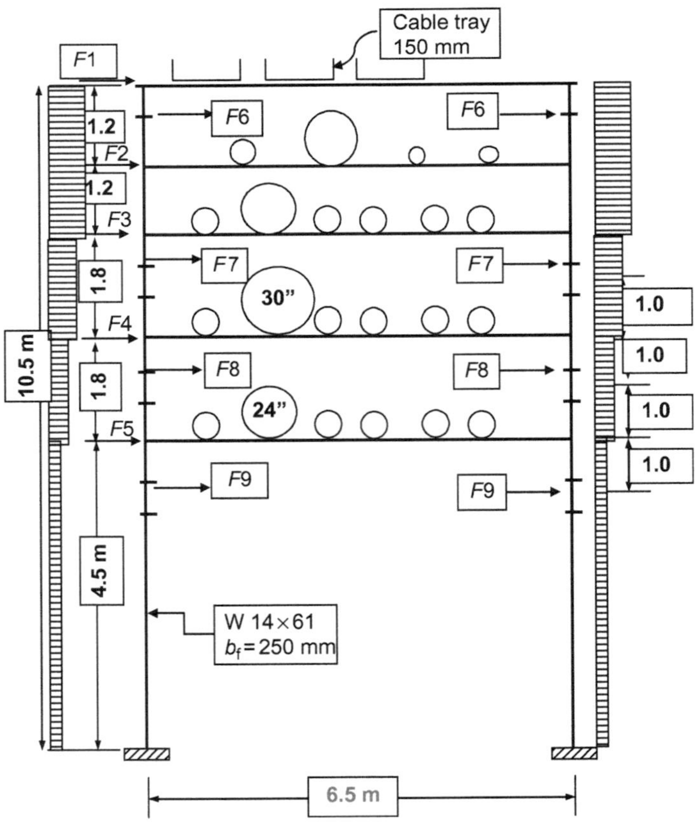

Figure 2.9 Sketch for Example 2 configuration.

$$F = q_z\ G\ C_f\ A_e \tag{2.6}$$

Design wind pressure, for 11 m elevation:

$$q_z = 1.65\ \text{kN}/m^2$$

Gust effect factor, **G = 0.85.**

- *Force Coefficients*

As per Section 2.2.3.2, the following are C_f values for different members:

For structural members	$C_f = 1.8$
For columns	$C_f = 2.0$
For pipes	$C_f = 0.7$
For cable trays	$C_f = 2.0$

- *Projected Area*

Projected area per foot of pipe rack, A_e = Largest pipe diameter or cable tray height + 10% of pipe rack width.

A—Piping and Cable Tray

The guidelines require the consideration of the piping or cable trays separately from the structural members. The following calculations are only for piping and cable trays without structural support members:

Force Calculation Force

F1 Cable tray 150 deep (at elevations of 11 m)

$C_f = 2.0$

$A_e = 0.15 + (10\% \times 6.5) = 0.8\ \text{m}^2/\text{m}$

$\mathbf{F1 = q_z\ G\ C_f\ A_e}$

F1 = [(1.65) (0.85) (2.0) (0.75 0.8)] × 5.5 m bent span.

F1 = 12.34 kN

*F*2 Pipe Level 10 m – 12″ Max. O.D.

$C_f = 0.7$

$A_e = 0.3 + (10\% \times 6.5) = 0.95\ \text{m}^2/\text{m}$

$F2 = q_z\ G\ C_f\ A_e$

$F2$ = [(1.65) (0.85) (0.7) (0.95)] × 5.5 m bent span. ***F2* = 5.13 kN**

*F*3 Pipe Level 8.0 m – 20″ Max. O.D.

$C_f = 0.7$

$A_e = 0.5 + (10\% \times 6.5) = 1.15\ \text{m}^2/\text{m}$

$F3 = q_z\ G\ C_f\ A_e$

F3 = [(1.5) (0.85) (0.7) (1.15)] × 5.5 m bent spa. ***F3* = 5.65 kN**

*F*4 Pipe Level 6.3 m – 30″ Max. O.D.

$C_f = 0.7$
$A_e = 0.75 + (10\% \times 6.5) = 1.4\ m^2/m$
$F4 = q_z\ G\ C_f\ A_e$
F4 $= [(1.45)\ (0.85)\ (0.7)\ (1.4)]$ 5.5 m bent spa. ***F4 = 6.64 kN***

F5 Pipe Level 4.5m—24″ Max. O.D.

$C_f = 0.7$
$A_e = 0.6 + (10\% \times 6.5) = 1.25\ m^2/m$
$\boldsymbol{F5 = q_z\ G\ C_f\ A_e}$
F5 $= [(1.3)\ (0.85)\ (0.7)\ (1.25)]$ 5.5 m bent spa. ***F5*** **= 5.32 kN**

B—Structural Members

For structural members, assume a 6.5-m-wide rack with bent spacing of 5.5-m centers, all stringers not shielded.

Stringers at elevations 11 m, 7.2 m, 5.4 m, and 4 m.

$C_f = 1.8$
$A_e = 0.24$ (beam depth) $\times$ 6.5 m (beam length) $= 1.56\ m^2$
$\boldsymbol{F = q_z\ G\ C_f\ A_e} = q_z \times 0.85 \times 1.56 = q_z \times 1.33$
F6 $= 1.59 \times 1.33 =$ **2.11 kN**
F7 $= 1.45 \times 1.33 =$ **1.93 kN**
F8 $= 1.3 \times 1.33 =$ **1.73 kN**
F9 $= 1.2 \times 1.33 = 1.6$ **kN**

Columns

$q_z =$ **1.59 kN** at elev. 11 m
Use $q_z =$ **1.59 kN** for all column (conservative)
$C_f = 2.0$
$A_e = 0.25$ (column width) $\times$ 1 m $= 25\ m^2/m$
$F = q_z\ G\ C_f\ A_e$
Force per column $\boldsymbol{F} = (1.59) \times 0.85 \times 2.0 \times 0.25 =$ **0.68 kN/m**

A wind directionality factor $\boldsymbol{K_d = 0.85}$ shall be applied to calculated q_z values above; otherwise, multiply the above calculated loads *F*1 through *F*9 (including column wind loads) by a factor of 0.85 when used with service and factored load combinations (Fig. 2.10).

Example 3: Wind Load on Open Frame Structure

The plan and elevation views of the structure are shown in the following Figs. 2.11–2.13. The structure considered is 10 m × 10 m × 18 m high, with three open frames in the direction of wind. The basic wind speed is $V = 190$ km/h. This is a 3-s gust wind speed with an annual probability of exceeding this speed of 0.02.

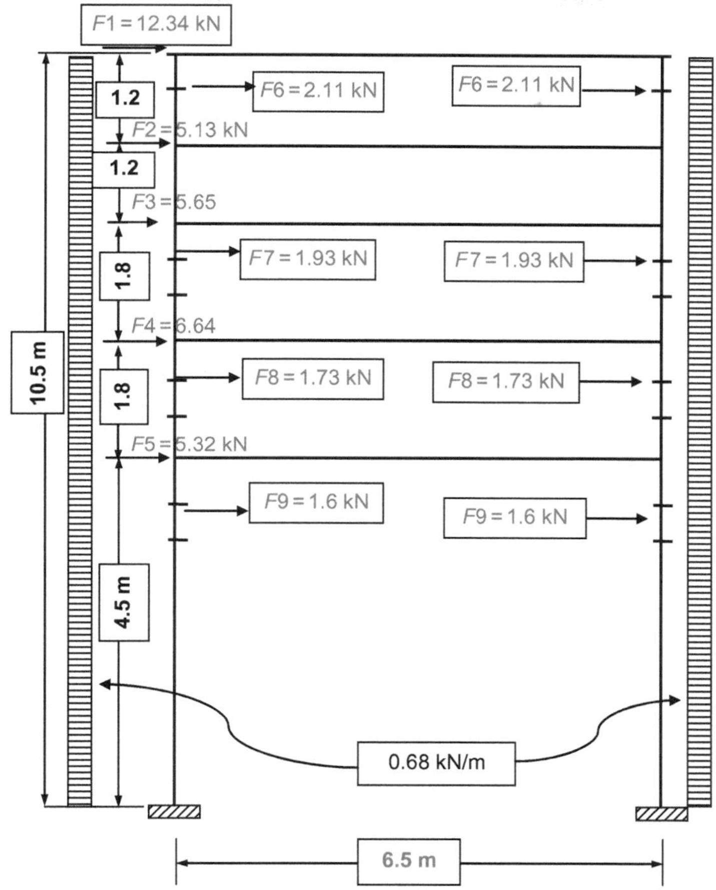

Figure 2.10 Sketch for example 2 configuration.

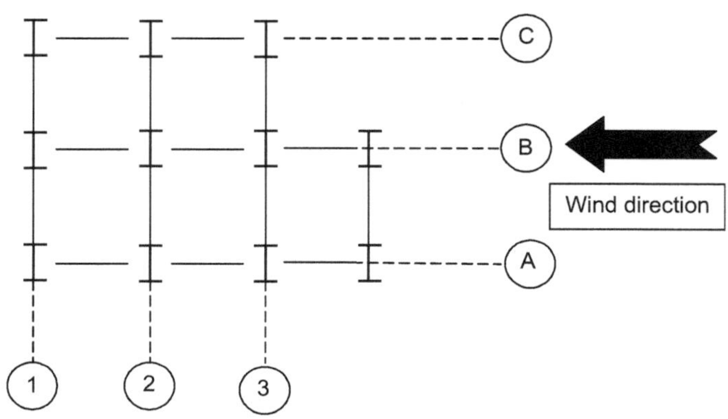

Figure 2.11 Sketch for example 3 plans for open frame structure.

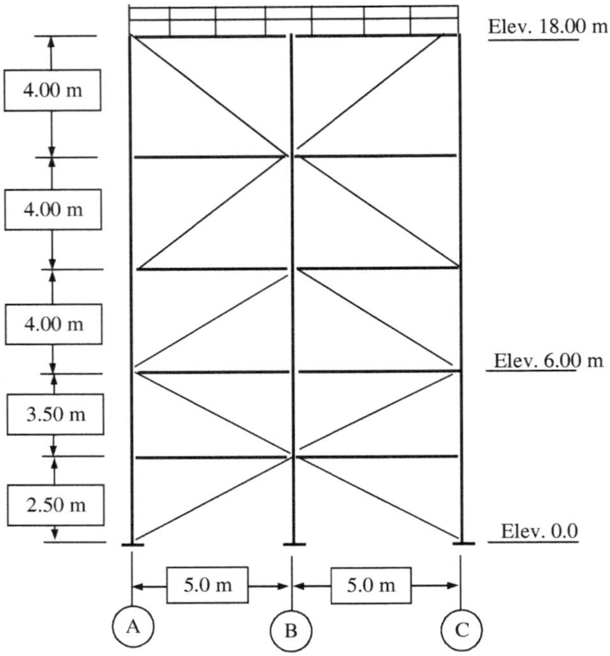

Figure 2.12 Example 3 view on line 3.

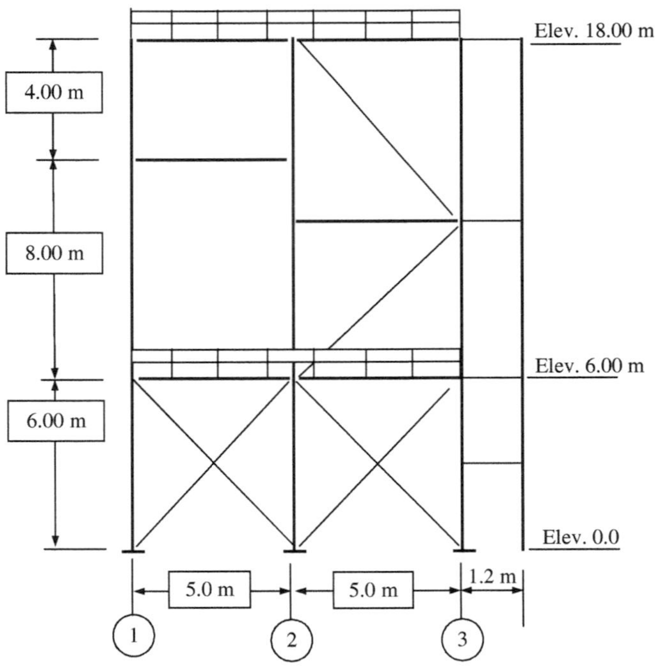

Figure 2.13 Example 3 view on axis (A).

Member sizes are assumed as follows:

Columns	HEB 300 (300 mm × 300 mm)
Beams at el. (6.0 m)	HEA 900
Beams at el. (18 m)	HEB 450
Braces	HEB 200
Intermediate beams	HEB 300

Solution:

$$F_S = q_z\ G\ C_f\ A_e$$

where F_s is the wind force on structural frame and appurtenance.

It is convenient to determine the velocity pressures at the mid-floor heights and at the top of the structure. The following are q_z values at different levels, with velocity pressure represented as q_z.

Height Above ground z	q_z (kN/m^2)
3 m	1.7
15 m	2.0

The gust effect factor is determined next. The ratio of height/least horizontal dimension $= 18/10\ \text{m} = 1.8 < 4$; therefore, the structure is not considered flexible. Use the gust effect factor, $G = 0.85$.

In order to calculate the force coefficient, the solidity ratio ε must first be computed from Eq. (2.3). The gross area (or envelope area) is the area within the outmost projections of the front face normal to the nominal wind direction. Note that the width used below is measured from outside column face to outside column face. For the wind direction shown in Fig. 2.11:

$$A_g = 18 \times 10.3 = 185.4\ \text{m}^2$$

The calculation of the effective solid area of the windward frame, A_s, is summation of the solid surface area of columns, beam, bracing, handrail, and stairs for this example as shown.

Since the middle and leeward frames (column lines 2 and 1, respectively) are similar to the windward frame (with the exception of not having stairs), the solid areas, and hence the solidity ratios for these two frames, will be less than the windward frame, so A_s is equal to the solid area of windward:

$$\varepsilon = A_s/A_g = 53.8/185.4 = 0.29$$

Next, the coefficient C_{Dg} is obtained from curves given in Fig. 2.1 as a function of the solidity ratio ε, the number of frames N, and the frame spacing ratio S_F/B. As defined in Fig. 2.1, $N = 3$ and $S_F/B = 5/10.3 = 0.49$. From Fig. 2.1, for $N = 3$ and extrapolating slightly for $\varepsilon = 0.29$:

$$C_{Dg} = 0.95$$

Next, the gross area force coefficient C_{Dg} is converted into a force coefficient compatible with ASCE 7:

$$C_f = C_{Dg}/\varepsilon = 0.95/0.29 = 3.28$$

The area of application of force A_e has already been determined per floor level during calculation of the solidity ratio. The wind force transmitted to each floor level may now be found.

The total force on the structural frame and appurtenances $\boldsymbol{F_S}$ is 265.90 kN. This is found by summing the forces at all levels, as shown in Table 2.9.

These forces are due to wind acting on the frames only. Note that the wind loads on vessels, equipment, and piping are not considered in this example.

Crosswind Force Calculations

The next step is to repeat the analysis for the nominal wind direction normal to column line A, as shown in the Fig. 2.13 nonwindward frame. The member sizes are the same on this elevation, except that the intermediate beams are HEA 260s and

Beams at el.	6 m	HEA 360
Beams at el.	18 m	HEB 400

The gross area of the windward face includes the stair tower on the right side of the structure.

$$A_g = (18 \times 10.3) + (1.2 \times 18) = 207 \text{ m}^2$$

Table 2.9 Total force—structural frame and appurtenances (F_S)

Floor level	q_z (kN/m²)	G	C_f	K_d	A_e (m²)	F (kN)
0	1.7	0.85	3.28	0.85	10	40.3
1	2.0	0.85	3.28	0.85	20	94.79
2	2.3	0.85	3.28	0.85	24	130.81
					$FS = \Sigma F =$	265.90

The solid areas for the windward frame are given below. The stairs column in the table includes areas of the stair column, struts, and handrails as shown in Table 2.10.

Since the solidity of neither the middle nor leeward frames (column lines B and C, respectively) exceeds that of the windward frame, $\boldsymbol{A_s}$ is equal to the solid area of windward frame, yielding:

$$\varepsilon = A_s/A_g = 59.6/207 = 0.29$$

The frame spacing ratio in this direction is $SF/B = 6/11.2 = 0.54$. Since the width is not uniform (the stair tower stops at the second floor), an average value of B is used. From Figs. 2.12 and 2.13 for $N = 3$ and $\varepsilon = 0.28$.

Therefore use $C_{Dg} = 0.9$:

$$C_f = C_{Dg}/\varepsilon = 0.9/0.29 = 3.1$$

The wind forces per floor level are shown in Table 2.11.

Open-Frame Example—Summary and Conclusion

The results thus far are summarized in the following table. The load combinations for design are application of F_T in one direction simultaneously with 0.5 F_s in the other.

These combinations are shown in Fig. 2.14 and Table 2.12.

In normal cases, these are a piping and vessel, so its wind load must be included. Here in the example, the load on the equipment as piping is only for illustration.

Table 2.10 Solid area (A_S)

Floor level	Tributary height (m)	Solid area (m²)						
		Columns	Beams	Intermediate beams	Bracing	Handrails	Stairs	Total
0	3	2.7	0	0	1.7	0	2.2	10.6
1	3–12	6.5	4.0	0.72	3.2	3.6	3.8	22
2	12–18	8.4	4.77	3.7	3.2	3.6	3.24	27
	Total solid area of windward frame (m²)							59.6

Table 2.11 Total force—structural frame and appurtenance ($\boldsymbol{F_S}$)

Floor Level	q_z (kN/m²)	G	C_f	A_e (m²)	K_d	F (kN)
0	1.7	0.85	3.1	10.6	0.85	40.36
1	2.0	0.85	3.1	22	0.85	98.55
2	2.3	0.85	3.1	27	0.85	139.1
					$F_s = \Sigma F =$	278.01

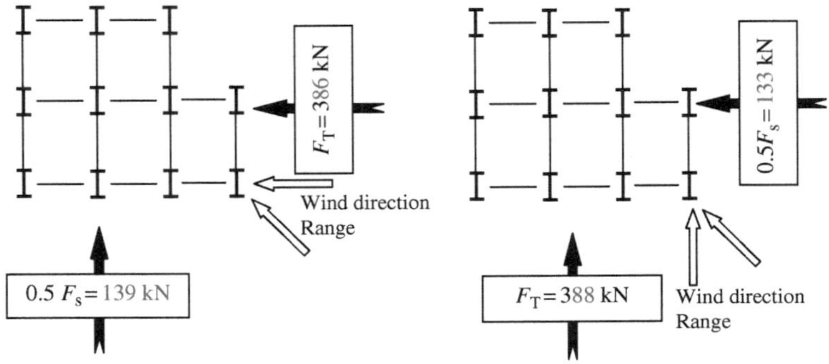

Figure 2.14 Sketch for design load cases in the open frame.

Table 2.12 Summary of load direction

	Load in wind direction 1 (kN)	**Load in wind direction 2 (kN)**
Wind load on structure frame, F_s	265.9	278.01
Wind load on piping, F_E	120	110
Total wind load, F_t on the structure	386	388

A wind directionality factor $K_d = 0.85$ shall be applied to calculated q_z values above, or otherwise multiply the above calculated loads (including column wind loads) by a factor of 0.85 when used with service and factored load combinations.

2.2.4 Earthquake Loads

An earthquake is a sudden and rapid shock that takes place in the Earth's crust. The earthquake take place when layers of the Earth's crust slide against each other; rocks are bent to adapt to the new situation, and when the rocks return to their original position in a sudden motion, the Earth's surface is shaken and quake waves spread from the quake center through the Earth's crust.

The earthquake's loads are very important and serious damage may be suffered by the structure as this load occurs at intervals of up to several years; when it happens, it only takes several seconds, but it has a significant impact on the building. In the past history, it was believed that there is an animal that holds the earth and when it weakens move from standing and shaking the ground. First definition of earthquake is known scholar "Ibn Sina" known as Style as the movement suffered a part of the land because underneath it.

It is worth mentioning that about 300,000 earthquakes occur annually throughout the world, but most are not strong, or do not occur close to residential areas. In the case of powerful earthquakes that occur near the areas where the facilities are located, landslides may occur partially or completely, according to the design of this building.

This code is based on the structure being designed to resist a minimum total lateral load V, which shall be assumed to act nonconcurrently in orthogonal directions parallel to the main axes of the structure.

The procedures and the limitations for the design of structures shall be determined considering seismic zoning, site characteristics, occupancy, configuration, structural system, and height in accordance with this section. Structures shall be designed with adequate strength to withstand the lateral displacements induced by the Design Basis Ground Motion, considering the inelastic response of the structure and the inherent redundancy, over strength, and ductility of the lateral-force-resisting system.

2.2.4.1 Design Spectral Response Acceleration Parameters

Design earthquake spectral response acceleration at short periods, $\boldsymbol{S_{DS}}$, and at a 1-s period, $\boldsymbol{S_{D1}}$ shall be calculated from the following equations:

$$S_{DS} = 2/3\ S_{MS} \tag{2.7}$$

$$S_{D1} = 2/3\ S_{M1} \tag{2.8}$$

The maximum considered earthquake spectral response acceleration for short periods **($\boldsymbol{S_{MS}}$)** and at a 1-s S_{M1}, adjusted for site class effects, shall be determined by Eqs. (2.9) and (2.10), respectively.

$$S_{MS} = Fa\ S_s \tag{2.9}$$

$$S_{M1} = Fv\ S_1 \tag{2.10}$$

where

S_1 = the mapped maximum considered earthquake spectral response acceleration at a period of 1-s.

S_S = the mapped maximum considered earthquake spectral response acceleration at short periods.

Table 2.13 Site classification with soil parameters

Site class definitions	Soil parameters
A	Hard rock with measured shear wave velocity, V_s, >1500 m/s
B	Rock with 760 m/s < V_s, ≤ 1500 m/s)
C	Very dense soil and soft rock with 370 m/s ≤ *Vs* ≤ 760 m/s) or *N* or *Nch* >50 or Su ≥ 100 kpa
D	Stiff soil with 180 ≤ *Vs* ≤ 370 m/s or with N and Nch >50 or 50 $kP_a \leq S_u \leq 100$ kP_a
E	A soil profile with V_s, <180 m/s or any profile with more than 3 m of soft clay. Soft clay is defined as soil with *PI* > 20, *w* ≥ 40%, and *su* <25 KP_a
F	Soils requiring site-specific evaluations: Soils vulnerable to potential failure or collapse under seismic loading such as liquefiable soils, quick and highly sensitive clays, collapsible weakly cemented soils

S_u, average undrained shear strength in top 100 ft (30 m); **ASTM** D2166-91 or ASTM D2850-87; ***N***, standard penetration resistance, **ASTM D 1536-84**; N_{ch}, Average field standard penetration resistance for cohesion less soil layer for the top **100** ft **(30** m).

Note that S_S and S_1 are determined based on the seismic map for the country that the project will be located in, and locate the plot plan on the map to define the seismic zone of the plant location.

The following website: http://earthquake.usgs.gov/earthquakes/world/seismicity_maps/ can provide the seismic map for the United States and the world but requires verification.

When the soil properties are not known in sufficient detail to determine the Site Class, Class (D) shall be used. Site Class (E) shall be used when the authority having jurisdiction determines that Site Class (E) is present at the site or in the event that Site (E) is established by geotechnical data (Tables 2.13, 2.14 and 2.15).

2.2.4.2 Architectural, Mechanical, and Electrical Components Systems

In industrial building structures there are always mechanical, electrical, and nonstructural systems, components, and elements which are permanently attached to structures including supporting structures and attachments.

The design criteria establish minimum equivalent static force levels and relative displacement demands for the design of components and their

Table 2.14 Value of F_a as a function of site class and mapped short period maximum considered earthquake spectral acceleration

Site class	Mapped maximum considered earthquake spectral response acceleration at short periods				
	$S_S \leq 0.25$	$S_S = 0.5$	$S_S = 0.75$	$S_S = 1.0$	$S_S \geq 1.25$
A	0.8	0.8	0.8	0.8	0.8
B	1.0	1.0	1.0	1.0	1.0
C	1.2	1.2	1.1	1.0	1.0
D	1.6	1.4	1.2	1.1	1.0
E	2.5	1.7	1.2	0.9	0.9
F	a	a	a	a	a

Note: Use straight-line interpolation for intermediate values of S_S.
[a]Site-specific geotechnical investigation and dynamic site response analyses shall be performed except that for structures with periods of vibration equal to or less than 0.5 s, values of $\boldsymbol{F_a}$ for liquefiable soils may be assumed equal to the values for the site class determined without regard to liquefaction.

Table 2.15 Values of F_v as a function of site class and mapped 1-S period maximum considered earthquake spectral acceleration

Site class	Mapped maximum considered earthquake spectral response acceleration at 1-s periods				
	$S_1 \leq 0.1$	$S_1 = 0.2$	$S_1 = 0.3$	$S_1 = 0.4$	$S_1 \geq 0.5$
A	0.8	0.8	0.8	0.8	0.8
B	1.0	1.0	1.0	1.0	1.0
C	1.7	1.6	1.5	1.4	1.3
D	2.4	2.0	1.8	1.6	1.5
E	3.5	3.2	2.8	2.4	2.4
F	a	a	a	a	a

Note: Use straight-line interpolation for intermediate values of S_1.
[a]Site-specific geotechnical investigation and dynamic site response analyses shall be performed except for structures with periods of vibration equal to or less than 0.5-s. Values of $\boldsymbol{F_v}$ for liquefiable soils may be assumed equal to the values for the site class determined without regard to liquefaction.

attachments to the structure, recognizing ground motion and structural amplification, component toughness and weight, and performance expectations.

The component's design shall be considered to have the same seismic design category as that of the structure that they occupy or to which they are attached in most cases.

Therefore, it needs to establish a minimum seismic design force requirement for nonbuilding structures that are supported by other structures where the weight of the nonbuilding structure is less than 25% of the combined weight of the nonbuilding structure and the supporting structure.

The seismic design force will be calculated from the following formula:

$$F_p = \frac{0.4a_p S_{DS} W_P}{R_P / I_P}\left(1 + 2\frac{z}{h}\right)$$

$\boldsymbol{F_p}$ is not required to be taken as greater than

$$F_p = 1.6\, S_{DS}\, I_P\, W_P$$

and F_p shall not be taken as less than

$$F_p = 0.3\, S_{DS}\, I_P\, W_P$$

where

F_p = Seismic design force centered at the component's center of gravity and distributed relative to the component's mass distribution.

S_{DS} = Spectral acceleration, short period

a_p = Component amplification factor that varies from 1.00 to 2.50

I_P = Component importance factor that varies as shown in Table 2.16

W_p = Component operating weight

R_p = Component response modification factor that varies from 1.50 to 5.00

z = Height in structure of point of attachment of component with respect to the base. For items at or below the base, z shall be taken equal to zero. The value of z/h need not exceed 1.0.

h = Average roof height of structure with respect to the base

The force (F) shall be applied independently longitudinally and laterally in combination with service loads associated with the component (Tables 2.16–2.18).

Table 2.16 The importance factor, I_p

Equipment criticality	I_P
Life safety component required to function after an earthquake (e.g., fire protection sprinkler system)	1.5
Component that contains hazardous content	1.5
Storage racks in structures open to the public (e.g., warehouse retails stores)	1.5
All other components	1.0
In addition, for structures in Seismic Use Group III:	1.5
All components needed for continued operation of the facility or whose failure could impair the continued operation of the facility	

Table 2.17 Architectural component coefficient

Architectural component or element	a_P	R_P
Interior nonstructural walls and partitions		
Plain (unreinforced) masonry walls	1	1.5
All other walls and partitions	1	2.5
Cantilever elements (unbraced or braced to structural frame below its center of mass)		
Parapets and cantilever interior nonstructural walls	2.5	2.5
Chimneys and stacks when laterally braced or supported by the structural frame	2.5	2.5
Cantilever elements (braced to structural frame above its center of mass)		
Parapets	1.0	2.5
Chimneys and stacks	1.0	2.5
Exterior nonstructural wall elements and connections		
Wall element		
Body of wall panel connections	1	2.5
Fasteners of the connecting system	1	2.5
Storage cabinets and laboratory equipment	1	2.5
Special access floors	1	2.5
All access floors	1	1.5
Signs and billboards	2.5	2.5
Other rigid components		
High deformability elements and attachments	1	3.5
Limited deformability elements and attachments	1	2.5
Low deformability materials and attachments	1	1.5
Other flexible components		
High deformability elements and attachments	2.5	3.5
Limited deformability elements and attachments	2.5	2.5
Low deformability materials and attachments	2.5	1.5

The seismic factor for architectural component is shown in Table 2.17.

2.2.4.2.1 Mechanical and Electrical Component Period

The seismic factor for mechanical and electrical component is shown in Table 2.18.

The mechanical and electrical equipment that are attached to the structure have a natural frequency. The natural frequency period of the mechanical and electrical component, T, shall be determined by the following equation, provided that the component and attachment can be

Table 2.18 Mechanical and electrical components seismic coefficients

Mechanical and electrical component or element	a_p	R_p
General mechanical equipment		
Boilers and furnaces	1.0	2.5
Pressure vessels on skirts and free-standing	2.5	2.5
Stack	2.5	2.5
Cantilevered chimneys	2.5	2.5
Other	1.0	2.5
Manufacturing and process machinery		
General	1.0	2.5
Conveyors (nonpersonnel)	2.5	2.5
Piping systems		
High deformability elements and attachment	1.0	3.5
Limited deformability elements and attachments	1.0	2.5
Low deformability elements and attachments	1.0	1.5
HVAC system		
Vibration isolated	2.5	2.5
Nonvibration isolated	1.0	2.5
Mounted in line with ductwork	1.0	2.5
Other	1.0	2.5
Elevator components	1.0	2.5
Escalator components	1.0	2.5
Trussed towers (freestanding or guyed)	2.5	2.5
General electrical		
Distribution systems (bus ducts, conduits, cable trays)	2.5	5.0
Equipment	1.0	2.5
Lighting fixtures	1.0	1.5

A lower value for a_p shall not be used unless justified by detailed dynamic analyses. The value for a_p shall not be less than 1.00. The value of $a_p = 1$ is for equipment generally regarded as rigid or rigidly attached. The value of $a_p = 2.5$ is for equipment generally regarded as flexible or flexibly attached.

Components mounted *on* vibration isolation systems shall have a bumper restraint or snubber in each horizontal direction. The design force shall be taken as $2F_p$ if the maximum clearance (air gap) between the equipment support frame and restraint is greater than 6 mm.

If the maximum clearance is specified on the construction documents to be not greater than 6 mm, the design force may be taken as F_p.

reasonably represented analytically by a simple spring-and-mass single degree of freedom system:

$$T = 2\pi\sqrt{\frac{W_P}{K_P g}}$$

where

T = component fundamental period
W = component operating weight
g = gravitational acceleration
K = stiffness of resilient support system of the component and attachment, determined in terms of load per unit deflection at the center of gravity of the component. Note that consistent units must be used.

Otherwise, determine the fundamental period of the component in seconds from experimental test data or by a properly substantiated analysis.

2.2.4.3 HVAC Ductwork

Attachments and supports for HVAC ductwork systems should be designed to meet the force and displacement requirement.

In addition to their attachments and supports, ductwork systems designated as having an $I_p = 1.5$ themselves shall be designed to meet the force and displacement, as the drift will not be more than 1/100 of the height level. Where HVAC ductwork runs between structures that could displace relative to one another, and for seismically isolated structures where the HVAC ductwork crosses the seismic isolation interface, the HVAC ductwork shall be designed to accommodate the seismic relative displacements.

Based on ASCE 7-02, the seismic restraints are not required for HVAC ducts with $\boldsymbol{I_p} = 1.0$ if either of the following conditions are met:

a. HVAC ducts are suspended from hangers 12 in. (305 mm) or less in length from the top of the duct to the supporting structure. The hangers shall be detailed to avoid significant bending of the hangers and their attachments,

or

b. HVAC ducts have a cross-sectional area of less than 0.557 m^2 (6 ft^2).

HVAC duct systems fabricated and installed in accordance with standards approved by the authority having jurisdiction shall be deemed to meet the lateral bracing requirements.

Equipment items installed in-line with the duct system such as fans, heat exchangers, and humidifiers weighing more than 75 lb (344 N) shall be supported and laterally braced independent of the duct system and shall also meet the force requirements.

In addition to that, the appurtenances such as dampers, louvers, and diffusers shall be positively attached with mechanical fasteners to the structure. Unbraced piping attached to in-line equipment shall be provided with adequate flexibility to accommodate differential displacements.

2.2.4.4 Piping Systems

The piping systems which are attachments and supports to the structure shall be designed to meet the force and displacement requirement to accommodate a safe operation.

In addition to their attachments and supports, piping systems designated as having $\boldsymbol{I_p} = 1.5$ shall be designed to meet the force and displacement requirement. Where piping systems are attached to structures that could displace relative to one another, and for seismically isolated structures where the piping system crosses the seismic isolation interface, the piping system shall be designed to accommodate the seismic relative displacements requirement to limit the drift to 1/100 of the level height.

Seismic effects that shall be considered in the design of a piping system include the dynamic effects of the piping system, its contents, and when appropriate, its supports. The interaction between the piping system and the supporting structures, including other mechanical and electrical equipment, shall also be considered.

Pressure piping systems designed and constructed in accordance with ASME B31, Code for Pressure Piping, shall be deemed to also meet specific force and displacement provisions provided in the ASME B31.

Fire protection sprinkler systems designed and constructed in accordance with NFPA 13 Standard for the Installation of Sprinkler Systems shall be deemed to meet the other requirements, and the force will be calculated as the force equation.

Other Piping Systems designated as having an $\boldsymbol{I_p} = 1.5$ but not designed and constructed in accordance with ASME B31 or NFPA 13 shall meet the criteria which is described in Section 2.2.4.6.

2.2.4.5 Boilers and Pressure Vessels

The seismic design of a boiler or pressure vessel shall include analysis of the following: the dynamic effects of the boiler or pressure vessel, its contents, and its supports; sloshing of liquid contents; loads from attached components such as piping; and the interaction between the boiler or pressure vessel and its support.

Boilers or pressure vessels designed in accordance with the ASME Boiler and Pressure Vessel Code shall be deemed to meet the force, displacement, and other requirements of this code.

2.2.4.6 General Precautions

Based on the ASCE 7-02 for the location of sites in seismic zones of equipment in general such as boilers, pressure vessels, mechanical equipment, and electrical equipment, the following design strength for seismic loads in combination with other service loads and appropriate environmental effects shall not exceed:

1. For equipment constructed with ductile materials (steel, aluminum, or copper), 90% of the material minimum specified yield strength.
2. For threaded connections in equipment or their supports constructed with ductile materials, 70% of the material minimum specified yield strength.
3. For equipment constructed with nonductile materials as plastic, cast iron, or ceramics, 25% of the material minimum specified tensile strength.
4. For threaded connections in equipment or their supports constructed with nonductile materials, 20% of the material minimum specified tensile strength.

2.2.4.7 Building and Nonbuilding Structure

The definition for the nonbuilding structures, includes all self-supporting structures that carry gravity loads and that may be required to resist the effects of earthquake, with the exception of buildings, vehicular and railroad bridges, nuclear power generation plants, offshore platforms, dams, and other structures. Nonbuilding structures supported by the Earth or supported by other structures shall be designed and detailed to resist the minimum lateral forces as specified.

For equipment in general such as machines, piping, pressure vessels and tanks, there are two cases; the first is during operation as the equipment will be full of liquid, and the other case is when the equipment is empty, which occurs during construction, maintenance, or in some cases shutdown. Whereas the weight of the equipment is different in these two cases, there will be two calculated values of earthquake loads as follows:

- E_o = Earthquake load considering the unfactored operating dead load and the applicable portion of the unfactored structure dead load.
- E_e = Earthquake load considering the unfactored empty dead load and the applicable portion of the unfactored structure dead load.

The design of nonbuilding structures shall provide sufficient stiffness, strength, and ductility consistent with the requirements specified herein for buildings to resist the effects of seismic ground motions as represented by these design forces:

1. Applicable strength and other design criteria shall be obtained from codes and standards which will apply to the project.
2. When applicable strength and other design criteria are not contained in project specifications, such criteria shall be obtained from approved national standards. Where approved national standards define acceptance criteria in terms of allowable stresses as opposed to strength, the design seismic forces shall be obtained and used in combination with other loads in the national standards.

2.2.4.7.1 Design Basis

Nonbuilding structures having specific seismic design criteria established in approved standards shall be designed using the standards as amended herein. In addition, nonbuilding structures shall be designed to resist minimum seismic lateral forces that are not less than the requirements of building structure with the following values of the response modification coefficient R as shown in Table 2.19. Apply the following equation:

$$Cs = 0.14\ S_{DS}\ I$$

and the minimum value specified shall be replaced by

$$Cs = 0.8 S_1 I / R$$

The over strength factor, Ω_o, shall be taken from the same Table 2.19 as the R factor. The importance factor I shall be as given in Table 2.20. The vertical distribution of the lateral seismic forces in nonbuilding structures shall be determined in accordance with an approved standard applicable to the specific nonbuilding structure.

For nonbuilding structural systems containing liquids, gases, and granular solids supported at the base, the minimum seismic design force shall not be less than that required by the approved standard for the specific system.

Table 2.19 Seismic coefficients for nonbuilding structures

Nonbuilding structure type	*R*	Ω	C_d
Nonbuilding frame systems			
Concentric braced frames of steel	6	2.5	5
Moment-resisting frame systems:			
Special moment frames of steel	8	3	5.5
Ordinary moment frames of steel	3.5	3	2.5
Special moment frames of concrete	8	3	5.5
Intermediate moment frames of concrete	5	5	4.5
Ordinary moment frames of concrete	3	3	2.5
Steel storage racks	4	2	3.5
Elevated tanks, vessels, bins, or hoppers			
On braced legs	3	2	2.5
On unbraced legs	3	2	2.5
Irregular braced legs single pedestal or skirt-supported	2	2	2
Welded steel	2	2	2
Concrete	2	2	2
Horizontal, saddle supported welded steel vessels	3	2	2.5
Tanks or vessels supported on structural towers similar to building	3	2	2
Flat-bottomed, ground-supported tanks, or vessels			
Anchored (welded or bolted steel)	3	2	2.5
Unanchored (welded or bolted steel)	2.5	2	2
Reinforced or prestressed concrete			
Tanks with reinforced nonsliding base	2	2	2
Tanks with anchored flexible base	3	2	2
Tanks with unanchored and unconstrained			
Flexible base	1.5	1.5	1.5
Other material	1.5	1.5	1.5
Cast-in-place concrete silos, stacks, and chimneys having walls continuous to the foundation	3	1.75	3
All other reinforced masonry structures not similar to buildings	3	2	2.5
All other nonreinforced masonry structures not similar to buildings	1.25	2	1.5
All other steel and reinforced concrete distributed-mass cantilever structures not covered herein including stacks, chimneys, silos, and skirt-supported vertical vessels that are not similar to buildings	3	2	2.5

(*Continued*)

Table 2.19 (Continued)

Nonbuilding structure type	*R*	Ω	C_d
Trussed towers (freestanding or guyed), guyed stacks, and chimneys	3	2	2.5
Cooling towers			
Concrete or steel	3.5	1.75	3
Telecommunication towers			
Truss: Steel	3	1.5	3
Pole: Steel	1.5	1.5	1.5
Wood	1.5	1.5	1.5
Concrete	1.5	1.5	1.5
Frame: Steel	3	1.5	1.5
Concrete	2	1.5	1.5
Inverted pendulum-type structures (except elevated tanks, vessels, bins, and hoppers)	2	2	2
Signs and billboards	3.5	1.75	3
All other self-supporting structures, tanks, or vessels not covered above or by approved standards that are similar to buildings	1.25	5	2.5

In cases of irregular structures in which the S_{DS} is greater than or equal to 0.50, and that cannot be modeled as a single mass, the modal analysis procedures shall be used.

2.2.4.7.2 Rigid Nonbuilding Structures

Nonbuilding structures that have a fundamental period T less than 0.06 s, including their anchorages, shall be designed for the lateral force obtained from the following:

$$V = 0.30 S_{DS} WI$$

where:

V is total design lateral seismic base shear force applied to a nonbuilding structure
S_{DS} is the site design response acceleration
W is nonbuilding structure operating weight
I is the importance factor, as determined from Table 2.20

Table 2.20 Importance factor *I* and seismic use group classification for nonbuilding structures

Importance factor	*I* = 1.0	*I* = 1.25	*I* = 1.5
Seismic use group	I	II	III
Hazard	H-I	H-II	H-III
Function	F-I	F-II	F-III

H-I: The nonbuilding structures that are not assigned to H-II or H-III.
H-II: The nonbuilding structures containing hazardous materials and classified as a Category III structure in Table 2.5.
H-IV: The nonbuilding structures containing extremely hazardous materials and classified as a Category IV structure in Table 2.5.
F-I: Nonbuilding structures not classified as F-II or F-III.
F-II: Nonbuilding structures classified as Category III structures in Table 2.5.
F-III: The nonbuilding structures classified as Category IV structures (essential facilities) in Table 2.5.

2.2.4.8 Flexibility of Piping Attachments

The piping systems connected to tanks and vessels is sensitive to the potential movement of the connection points during earthquakes, so it requires sufficient flexibility to avoid release of the product by failure of the piping system. Forces and deformations imparted by piping systems or other attachments shall not exceed the designated limitations at points of attachment to the tank or vessel shell. Mechanical devices which add flexibility, such as bellows, expansion joints, and other flexible apparatus, may be used when they are designed for seismic loads and displacements.

Unless otherwise calculated, the minimum displacements in Table 2.21 shall be assumed.

For attachment points located above the support or foundation elevation, the displacements in Table 2.21 shall be increased to account for drift of the tank or vessel relative to the base of support.

When the elastic deformations are calculated, the minimum design displacements for piping attachments shall be the calculated displacements at the point of attachment increased by the amplification factor C_d.

Table 2.21 does not include the influence of relative movements of the foundation and piping anchorage points due to foundation movements (e.g., settlement or seismic displacements). The effects of the foundation movements shall be included in the piping system design, including the determination of the mechanical loading on the tank or vessel and the total displacement capacity of the mechanical devices intended to add flexibility.

2.2.4.9 Design Review for Seismic Load

In the case of areas of high seismic zones or very special plats, such as LNG and NGL gas plants, a design review of the seismic force-resisting system

Table 2.21 Minimum displacement for piping attachments

Tanks or vessel conditions	Displacements (mm)
Anchored tanks or vessels	
Vertical displacement relative to support or foundation	50
Horizontal (radial and tangential) relative to support or foundation	12
Unanchored tanks or vessel (at grade)	
Vertical displacement relative to support or foundation if designed to meet approved standard	150
If designed for seismic loads per these provisions but not covered by an approved standard	300
For tanks and vessels with a diameter <40 ft, horizontal (radial and tangential) relative to support or foundation	200

and the structural analysis shall be performed by an independent team of registered design professionals in the appropriate disciplines and others experienced in seismic analysis methods and the theory and application of nonlinear seismic analysis and structural behavior under extreme cyclic loads. The design review shall include, but not be limited to, the following:

1. Review of any site-specific seismic criteria employed in the analysis including the development of site-specific spectra and ground motion time histories.
2. Review of acceptance criteria used to demonstrate the adequacy of structural elements and systems to withstand the calculated force and deformation demands, together with laboratory data and other data used to substantiate these criteria.
3. Review of the preliminary design including the selection of structural systems and the configuration of structural elements.
4. Review of the final design of the entire structural system and all supporting analyses.

2.2.5 Impact Loads

Impact loads shall be in accordance with *SEI/ASCE 7*. Impact loads for davits shall be the same as those for monorail cranes (powered).

Lifting lugs or pad eyes and internal members (including both end connections) framed into the joint where the lifting lug or pad eye is located shall be designed for 100% impact.

All other structural members transmitting lifting forces shall be designed for 15% impact.

The allowable stresses shall not be increased when combining impact with dead load.

2.2.6 Thermal Loads

In this book, the thermal loads are designated by the following nomenclature:

$$T_p,\ T,\ A_f,\ \text{and}\ F_f$$

where

T_p = Forces on vertical vessels, horizontal vessels, or heat exchangers caused by the thermal expansion of the pipe attached to the vessel

T = Self-straining forces caused by the restrained thermal expansion of structural steel in pipe racks or horizontal vessels and heat exchangers

A_f = Pipe anchor and guide forces

F_f = Pipe rack friction forces caused by the sliding of pipes or friction forces caused by the sliding of horizontal vessels or heat exchangers on their supports, in response to thermal expansion

All support structures and elements thereof shall be designed to accommodate the loads or effects produced by thermal expansion and contraction of equipment and piping.

Thermal loads shall be included with operating loads in the appropriate load combinations. Thermal loads shall have the same load factor as dead loads.

Thermal loads and displacements shall be calculated on the basis of the difference between ambient or equipment design temperature and installed temperature. To account for the significant increase in temperatures of steel exposed to sunlight, 35°F (20°C) shall be added to the maximum ambient temperature.

Friction loads caused by thermal expansion shall be determined using the appropriate static coefficient of friction. Coefficients of friction shall be in accordance with Table 2.22.

Table 2.22 Coefficient of friction

Steel to steel	0.4
Steel to concrete	0.6
Special sliding surfaces such as Teflon	According to manufacturer's instruction

The friction loads shall be considered temporary and shall not be combined with wind or earthquake loads. However, anchor and guide loads (excluding their friction component) shall be combined with wind or earthquake loads.

For pipe racks supporting multiple pipes, 10% of the total piping weight shall be used as an estimated horizontal friction load applied only to local supporting beams. However, an estimated friction load equal to 5% of the total piping weight shall be accumulated and carried into pipe rack struts, columns, braced anchor frames, and foundations.

In case of normal loading conditions with multiple pipes, tensional effects on the local beam need not be considered because the pipes supported by the beam limit the rotation of the beam to the extent that the torsion stresses are minimal. Under certain circumstances, engineering judgment shall be applied to determine whether a higher friction load and/or torsion effects should be used.

Pipe anchor and guide loads shall have the same load factor as dead loads. The internal pressure and surge shall be taken into consideration for pipe anchor and guide loads.

During the design stage of the beams, struts, columns, braced anchor frames, and foundations, they should be capable of resisting actual pipe anchor and guide loads.

For local beam design, only the top flange shall be considered effective for horizontal bending unless the pipe anchor engages both flanges of the beam.

2.2.7 Bundle Pull Load (Bp) for Heat Exchanger

Structures and foundations supporting heat exchangers subject to bundle pulling shall be designed for a horizontal load equal to 1.0 times the weight of the removable tube bundle but not less than 2000 lb (9.0 kN). If the total weight of the exchanger is less than 2000 lb (9.0 kN), the bundle pull load is permitted to be taken as the total weight of the exchanger. The bundle pull load shall be applied at the center of the bundle. It is clearly described in Chapter 3, Static Equipment Foundation Design for heat exchanger foundation design.

If it can be assured that the bundles will be removed strictly by the use of a bundle extractor attached directly to the exchanger (such that the bundle pull force is not transferred to the structure or foundation), the structure or foundation need not be designed for the bundle pull force. Such assurance would typically require the addition of a sign posted on the exchanger to indicate bundle removal by an extractor only.

The portion of the bundle pull load at the sliding end support shall equal the friction force or half the total bundle pull load, whichever is less. The remainder of the bundle pull load shall be resisted at the anchor end support.

Considering that the sustained thermal loads not relieved by sliding caused by vessel or exchanger expansion shall be considered in operating load combinations with wind or earthquakes. Thrust forces caused by thermal expansion of piping shall be included in the calculations for operating load combinations, if deemed advisable. The pipe stress engineer shall be consulted for any thermal loads that are to be considered. The same load factor used for dead load shall be used.

2.2.8 ICE Loads (S)

Atmospheric ice loads due to freezing rain, snow, and ice shall be considered in the design of ice-sensitive structures. In areas where records or experience indicate that snow or in-cloud icing produces larger loads than freezing rain, site-specific studies shall be used. Structural loads due to hoarfrost are not a design consideration.

Mountainous terrain and gorges shall be examined for unusual icing conditions. Site specific studies shall be used to determine the 50-year mean recurrence interval ice thickness and concurrent wind speed in:

1. Alaska
2. Areas where records or experience indicate that snow or in-cloud icing produces larger loads than freezing rain
3. Special icing regions identified through an official map
4. Mountainous terrain and gorges where examination indicates unusual icing conditions exist

Site-specific studies shall be subject to review and approval by the authority having jurisdiction.

In lieu of using the mapped values, it shall be permitted to determine the ice thickness and the concurrent wind speed for a structure from local meteorological data based on a 50-year mean recurrence interval provided that:

1. The quality of the data for wind and type and amount of precipitation has been taken into account
2. A robust ice accretion algorithm has been used to estimate uniform ice thicknesses and concurrent wind speeds from the data

3. Extreme-value statistical analysis procedures acceptable to the authority having jurisdiction have been employed in analyzing the ice thickness and concurrent wind speed data
4. The length of record and sampling error have been taken into account.

The ice load shall be determined using the weight of glaze ice formed on all exposed surfaces of structural members, guys, components, appurtenances, and cable systems. On structural shapes, prismatic members, and other similar shapes, the cross-sectional area of ice shall be determined by:

$$A_i = \pi t_d (D_c + t_d)$$

D_c is shown for a variety of cross-sectional shapes.

On flat plates and large three-dimensional objects such as domes and spheres, the volume of ice shall be determined by:

$$V_i = \pi \, t_d \, A_s$$

where:

$$A_s = \pi r^2$$

where:

r is the radius of sphere or dome.

It is acceptable to multiply by 0.6 for horizontal plates and by 0.8 for vertical plates. The ice density shall be not less than 56 pcf (900 kg/m^3).

Nominal ice thickness, either from the weather data of the site location or through its maps, show the equivalent uniform radial thicknesses t of ice due to freezing rain at a height of 33 ft (10 m) over the contiguous United States for a 50-year mean recurrence interval. Also shown are concurrent 3-s gust wind speeds. Thicknesses for Alaska and Hawaii, as well as for ice accretions due to other sources in all regions, shall be obtained from local meteorological studies.

The height factor f_z used to increase the radial thickness of ice for height above ground z shall be determined by:

$$f_z = \left(\frac{z}{10}\right)^{0.1} \quad \text{for } 0 < z \leq 275 \text{ m}$$

$$f_z = 1.4 \quad \text{for } z > 275 \text{ m}$$

Importance factors to be applied to the radial ice thickness and wind pressure according to the structure classifications are shown in Table 2.5.

The importance factor multiplier I must be on the ice thickness and not the ice weight because the ice weight is not a linear function of thickness.

Both the ice thickness and concurrent wind speed for structures on hills, ridges, and escarpments are higher than those on level terrain because of wind speed-up effects.

The design ice thickness t_d shall be calculated by the following equation:

$$t_d = 2.0tI_if_z(K_{zt})^{0.35} \tag{2.11}$$

In the case of wind on ice-covered chimneys, tanks, and similar structures, the force coefficients for structures with square, hexagonal, and octagonal cross sections shall be as given in Table 2.7. Force coefficients for structures with round cross sections shall be as given in Table 2.7 for round cross sections with $D\,(q_z)^{0.5} \leq 2.5$ for all ice thicknesses, wind speeds, and structure diameters.

Ice accreted on structural members, components, and appurtenances increases the projected area of the structure exposed to wind. The projected area shall be increased by adding t_d to all free edges of the projected area. Wind loads on this increased projected area shall be used in the design of ice-sensitive structures. The wind load coefficient area should be increased due to the ice thickness.

2.3 LOAD COMBINATIONS

Buildings, structures, equipment, vessels, tanks, and foundations shall be designed for the following:

- Appropriate load combinations which are usually preferred to be considered from *SEI/ASCE* 7 except as otherwise specified by the owner
- Local building codes
- Any other applicable design codes and standards
- Any other probable and realistic combination of loads

Load combinations are provided to cover all cases of probability of combining for any types of loads to the structure at the same time during its lifetime. For industrial projects, the special combination should be studied carefully and review in every projects as it is different from mode of operation to another, the way of testing the vessel and the piping and the construction procedure itself. In this book, the load combination will be according to ASCE 7-05 if the owner has not specified any another code; it is preferred to use this standard, and the load combination in this section is considered as a guideline.

These load combinations are for specific types of structures in both allowable stress design (ASD) and load resistance.

1. ASD
   a. The noncomprehensive list of typical load combinations for each type of structure provided here shall be considered and used as applicable.
   b. Engineering judgment shall be used in establishing all appropriate load combinations.
   c. The use of a one-third stress increase for load combinations including wind or earthquake loads shall not be allowed for designs using the AISC ASD.
   d. Steel structures in seismic design category (*D*) or higher shall use factored load combinations as specified in ANSI/AISC 341-02, Part III (Allowable Stress Design Alternative).

   The dead load factor used for the seismic uplift ASD load combinations is generally taken as 0.9.

   This factor is greater than the 0.6 dead load factor used in the ASD load combinations of *SEI/ASCE 7-02*, Section 2.2, because the dead loads of nonbuilding structures are known to a higher degree of accuracy than are the corresponding dead loads of buildings. A dead load factor of 0.9 instead of 1.0 is used to account for the effect of vertical seismic forces.

   The use of this reduction is necessary because foundations sized using ASD loads, except for foundations for ground-supported storage tanks, are generally not required to consider the effect of vertical seismic forces. A dead load factor of 1.0 is used for the wind uplift ASD load combinations because of the higher accuracy of dead loads of nonbuilding structures.
2. Load resistance factor design
   a. The noncomprehensive list of typical factored load combinations for each type of structure provided shall be considered and used as applicable.
   b. Engineering judgment shall be used in establishing all appropriate load combinations.
   c. The following load combinations are appropriate for use with the strength design provisions of either AISC LRFD (third edition or later) or ACI 318 (2002 edition or later).

For all the load combinations the value of structure dead load (D_S), which is the weight of material forming the structure, will be included in other values of the dead load; in most applicable cases only the operating load and empty load will be available.

2.3.1 Vertical Vessels

The load combination that affects the vertical vessel will be as shown in Tables 2.23 and 2.24 for allowable stress design (ASD) and Load resistance

Table 2.23 Loading combinations ASD for vertical vessel

Load comb. no.	Load combination	Allowable stress multiplier	Description
1	$D_o + L$	1.00	Operating weight + live load
2	$D_o + (W \text{ or } 0.7\ E_o{}^{a})$	1.00	Operating weight + wind or earthquake
3	$D_e + W$	1.00	Empty weight + wind (wind uplift case)
4a	$0.9\ (D_o) + 0.7\ E_o{}^{a}$	1.00	Operating weight + earthquake (earthquake uplift case)
4b	$0.9\ (D_e) + 0.7\ E_e{}^{a}$	1.00	Empty weight + earthquake (earthquake uplift case)
5	$D_f + W_p$	1.00	Erection weight + partial wind[b] (wind uplift case)
6	$D_t + W_p$	1.20	Test weight + partial wind

a. For skirt-supported vertical vessels and skirt-supported elevated tanks classified as Occupancy Category IV in accordance with *ASCE/SEI 7-05*, the critical earthquake provisions and implied load combination of *ASCE/SEI 7-05*, shall be followed.

b. Erection weight + partial wind is required only if the erection weight of the vessel is significantly less than the empty weight of the vessel. This case only applies if the erection weight of the vessel is significantly less than the empty weight of the vessel.

c. Thrust forces caused by thermal expansion of piping shall be included in the calculations for operating load combinations, if deemed advisable. The pipe stress engineer shall be consulted for any thermal loads that are to be considered. The same load factor used for dead load shall be used.

factor design (LRFD) respectively. A guide for the design of the vertical vessel foundation is illustrated in Chapter 3.

Table 2.24 Loading combinations For LRFD for vertical vessel

Load comb. no.	Load combination	Description
1	1.4 (D_o)	Operating weight
2	1.2 (D_o) + 1.6 L	Operating weight + live load
3	1.2 (D_o) + (1.6 W or 1.0 E_o[a])	Operating weight + wind or earthquake
4	0.9 (D_e) + 1.6 W	Empty weight + wind (wind uplift case)
5a	0.9 (D_o) + 1.0 E_o[a]	Operating weight + earthquake (earthquake uplift case)
5b	0.9 (D_e) + 1.0 E_e[a]	Empty weight + earthquake (earthquake uplift case)
6	0.9 (D_f) + 1.6 W_p	Erection weight + partial wind[b] (wind uplift case)
7	1.4 (D_t)	Test weight
8	1.2 (D_t) + 1.6 W_p	Test weight + partial wind

a. For skirt-supported vertical vessels and skirt-supported elevated tanks classified as Occupancy Category IV in accordance with *ASCE/SEI 7-05*, the critical earthquake provisions and implied load combination of *ASCE/SEI 7-05*, shall be followed.
b. Erection weight + partial wind is required only if the erection weight of the vessel is significantly less than the empty weight of the vessel. This case only applies if the erection weight of the vessel is significantly less than the empty weight of the vessel.
c. Thrust forces caused by thermal expansion of piping shall be included in the calculations for operating load combinations, if deemed advisable. The pipe stress engineer shall be consulted for any thermal loads that are to be considered. The same load factor used for dead load shall be used.

2.3.2 Horizontal Vessels and Heat Exchangers

The design of the exchanger and horizontal vessel such as the separator knockout drums and other vessels shall use the load combination for ASD and LRFD as presented in Tables 2.25 and 2.26.

Table 2.25 Load combination for ASD for horizontal pressurer vessel

Load comb. no.	Load combination	Allowable stress multiplier	Description
1	$D_o + (T \text{ or } F_f)^b$	1.00	Operating weight + thermal expansion or friction force
2	$D_o + L + (T \text{ or } F_f)^b$	1.00	Operating weight + live load + thermal expansion or friction force
3	$D_o + (W^a \text{ or } 0.7\ E_o)$	1.00	Operating weight + wind or earthquake
4	$D_e + W^a$	1.00	Empty weight + wind (wind uplift case)
5a	$0.9\ (D_o) + 0.7\ E_o$	1.00	Operating weight + earthquake (earthquake uplift case)
5b	$0.9\ (D_e) + 0.7\ E_e$	1.00	Empty weight + earthquake (earthquake uplift case)
6	$D_f^c + W_p$	1.00	Erection weight + partial wind (wind uplift case)
7	$D_t + W_p$	1:20	Test weight + partial wind (for horizontal vessels only)
8	$D_e^d + B_p$	1.00	Empty weight + bundle pull (for heat exchangers only)

a. Wind and earthquake forces shall be applied in both transverse and longitudinal directions, but shall not necessarily be applied simultaneously.
b. The design thermal force for horizontal vessels and heat exchangers shall be the lesser of T or F_f.
c. Erection weight + partial wind is required only if the erection weight of the vessel or exchanger is significantly less than the empty weight of the vessel or exchanger.
d. Heat exchanger empty dead load will be reduced during bundle pull because of the removal of the exchanger head.

Sustained thermal loads not relieved by sliding from vessel or exchanger expansion shall be considered in operating load combinations with wind or earthquakes. The thrust forces caused by thermal expansion of piping shall be included in the calculations for operating load combinations, if deemed advisable. The pipe stress engineer shall be consulted for any thermal loads that are to be considered. The same load factor used for dead load shall be used.

Table 2.26 Loading combinations and load factors—strength design

Load comb. no.	Load combination	Description
1	$1.4\ (D_o) + 1.4\ (T \text{ or } F_f)^b$	Operating weight + thermal expansion or friction force
2	$1.2\ (D_o) + 1.6\ L + 1.2\ (T \text{ or } F_f)^b$	Operating weight + live load + thermal expansion or friction force
3	$1.2\ (D_o) + (1.6\ W^a \text{ or } 1.0\ E_o)$	Operating weight + wind or earthquake
4	$0.9\ (D_e) + 1.6\ W^a$	Empty weight + wind (wind uplift case)
5a	$0.9\ (D_o) + 1.0\ E_o$	Operating weight + earthquake (earthquake uplift case)
5b	$0.9\ (D_e) + 1.0\ E_e$	Empty weight + earthquake (earthquake uplift case)
6	$0.9\ (D_f^c) + 1.6\ W_p$	Erection weight + partial wind (wind uplift case)
7	$1.4\ (D_t)$	Test weight (for horizontal vessels only)
8	$1.2\ (D_t) + 1.6\ W_p$	Test weight + partial wind (for horizontal vessels only)
9	$1.2\ (D_e^d) + 1.6\ B_p$	Empty weight + bundle pull (for heat exchangers only)
10	$0.9\ (D_e^d) + 1.6\ B_p$	Empty weight + bundle pull (for heat exchangers only) (bundle pull uplift case)

a. Wind and earthquake forces shall be applied in both transverse and longitudinal directions, but shall not necessarily be applied simultaneously.
b. The design thermal force for horizontal vessels and heat exchangers shall be the lesser of T or F.
c. This case only applies if the erection weight of the vessel is significantly less than the empty weight of the vessel.
d. Heat exchanger empty dead load will be reduced during bundle pull because of the removal of the exchanger head.

2.3.3 Pipe Rack and Pipe Bridge Design

See Tables 2.27 and 2.28.

Table 2.27 Load combinations for ASD for pipe rack and pipe bridge design

Load comb. no.	Load combination	Allowable stress multiplier	Description
1	$D_s + D_o + F_f + T + A_f$	1.00	Operating weight + friction force + thermal expansion + anchor force
2	$D_s + D_o + A_f +$ $(W^c$ or $0.7\ E_o^{\ b})$	1.00	Operating weight + anchor + wind or earthquake
3	$D_s + D_e^{\ c} + W^c$	1.00	Empty weight + wind (wind uplift case)
4a	$0.9\ (D_s) + 0.6\ (D_o) +$ $A_f + 0.7\ E_o^{\ d}$	1.00	Operating weight + earthquake (earthquake uplift case)
4b	$0.9\ (D_s + D_e^{\ c}) + 0.7\ E_e$	1.00	Empty weight + earthquake (earthquake uplift case)
5	$D_s + D_t + W_p$	1.20	Test weight + partial wind[e]

a. Considerations of wind forces are normally not necessary in the longitudinal direction because friction and anchor loads will normally govern.
b. Earthquake forces shall be applied in both transverse and longitudinal directions, but shall not necessarily be applied simultaneously.
c. $0.6\ D_o$ is used as a close approximation of the empty pipe condition D_e.
d. Full $D_s + D_o$ value shall be used for the calculation of E_o in load combination 4a.
e. Test weight + partial wind is normally required only for local member designs because the test is not typically performed on all pipes simultaneously.

2.3.4 Ground-Supported Storage Tank Load Combinations

Load combinations for ground-supported storage tanks shall be taken from API Standard 650. Load combinations from API Standard 650 and modified for use with *SEI/ASCE* 7 loads and Process Industrial Practice (PIP) nomenclature as presented in Table 2.29.

It is worth to mention that load combination for static machinery, skid and modular equipment, filters, and other elements shall be similar to the load combinations for vertical vessels.

Table 2.28 Load combinations for LRFD for pipe rack and pipe bridge design

Load comb. no.	Load combination	Description
1	$1.4\ (D_o + F_f + T + A_f)$	Operating weight + friction force + thermal expansion + anchor
2	$1.2\ (D_o + A_f) + (1.6\ W^a \text{ or } 1.0\ E_o{}^b)$	Operating weight + anchor + wind or earthquake
3	$0.9\ (D_e{}^c) + 1.6\ W^a$	Empty weight + wind (wind uplift case)
4a	$0.9\ (D_o) + 1.2\ (A_f) + 1.0\ E_o$	Operating weight + earthquake (earthquake uplift case)
4b	$0.9\ (D_e{}^c) + 1.0\ E_e$	Empty weight + earthquake (earthquake uplift case)
5	$1.4\ (D_t)$	Test weight
6	$1.2\ (D_t) + 1.6\ W_p$	Test weight + partial wind[d]

a. Considerations of wind forces are normally not necessary in the longitudinal direction because friction and anchor loads will normally govern.
b. Earthquake forces shall be applied in both transverse and longitudinal directions, but shall not necessarily be applied simultaneously.
c. $0.6D_o$ is used as a close approximation of the empty pipe condition D_e.
d. Test weight and partial wind normally is required only for local member designs because the test is not typically performed on all pipes simultaneously.

Table 2.29 Load combination for ASD for Storage tank

Load comb. no.	Load combination	Description
1	$D_o + P_i$	Operating weight + internal pressure[a]
2	$D_t + P_t$	Test weight + test pressure
3	$(D_e \text{ or } D_o) + W + 0.4\ P_i{}^b$	Empty or operating weight + wind + internal pressure[a]
4	$(D_e \text{ or } D_o) + W + 0.4\ P_e{}^b$	Empty or operating weight + wind + external pressure
5	$D_o + (L \text{ or } S) + 0.4\ P_e{}^b$	Operating weight + live or snow + external pressure
6	$(D_e \text{ or } D_o) + 0.4\ (L \text{ or } S) + P_e$	Empty or operating weight + live or snow + external pressure
7	$D_o + 0.1\ S + E_o{}^c + 0.4\ P_i{}^b$	Operating weight + snow + earthquake + internal pressure[a] (earthquake uplift case)
8	$D_o + 0.1\ S + E_o{}^c$	Operating weight + snow + earthquake

[a]For internal pressures sufficient to lift the tank shell according to the rules of *API Standard 650*, tank, anchor bolts, and foundation shall be designed to the additional requirements of API Standard 650 Appendix F.7.
[b]If the ratio of operating pressure to design pressure exceeds 0.4, the owner shall consider specifying a higher factor on design pressure in load combinations 3, 4, 5, and 7 of Table 3.26.
[c]Earthquake loads for API Standard 650 tanks taken from *ASCE/SEI* 7 "bridging equations" or from API Standard 650 already include the 0.7 ASD seismic load factor.

2.3.5 Test Load Combinations

Engineering judgment shall be used in establishing the appropriate application of test load combinations to adequately address actual test conditions in accordance with project and code requirements while avoiding over conservative design.

Consideration shall be given to the sequence and combination of testing for various equipment, vessels, tanks, and/or piping systems supported on common structures, pipe racks, or foundations.

Full wind and earthquake loads are typically not combined with test loads unless an unusually long test duration is planned (i.e., if a significant probability exists that the "partial wind velocity" will be exceeded or that an earthquake event may occur).

Additional loading shall be included with test loads if specified in the contract documents.

For allowable stress design, a 20% allowable stress increase shall be permitted for any test load combination.

For ultimate strength/limit state design, no load factor reduction shall be permitted for any test load combination.

REFERENCES

ACI318-02, Building Code Requirements for Structural Concrete, ACI Manual of Concrete Practice, 2002.

AISC Manual of Steel Construction—Allowable Stress Design (ASD), 2000.

American Institute of Steel Construction (AISC), Seismic Provisions for Structural Steel Buildings, Including Supplement No. 1 (February 15, 1999, July 1997, Parts I and II).

ASCE, Wind Load and Anchor Bolt Design for Petrochemical Facilities, Task committee in wind induced forces and task committee in anchor bolt design for petrochemical committee of the energy division of ASCE, 1997.

ASCE 7 −02 Minimum Design Loads for Buildings and Other Structures, 2002.

ASCE 7 −05 Minimum Design Loads for Buildings and Other Structures, 2005.

ASME. ASME B31.8, 1995. Gas Transmission and Distribution Piping Systems, 1995.

ASME, 2000. Boiler and Pressure Vessel Code, Including Addenda through 2000.

ASTM. Test Method for Unconsolidated, Undrained Compressive Strength of Cohesive Soils in Triaxial Compression. ASTM 02850-87,1987.

ASTM. Test Method for Unconfined Compressive Strength of Cohesive Soil. ASTM 021 66-91, 1991.

ASTM. Test Method for Laboratory Detednation of Water (Moisture) Content of Soil and Rock. ASTM 02216-92, 1992.

CHAPTER 3

Static Equipment Foundation Design

3.1 INTRODUCTION

In general, the equipment in oil and gas plants consists of two categories: the static equipment and the rotating equipment. In this chapter, we will focus on the static equipment as the separator vertical or horizontal, the knockout drum, the vertical desalination towers, and other equipment which are called also as the pressure vessel as most of these static equipment contained a fluid under pressure.

This chapter establishes guidelines and recommended procedures for use by engineers analyzing and designing the foundation under the static equipment. The two most well-known pieces of equipment are the heat exchanger and horizontal vessel foundations; these should be used where applicable unless otherwise specified.

3.2 DESIGN PROCEDURE

The engineer should review the project design criteria to determine wind and earthquake loadings, corrosion allowances for anchor bolts, anchor bolt types, and any special requirements dictated by the owner.

The engineer should verify that the design is based on applicable codes in existence when foundation drawings are issued.

3.2.1 Dead Loads

The dead load will consist of the following items if you apply working-strength design or limit-state design. These loads represent all the types loads of that will affect the pressure vessel during its lifetime.

a. Structure dead load (D_s), which is the weight of the foundation and soil above the part of the foundation that resists uplift.

b. Erection dead load (D_r), which is the fabricated weight of the exchanger or vessel, generally taken from certified exchanger or vessel drawings.

Onshore Structural Design Calculations.
DOI: http://dx.doi.org/10.1016/B978-0-08-101944-3.00003-6

c. Empty dead load (D_e), which is the empty weight of the exchanger or vessel including all attachments, trays, internals, bundle, insulation, fireproofing, agitators, piping, ladders, platforms, and others. The eccentric load should also be added to the empty dead load weight.

d. Operating dead load (D_o), which is the load during the operation. This load will be the empty dead load of the exchanger or vessel in addition to the maximum weight of contents during normal operation. The eccentric load should also be added to the operating dead load weight.

e. Test dead load (D_t), which is the load during the test of the vessel. This load is a dead load in cases when the vessel is empty with the weight of the test medium contained in the system. The fluid that will be used in the test should be as specified in the contract documents or as specified by the owner. Unless otherwise specified, a minimum specific gravity of 1.0 should be used for the test medium.

The load due to cleaning should be used instead of test dead load if cleaning fluid has a specific gravity higher than 1. Whether test or cleaning will actually be done in the field should be determined.

It is generally desirable to design for test dead load because unforeseen circumstances may occur. The eccentric load should also be added to the test dead load weight.

Use of the following guidelines is recommended if more exact information about piping supported on the exchanger or horizontal vessel is not available, and should be verified after obtaining the actual data:

1. In case of heat exchangers

   a. A load of an additional 20% of the applicable weight (empty or operating) for heat exchangers with diameters less than 600 mm (24 in.).

   b. A load of an additional 10% of the applicable weight (empty or operating) for heat exchangers with diameters equal to or greater than 600 mm (24 in.).

2. In case of vessels

   c. A load of an additional 10% of the applicable weight (empty, operating, or test) for horizontal vessels.

This additional load should be applied at a perpendicular horizontal distance of $D/2$ plus 450 mm (18 in.) from the longitudinal centerline of the vessel.

D is the basic diameter and calculated from the following equation.

Basic diameter (D) = vessel I.D. + 2 times the wall thickness + 2 times the insulation thickness.

This additional eccentric load (vertical load and moment caused by eccentricity) should be distributed to each pedestal in proportion to the distribution of the operating load to each pedestal.

In case of stacked exchangers, the weight of only the largest exchanger should be used to estimate the eccentric load.

In case of heat exchangers, the load distribution for most common shell and tube heat exchangers, vertical dead loads should normally be distributed with 60% to the channel end support and 40% to the shell end support.

The actual exchanger shape and support configuration should be reviewed when determining weight distribution because in many cases load distribution may vary.

3.2.2 Live Loads

The live load (L) data will be as described in Chapter 2 noting that the live load data presents the case of maintenance of the valves, measurement, and other activities, but it will be less load effect with respect to the other loads in the case of the pressure vessels.

3.2.3 Wind Loads

The engineering office is responsible for determining wind loads (W) used for the foundation design. Wind loads from vendors or other engineering disciplines should not be accepted without verification.

Transverse wind—The wind pressure on the projected area of the side of the exchanger or vessel should be applied as a horizontal shear at the center of the exchanger or vessel. Including the wind loading on projections such as piping, manways, insulation, and platforms during the wind analysis is important. The saddle-to-pier connection should be considered fixed for transverse loads.

Longitudinal wind—The wind pressure on the end of the exchanger or vessel should be applied as a horizontal shear at the center of the exchanger or vessel. The flat surface wind pressure on the exposed area of both piers or both columns should also be included, applied as a horizontal shear at the centroid of the exposed area. The saddle-to-pier connection will be considered pinned for longitudinal loads unless more than one row of anchor bolts exists.

Shielding—in normal cases, no allowance should be made for shielding from wind by nearby equipment or structures except under unusual conditions. Industrial facilities usually relocate and change their equipment.

3.2.4 Earthquake Loads

Earthquake loads (E) should be calculated in accordance with the relevant code of the country which the plant is located in; but in general, UBC is the most popular code for earthquake load for the building, and ASCE07 is preferred for industrial structures as this code focuses on industrial structures.

The engineer is responsible for determining the earthquake loads used for the foundation design. Earthquake loads from vendors or other engineering disciplines should not be accepted without verification.

For low-friction slide plates ($\mu \leq 0.2$), all the longitudinal earthquake loads should be applied at the fixed pier. In case of higher friction slide plates ($\mu > 0.2$), 70% of the earthquake loads should be applied at the fixed pier. The earthquake loads in transverse and vertical should be distributed in proportion to the vertical load applied to both piers.

The piers are normally designed for the fixed end, and then the pier for the sliding end is made identically in order to avoid potential errors in construction and to reduce engineering time. If this proves to be uneconomical, the sliding end should be designed for 30% of the longitudinal earthquake load if using low-friction slide plates, and for 50% of the longitudinal earthquake load if using higher friction slide plates.

The horizontal loads should be divided equally between piers unless otherwise required by the engineering judgment.

3.2.5 Bundle Pull Load (Bp) (Exchangers)

This load only exists in the case of the heat exchanger, as during operation it needs to remove the bundle tubes inside the heat exchanger for cleaning; it should be defined by the operation group and from the data sheet, as there is some sort of cleaning without removing the bundle.

In this calculation, reducing the empty weight of the exchanger should be considered due to the removal of the exchanger head (channel) to pull the bundle. The weight of the exchanger head (channel) is typically within the range of 8–15% of the empty weight of the exchanger.

Table 3.1 gives the approximate weight of standard heat exchangers, all in tons.

The tube material is 14-gage steel (Table 3.2). For the weights of heat exchangers with other tube lengths, multiply by the following factors:

Table 3.3 gives the approximate weight of standard tube bundles, all in tons. The tubes are (3/4 in.), 14 gage, and 4.8 m (192 in.) long. The tubes are two pass on a square pitch. The baffle spacing ranges from 200 mm on the 375 mm (15 in.) exchanger to 400 mm on the 1200 mm (48 in.) exchanger.

Table 3.1 Approximate heat exchanger weight (tons)

Heat exchanger diameter (in.)	Heat exchanger pressure class		
	150	300	450
15	2.2	2.5	2.9
18	2.8	3.2	3.8
20	3.4	3.9	4.6
22	4.0	4.4	5.4
24	4.8	5.4	6.2
26	5.6	6.4	7.2
28	6.4	7.2	8.2
30	7.2	8.3	8.4
32	8.2	9.4	10.8
34	9.2	10.6	12.3
36	10.4	12.0	13.9
38	11.6	13.6	15.4
40	12.8	14.6	17.0
42	14.2	16.2	18.8
44	15.4	17.8	
46	16.8	19.5	
48	18.5	21	

Table 3.2 Modification factor for different tube length

Tube length (m)	Modification factor
2.4	0.80
3.0	0.85
3.6	0.90
4.2	0.95
4.8	1.00
6.0	1.10

Table 3.3 Approximate tube bundle weight

Heat exchanger diameter (in.)	Heat exchanger pressure class			
	150	300	450	600
15	0.9	0.9	0.9	1.0
18	1.2	1.3	1.3	1.4
20	1.5	1.6	1.6	1.7
22	1.8	2.0	2.0	2.1
24	2.3	2.4	2.5	2.7
26	2.7	2.8	2.9	3.0
28	3.2	3.3	3.4	3.6
30	3.7	3.9	4.0	4.2
32	4.2	4.45	4.55	4.8
34	4.8	5.0	5.25	5.45
36	5.4	5.7	6.0	6.15
38	6.1	6.4	6.75	6.9
40	6.7	7.0	7.3	7.6
42	7.4	7.7	8.0	8.4
44	8.1	8.4	8.95	9.3
46	8.9	9.2	9.75	10.2
48	9.6	10.0	10.5	11.0

3.2.6 Thermal Force

By knowing the calculated maximum design temperature, we can calculate the thermal growth using the thermal coefficients as shown in Table 3.4.

The thermal force used for design should be the smaller value resulting from the following two calculations:

a. The force required to overcome static friction between the exchanger or vessel support and the slide plate:

$$F_f = \mu(P_o) \tag{3.1}$$

where,

F_f = static friction force
μ = coefficient of friction
P_o = nominal operating compression dead load on slide plate

b. The force required to deflect the pier or column an amount equal to half of the thermal expansion between exchanger or vessel saddles:

$$T = \frac{3\Delta EI}{2H^3} \tag{3.2}$$

Table 3.4 Total linear xpansion between 70°F and indicated temperature (mm/m) for different material

Temperature (°F)	Shell material							
	Carbon steel Carbon-Moly Low-Chrome (through 3 Cr Mo)	5 Cr Mo through 9 Cr Mo	Austenitic stainless steels 18Cr 8 Ni	12 Cr 17 Cr 27 Cr	12 Cr 20 Ni	Monel 67 Ni 30 Cr	3-1/2 Nickel	Ni–Fe–Cr
70	0	0	0	0	0	0	0	0
100	0.18	0.28	0.17	0.27	0.23	0.19	0.23	0.18
125	0.33	0.52	0.30	0.48	0.43	0.35	0.43	0.33
150	0.48	0.75	0.44	0.70	0.63	0.51	0.63	0.48
175	0.63	0.98	0.58	0.92	0.83	0.68	0.83	0.63
200	0.78	1.22	0.72	1.14	1.02	0.84	1.03	0.78
225	0.94	1.46	0.86	1.37	1.22	1.01	1.24	0.94
250	1.11	1.69	1.01	1.59	1.43	1.18	1.47	1.11
275	1.27	1.93	1.15	1.82	1.63	1.36	1.69	1.27
300	1.43	2.18	1.30	2.04	1.84	1.53	1.92	1.43
325	1.58	2.42	1.45	2.27	2.03	1.71	2.16	1.58
350	1.75	2.67	1.61	2.49	2.23	1.88	2.40	1.75
375	1.92	2.92	1.76	2.72	2.43	2.06	2.65	1.92
400	2.08	3.17	1.92	2.94	2.71	2.24	2.90	2.08
425	2.27	3.42	2.08	3.17	2.93	2.43	3.13	2.27
450	2.44	3.68	2.24	3.39	3.16	2.61	3.37	2.44
475	2.62	3.93	2.41	3.62	3.38	2.79	3.59	2.62
500	2.79	4.18	2.57	3.84	3.61	2.98	3.83	2.79
525	2.98	4.43	2.73	4.07	3.84	3.18	4.06	2.98
550	3.17	4.68	2.91	4.29	4.08	3.37	4.30	3.17

(Continued)

Table 3.4 (Continued)

Temperature (°F)	Shell material							
	Carbon steel Carbon-Moly Low-Chrome (through 3 Cr Mo)	5 Cr Mo through 9 Cr Mo	Austenitic stainless steels 18Cr 8 Ni	12 Cr 17 Cr 27 Cr	12 Cr 20 Ni	Monel 67 Ni 30 Cr	3-1/2 Nickel	Ni–Fe–Cr
575	3.35	4.94	3.08	4.52	4.32	3.56	4.53	3.35
600	3.53	5.20	3.25	4.74	4.55	3.75	4.77	3.53
625	3.73	5.46	3.42	4.97	4.79	3.95	5.01	3.73
650	3.91	5.73	3.59	5.19	5.04	4.15	5.25	3.91
675	4.10	5.98	3.77	5.42	5.28	4.35	5.48	4.10
700	4.28	6.25	3.94	5.64	5.53	4.55	5.73	4.28
725	4.48	6.52	4.12	5.87	5.78	4.75	5.98	4.48
750	4.68	6.79	4.30	6.09	6.04	4.95	6.23	4.68
775	4.88	7.06	4.48	6.32	6.29	5.15	6.47	4.88
800	5.08	7.33	4.67	6.54	6.54	5.36	6.72	5.08
825	5.28	7.61	4.85	6.79	6.80	5.57	6.96	5.28
850	5.49	7.88	5.04	7.04	7.07	5.78	7.22	5.49
875	5.69	8.16	5.23	7.29	7.33	5.98	7.46	5.69
900	5.89	8.43	5.41	7.54	7.60	6.19	7.72	5.89
925	6.09	8.72	5.59	7.79	7.87	6.40	7.97	6.09
950	6.30	9.00	5.78	8.04	8.14	6.61	8.23	6.30
975	6.51	9.28	5.98	8.29	8.41	6.81	8.48	6.51
1000	6.72	9.57	6.17	8.54	8.68	7.01	8.74	6.72
1025	6.92	9.85	6.35	8.79	8.96	0.00	9.00	6.92
1050	7.13	10.13	6.63	9.04	9.24	0.00	9.26	7.13

1075	7.33	10.42	6.82	9.29	9.53	0.00	9.52	7.33
1100	7.54	10.70	6.93	9.54	9.81	0.00	9.78	7.54
1125	7.73	10.98	7.11	9.82	10.09	0.00	10.04	7.73
1150	7.93	11.27	7.30	10.09	10.39	0.00	10.32	7.93
1175	8.13	11.55	7.48	10.37	10.68	0.00	10.58	8.13
1200	8.33	11.83	7.67	10.64	10.96	0.00	10.85	8.33
1225	8.55	12.12	7.85	10.92	11.25	0.00	11.13	8.55
1250	8.78	12.40	8.04	11.19	11.55	0.00	11.43	8.78
1275	8.99	12.68	8.23	11.47	11.85	0.00	11.70	8.99
1300	9.22	12.97	8.43	11.74	12.15	0.00	11.99	9.22
1325	9.42	13.25	8.61	11.99	12.45	0.00	12.28	9.42
1350	9.63	13.53	8.80	12.24	12.75	0.00	12.58	9.63
1375	9.83	13.82	8.98	12.49	13.05	0.00	12.87	9.83
1400	10.04	14.10	9.18	12.74	13.35	0.00	13.17	10.04
1425	0.00	14.42	0.00	0.00	0.00	0.00	13.47	0.00
1450	0.00	14.74	0.00	0.00	0.00	0.00	13.78	0.00
1475	0.00	15.07	0.00	0.00	0.00	0.00	14.07	0.00
1500	0.00	15.39	0.00	0.00	0.00	0.00	14.38	0.00

where

T = force from thermal expansion required to deflect pier or column

Δ = total growth between exchanger/vessel saddles = $\varepsilon\ L$

ε = thermal expansion coefficient in accordance with Table 3.4

L = length of exchanger/vessel between saddles

E = modulus of elasticity of concrete pier

I = pier moment of inertia

H = pier height

The thermal force should be applied at the top of the piers.

The horizontal loads should be divided equally between piers unless otherwise required by vendor.

3.2.7 Load Combinations

Heat exchangers and horizontal vessel foundations should be designed using load combinations. The piping thermal loads should be included in combinations when deemed advisable and should be considered as dead loads when applying load factors.

3.3 ANCHOR BOLTS

Friction force at the bottom of the saddle should be overcome before lateral load is assumed to produce shear in the anchor bolts.

For earthquake loads, horizontal shear forces should be applied to the anchor bolts, assuming no frictional resistance.

3.4 SLIDE PLATES

A steel slide plate or low-friction slide plate assembly should typically be provided at the sliding end of every exchanger or vessel regardless of the flexibility inherent in the structural support. Small, lightly loaded exchangers or vessels may not require slide plates. The saddle with sliding plate configuration is shown in Fig. 3.1.

Low-friction manufactured slide plate assemblies should be used to reduce high-frictional resistance, especially in the case of heavy exchangers or for exchangers with significant thermal growth.

For exchangers with bundle pull, steel slide plates may be more cost efficient than low-friction slide plate assemblies.

Typically, a low-friction slide plate assembly consists of multiple individual slide plate components spaced out along the length of the saddle.

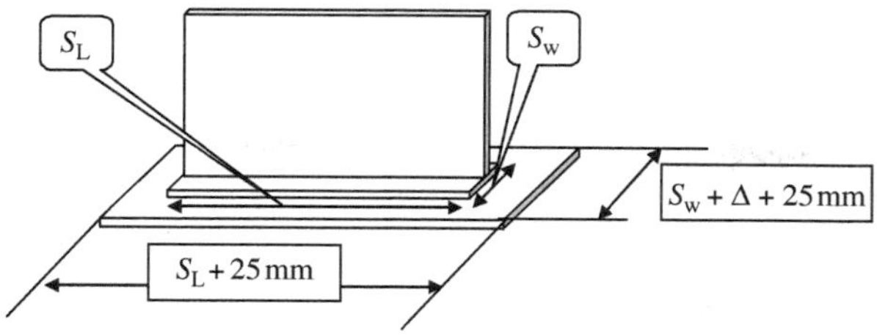

Figure 3.1 Steel sliding plate size.

Each slide plate component consists of an upper element and a lower element, and the sliding surface is at the interface of the upper and lower elements. The elements should be fabricated with a carbon steel backer plate attached to the elements to facilitate welding of the upper elements to the saddles and the lower elements to the steel bearing plate.

Typical coefficients of friction are as follows. For low-friction slide plate assemblies, the manufacturer's data sheet should be read carefully, as the coefficients of friction vary with slide plate material, temperature, and pressure. The different coefficient of friction values will be as follow:

a. No slide plate (steel support on concrete) 0.60
b. Steel slide plate 0.40
c. Low-friction slide plate assemblies 0.05–0.20

The suggested criteria for sizing low-friction slide plate elements are as follows.

The manufacturer's data sheet and specifications should be consulted for temperature restrictions, pressure limitations, and other requirements that may affect the size and types of materials used for the slide plate elements, and these data should be available to the mechanical department, which is always responsible to deliver this data.

Considering that Δ is the total thermal growth between exchanger or vessel saddles, the element widths will be defined by the following limits:

a. Upper element = saddle width + 25 mm minimum to allow for down hand welding on the element-to-saddle weld (larger upper element width may be required for exchangers or vessels with large Δ values).
b. Lower element = upper element width − 2 (Δ) − 25 mm (minimum of 25 mm narrower than upper element)

Element lengths (use 450 mm maximum clear distance between lower elements):

a. Lower element = based on allowable contact pressure in accordance with the manufacturer's literature and lower element width

b. Upper element = lower element length + 25 mm
Plates should be aligned with saddle stiffeners where practical.

A continuous steel bearing plate should be provided under the lower elements so that lower elements can be welded to the bearing plate.

The minimum width of the bearing plate should be 25 mm larger than the width of the lower elements. The minimum length of the bearing plate should be 25 mm larger than the saddle length. Bearing stress on concrete should be checked in accordance with *ACI 318*.

In summary, the suggested criteria for sizing steel slide plates are as follows:

a. Minimum width = saddle width + 2 (Δ) + 25 mm
b. Minimum length = saddle length + 25 mm
Bearing stress on concrete should be checked in accordance with *ACI 318*.

3.5 PIER DESIGN

The pier dimensions should be sized on the basis of standard available forms for the project. When form information is not available, pier dimensions should be sized in 2-in or 50 mm increments to allow for the use of standard contractor wood forms.

Minimum pier dimensions should equal the maximum of the saddle, bearing plate, or steel slide plate dimensions plus 100 mm, and should be sized to provide adequate anchor bolt edge distance based on the anchor bolt design section. Minimum pier width should be no less than 250 mm or 10% of the pier height.

3.5.1 Anchorage Considerations

It is normally desirable to make the pier high enough to contain the anchor bolts and to keep them out of the footing. Consideration must be given to anchor bolt development and foundation depth requirements.

3.5.2 Reinforcement for Piers

In general, the piers should normally be designed as cantilever beams with two layers of reinforcement. If the pier is a compression-controlled member, the pier should be designed as a column.

In any case, the size and reinforcement for each pier should normally be the same. Dowel splices are not required if the vertical pier reinforcing projection is less than the larger of 1.8 m or the rebar size in feet above the top of footing. For example, #8 (25 mm) rebar can extend up to 8 ft (2.5 m) above the mat without dowel splices. For cases that exceed this limit, dowels with minimum projections required for tension splices should be used in accordance with *ACI 318* or BS 8110.

The vertical reinforcement in the piers may need to be increased to account for shear friction. The following formula should be used to calculate the area of reinforcement required for shear friction:

$$A_{vf} = [V_u/(\mu\varphi) - P_u\text{pier}]/f_y \qquad (3.3)$$

V_u = Strength design factored shear force at bottom of pier

3.6 FOUNDATION DESIGN

The soil bearing pressure should be computed for footing design and checked against the allowable bearing capacity which is available from the geotechnical report based on the following equation (the total footing area will be under compression in the case of ($e \leq b/6$)):

$$e = M/P$$

$$q = \frac{P}{A}\left(1 \pm \left(\frac{6e}{b}\right)\right) \qquad (3.4)$$

The footing under eccentricity in one direction and in two directions are presented in Figs. 3.2 and 3.3, respectively.

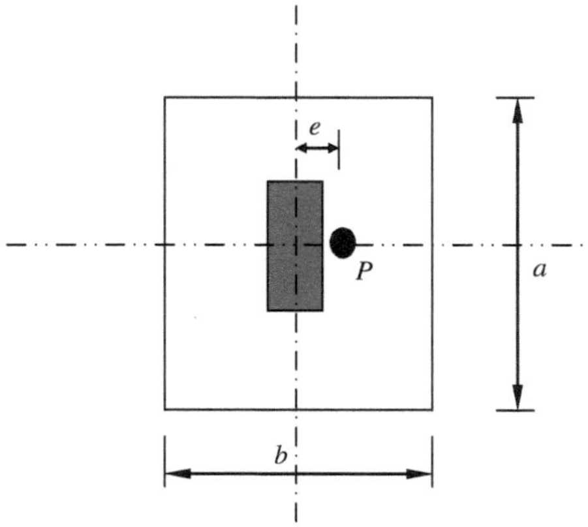

Figure 3.2 Sketch of the load eccentricity.

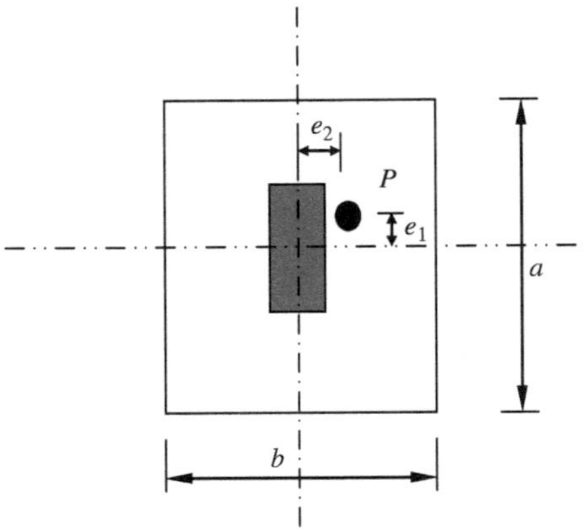

Figure 3.3 Sketch of the load eccentricity in two directions.

In the case of ($e > b/6$) the total footing area is not in compression pressure, and some of the area is under tension.

In Eq. (3.4):

e = eccentricity of vertical service load caused by horizontal service load

a = size of footing perpendicular to direction of horizontal load

b = size of footing parallel to direction of horizontal load

P = total vertical service load (exchanger or vessel, pier, footing, and soil)

A = area of footing

In some cases, the eccentricity exists in both directions; in this situation, these equations cannot apply.

The numerical solutions can be found in many soil mechanics textbooks, and commercial software is also available for such calculations.

The following principal equation will be applied as follows:

$$e_1 = \frac{M_1}{p} \; \textit{and} \; e_2 = \frac{m_2}{p} \quad q_{\max} = \frac{p}{a} + \frac{6p.e_1}{ba^2} + \frac{6pe_2}{ab^2} \quad q_{\min} = \frac{q}{a} - \frac{6pe_1}{ba^2} - \frac{6pe_2}{ab^2}$$

3.6.1 Foundation Reinforcement

The strength design factored moment and shear should be figured on a unit width strip, assuming a simple cantilever. The critical section for moment and diagonal tension shear should be taken at the pier or column face. If shear is excessive, the strength design factored shear should be rechecked using the critical section for shear specified in *ACI 318* or any codes you specified in your project.

The resulting reinforcing steel should be placed continuously across the entire footing. The minimum amount of bottom reinforcement is 16 mm at every 300 mm based on ACI.

In this case, this is a normal design for concrete section procedure, so it will be according to the code or standard that you are using in design; in this book we will apply ACI 318.

Except where seismic effects create tensile stresses, top reinforcement in the footing is not necessary if the factored tensile stress at the upper face of the footing does not exceed the flexural strength of structural plain concrete, as follows:

$$f_t' = 0.4167\varphi(f_c')^{1/2} \tag{3.5}$$

where

f_t' = flexural strength of structural plain concrete, MPa

f_c' = compressive strength of concrete, MPa

φ = strength reduction factor for structural plain concrete = 0.55

The actual thickness (t) that is specified in the drawing should be higher than the calculated thickness of the footing for tensile stress calculations by about 50 mm to allow for unevenness of excavation and contamination of the concrete adjacent to soil.

For footings cast against a plain concrete, the actual thickness of the footing may be used for the calculated depth. If the factored tensile stress exceeds the flexural strength of structural plain concrete, top reinforcement should be used if an increase in the footing thickness is not feasible.

The following formulas are used for calculating the required footing thicknesses with no top reinforcing steel:

For footings cast against soil:

$$t = t_c + 50 \text{ mm} \tag{3.6}$$

with t_c calculated as follows:

$$t_c = \left(6M_u/f_t'\right)^{1/2} \tag{3.7}$$

where,

t = required footing thickness with no top reinforcing steel, mm

t_c = calculated footing thickness, mm

M_u = strength design factored moment caused by the weight of soil and concrete acting on a 1-m strip in the footing at the face of the pier, m.t/m', using a load factor of 1.4

$$f_t = \text{flexural strength of structural plain concrete, MPa} \tag{3.8}$$

Table 3.5 Modification factor for different tube length

Tube length (m)	Modification factor
3.0	0.70
3.6	0.80
4.2	0.90
4.8	1.00
6.0	1.20

If tensile stress in the upper face of the footing exceeds ACI plain concrete design requirements, top steel should be used if increasing the footing thickness is unfeasible. If top reinforcement is required, minimum reinforcement is 12 mm diameter every 300 mm.

For the weight of bundles with other lengths (Table 3.5), multiply by the following factors:

3.7 EXAMPLE OF HEAT EXCHANGER DATA

Design data

Empty weight = 15 tons
Operating weight = 22 tons
Bundle weight = 8 tons
Channel weight = 2 tons
Basic diameter = 1.2 m
Max. design temperature = 550°F
Exchanger material: Carbon steel
Bolts: 2 1-1/4-in. diameter *ASTM F1554*, Grade 36 (galvanized) per pier
Bolt spacing = 80 mm
Saddle: 1000 mm by 300 mm
Assume load distribution: 60% at channel end, 40% at shell end

The heat exchanger dimensions and the foundations dimensions is presented in Figs. 3.4 and 3.5 respectively.

Design Criteria

Concrete: $f_c' = 40$ MPa
Reinforcing: $f_y = 248$ Mpa
Soil unit weight: $\gamma = 2.0$ tons/m^3
Allowable net soil bearing pressure at 1.2 m depth: $q_{all} = 15$ tons/m^2
Wind load: *ASCE/SEI 7-05*
Earthquake load: *ASCE/SEI 7-05*

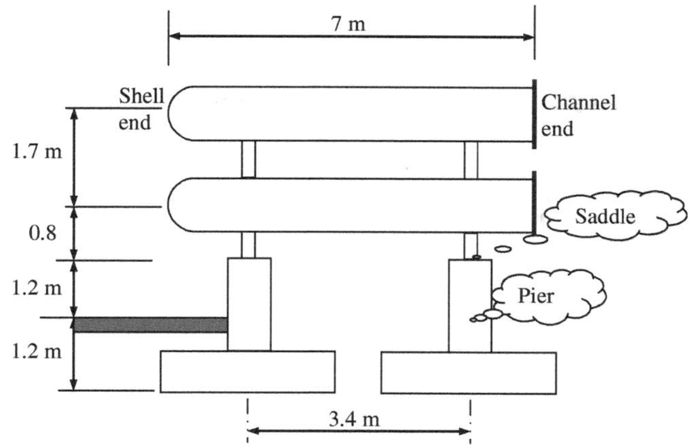

Figure 3.4 Sketch of heat exchanger example.

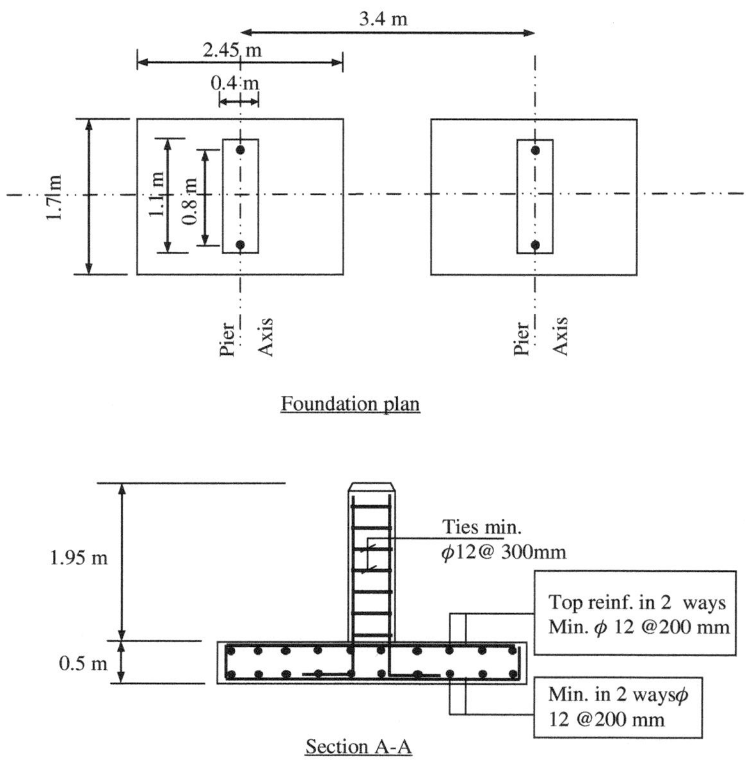

Figure 3.5 Sketch of foundation plan and sections.

Solution:

(A) Load calculation

Empty and Operating Loads

Exchanger weight supplied by outside manufacturers does not include the weight of attached accessories such as pipes and insulation. It is recommended to increase the exchanger weight by 10% of the larger exchanger to account for these attached items as the basic diameter is larger than 600 mm.

Empty dead load (D_e) = 15 + 15 (1.1) = 31.5 tons

Operating dead load (D_o) = 21 +21 (1.1) = 44.1 tons

	40% at shell/fixed end	60% at channel/sliding end
Empty dead load	12.6 tons	18.9 tons
Operating dead load	17.64 tons	26.46 tons

Transverse Moment From Pipe Eccentricity

Eccentricity = (basic diameter)/2 + (0.45 m) = (1.2)/2 + 0.45 = 1.05 m

Empty M_{Te} (channel end) = 15 (0.1) (0.6 channel end)(1.05) = 0.945 m.t

Empty M_{Te} (shell end) = (15)(0.1)(0.4 shell end)(1.05) = 0.63 m.t

Operating M_{To} (channel end) = (22)(0.1) (0.6 channel end)(1.05) = 1.39 m.t

Operating M_{To} (shell end) = (22)(0.1)(0.4 shell end)(1.05) = 0.92 m.t

Wind Loads

Wind load will be calculated based on the load calculation in Chapter 2 and this data will be obtained from the vendor data.

The lateral force due to wind load will affect the exchanger at the center of each exchanger.

Transverse wind: H_w = 0.58 ton (per exchanger)

Longitudinal wind: H_w = 0.11 ton (per exchanger)

Transverse or longitudinal wind on each pier: H_w = 1.87 kN/m^2

Earthquake Loads

The lateral force due to earthquakes will also be calculated based on the methodology described in Chapter 2.

The strength design load will be as depicted in the following table:

	Empty (per exchanger)	Operating (per exchanger)	Pier
Transverse	2.5 tons	3.4 tons	0.154 W*
Longitudinal	4 tons	5.5 tons	0.250 W

*W, weight of the pier.

For calculations based on allowable stress design (service loads), the strength design loads shown in the preceding table should be converted to service loads by multiplying by 0.7.

	Empty (per exchanger)	Operating (per exchanger)	Pier
Transverse	1.72 tons	2.37 tons	0.108 W
Longitudinal	2.8 tons	3.84 tons	0.175 W

Bundle Pull

The bundle weight which is applied at the centerline of the top exchanger will be as follows:

$$V_B = 8.62 \text{ ton}$$

Note that a reduction in the empty load of the exchanger due to the removal of the exchanger head (channel) to pull the bundle is not included in this foundation calculation because the reduction in the empty load is not considered to have a significant effect on the design.

Thermal Force

1. Compute sliding force (assume that a steel slide plate is used):
   Then the coefficient of friction:
   $\mu = 0.40$
   Operating load:
   $F_f = \mu\ (P_o) = 0.4\ (26.5) = 10.6$ tons (at channel end)
2. Compute force required to deflect pier:
   Assume pier is 1.05 m long by 0.4 m wide by 1.95 m height.
   Moment of inertia:
   $I = b(h)^3/12 = (1.05)(0.4)^3/12 = 0.0056\ \text{m}^4$

Based on *ACI 318-05*, Section 8.5.1., the modulus of elasticity will be calculated.

$$E_C = 4700\sqrt{f'_c}\ \text{Mpa}$$

$E_c = 4700\ (f'_c)^{0.5} = 4700\ (40)^{0.5} = 29725$ MPa

Using Table 6.1 or the vessel data sheet as a guideline, we can obtain the thermal expansion coefficient.

Thermal expansion coefficient for carbon steel at 550°F:

$\varepsilon = 3.4$ mm/m

Thermal growth between saddles:

$\Delta = (\varepsilon)(L) = (3.4 \text{ mm/m})(3.35) = 11.48$ mm

$T = 3\ \Delta\ E\ I/2\ H^3 = 3\ (11.48/1000)(29725 \times 10^6)(0.0056)/2\ (1.95)^3 =$ 38,6579.8 $N = 38.7$ tons

Because $F_f < T$ and because a lower friction factor will not help the distribution of earthquake and bundle pull loads, use a steel slide plate.

(B) Design element

1. *Size steel slide plate*

   Width = (saddle width) + 2(Δ) + 25 = 300 + 2 (11.48) + 25 = 348 mm. Take it 350 mm

   Length = (saddle length) + 25 mm = 1000 + 25 = 1025 mm

   Check bearing stress (operating and longitudinal earthquake):

   $P_{Eo} = 5.5\ (0.85 + 2.5)/3.4 = 5.5$ tons

   Based on Table 2.26 in Chapter 2 for Load Combination 3.

   $P_u = 1.2\ (P_o) + 1.0\ (P_{Eo})$

   $P_u = 1.2(26.46) + 1.0(5.5) = 37.3$ tons

   Based on *ACI 318-08*, Section 10.17.

   $P_n = \varphi\ 0.85\ f_c'\ A1 = (0.65)(0.85)(40 \text{ MPa})(350 \text{ mm})(1025 \text{ mm}) =$ 7928357 N = 792.8 tons $> P_u$

   - Choose a steel slide plate that is 1025 mm by 350 mm, with 10 mm thickness

2. *Pier size*
   - Pier length
     1. (c/c bolts) + (2)(125 mm minimum anchor bolt edge distance) = (800) + 2 (125 mm) = 1050 mm
     2. (steel slide plate length) + (100 mm) = (1025) + (100) = 1125 mm

     So, choose pier length not less than 1150 mm
   - Pier width

     Choose the maximum from:

     a. 250 mm

     b. 10% of pier height = (0.10)(1.95 mm) = 0.195 mm (assumed pier height is assumed)

c. (2)(125 mmminimum anchor bolt edge distance) = 2 (125 mm) = 250 mm

- (steel slide plate width) + (100 mm) = (350) + (100) = 450 mm ← controls

The wooden form should be in increments of 50 mm, therefore the pier size will be:

1.15 m length by 0.45 m width

For the anchor bolt design in this example, assume using two 25 mm diameter, *ASTM F1554*, Grade 36 anchor bolts per pier.

3. *Pier design*

Assume the footing thickness is equal to 0.5 m:

Pier height = 2.45 − 0.5 = 1.95 m

Noting that the pier height information is constrained by the piping that is attached to the vessel.

Pier weight = 2.5(0.45) (1.15) (1.95) = 2.5 tons

1. *Operating and longitudinal earthquake at fixed end*

Apply 70% of exchanger earthquake loads at fixed end.

Horizontal load at fixed end:

$V_{FX} = 1.0\ [(0.7)(5.5)(2\ \text{exchangers}) + (0.25)(2.5)] = 7.7 + 0.63 = 8.33$ tons

(based on Chapter 2, Table 2.26, Load Combination 3)

Therefore, shear and moment at bottom of pier will be:

$V_{FX} = 8.33$ tons

$M_{FX} = 7.7\ (1.95)\ + 1.14\ (1.95/2) = 16.13$ m.t

2. *Empty and bundle pull at fixed end*

Bundle pull force,

$V_B = 1.6(8.6) = 13.76$ tons [Chapter 2, Table 2.26, Load Combination 9 or 10]

Vertical load at top of pier due to bundle pull on top exchanger,

$P_B = 13.76\ (0.85 + 1.68)/3.4 = 10.24$ tons

Net vertical load on sliding pier pushing top bundle in use (0.9D) for load factor,

Chapter 2, Table 2.26, Load Combination 10,

$P_{SL} = (0.9)(18.9) - (10.24) = 6.77$ tons

Horizontal load at sliding end:

$V_{SL} = \mu\ (P_{SL}) = 0.40\ (6.33) = 2.53$ tons $<$ 1/2 bundle pull force (V_B)

Horizontal load at fixed end:

$V_{FX} = (V_B) - (V_{SL}) = (13.76) - (2.53) = 11.23$ tons

Shear and moment at bottom of pier:

$V_{FX} = 11.23$ tons

$M_{FX} = (11.23)(1.95 \text{ m}) = 21.9$ m.t

3. *Operating and thermal at fixed end*

a. Thermal force

$V_T = 1.4(10.6 \text{ kips}) = 14.84$ tons

$V_T =$ horizontal force due to thermal effect.

Horizontal load at fixed end:

$V_{FX} = V_T = 14.84$ tons

Shear and moment at bottom of pier:

$V_{FX} = 14.84$ tons

$M_{FX} = (14.84)(1.95) = 28.94$ m.t

Therefore, the pier will be designed on:

Shear force, $V = 14.84$ tons

Bending moment, $M = 28.94$ m.t

As this is the maximum load case, proceed with the design based on the applicable code for the project; a guide to the design by ACI is provided later in this chapter.

Check diagonal tension shear:

$d = 450 - (50 \text{ clear}) - (12 \text{ ties}) - (16 \text{ bar thickness}) = 372$ mm

Based on *ACI 318-08M:*

$\varphi\ V_c = \varphi\ (f_c')^{0.5}\ b_w\ d/6$

$= 0.75\ (40)^{0.5}\ (372)(1100)/6/10000 = 32.35 \text{ tons} > V = 14.1$ tons

$0.5\ \varphi\ V_c = (0.5)(32.35) = 16.2 \text{ tons} < V = 14.1$ tons

Minimum tie requirements from practical point of view is 12 mm ties at 300 mm spacing.

However, because $V > 0.5\ \varphi\ V_c$, spacing requirement should be checked for ϕ12 mm ties to meet minimum shear reinforcement requirements of *ACI 318-08M*, Section 11.5.5:

$A_v = (1/16)\ (f_c')^{0.5}\ b_w\ s/f_y$ but not less than $0.33\ b_w\ s/f_y$

Where b_w, and s in mm

$s = A_v f_y/(1/16)(f_c')^{0.5} b_w = (113)(2)(248)/(1/16)\ (40)^{0.5}\ (1100)$

$= 128$ mm but not more than $A_v f_y/0.33\ b_w$

$= (113)(2)(248)/(0.33)(1100) = 154$ mm

Choose ϕ12 mm ties at 15 mm spacing.

Design for moment:

After designing the section by ACI, choose 6ϕ 25 mm at each face.

4. *Footing size*

Determine minimum footing thickness to develop standard hook for ϕ 25 mm and by applying the development length equation, the foot by thickness will not be less than 45 mm pier reinforcing:

$q_{all} = (15\ t/m^2\ net)$.

- Try a 2.8 m by 2.1 m footing, 0.6 m thickness.

Total pier weights, $P_s = 2.5\ (1.15)(0.45)\ (1.95) = 2.52$ tons

Soil-bearing and stability ratio checks:

Use load combinations for allowable stress design (service loads) from Chapter 2, Table 2.25

1. *Check operating and thermal and eccentric (channel/sliding end)*:

Thermal force at top of pier,

$V_T = 10.08$ tons

Maximum axial load at bottom of footing,

$$P_{max} = P_s + P_o = (2.52) + (26.46) = 29\ \text{ton} \tag{3.9}$$

(Chapter 2, Table 2.25, Load Combination 1)

Moments at bottom of footing:

$M_L = (10.08)(2.45) = 24.2\ 24.7$ m.t

M_{To} (from pipe eccentricity) = 1.39 m.t

Check soil bearing using maximum axial load:

$P = 29$ tons

$M_L = 24.7$ m.t

$M_{to} = 1.39$ m.t

$$q = \frac{29}{(2.8)(2.1)} + \frac{6(24.7)}{2.1(2.8)^2} + \frac{6(1.39)}{2.8(2.1)^2} = 14.6\ t/m^2$$

$q < q_{all}$

2. *Check operating and longitudinal earthquake and eccentric (shell/fixed end)*:

Longitudinal operating earthquake load on exchangers:

$0.7\ V_{LEo} = 3.84$ tons applied at the center of each exchanger

Vertical load at top of piers from longitudinal operating earthquake load on exchangers

(owing to overturning moment),

$0.7\ P_{Eo} = (3.84)(0.85 + 2.5)/(3.4) = \pm 3.78$ tons

Axial loads at bottom of footing:

$P_{max} = P_s + P_o + 0.7 \quad P_{Eo} = (2.52) + (17.64) + (3.78) = 23.94$ tons

(Chapter 2, Table 2.25, Load Combination 3)

$P_{min} = 0.9\ (P_s + P_o) - 0.7\ P_{Eo} = (0.9)(2.52 + 17.64) - (3.78) = 14.36$ tons

(Chapter 2, Table 2.25, Load Combination 5a)

Moments at bottom of footing:

$M_L = (0.7$ at fixed end$)(3.78)(2$ exchangers$)(2.45) + (0.175)(2.15)(1.95/2 + 0.5)$

$= 13.52$ m.t

M_{To} (from pipe eccentricity) $= 0.78$ m.t

Check the soil-bearing capacity:

$q = \frac{23.94}{(2.8)(2.1)} + \frac{6(13.52)}{2.1(2.8)^2} + \frac{6(0.78)}{2.8(2.1)^2} = 9.36\ \text{t/m}^2$

$q < q_{all}$

Stability ratio check using minimum axial load:

Overturning moment $= M_{OT} = [M_L + (0.7P_{Eo})(a/2)] = [(13.52) + (3.84)(2.45)/2] = 18.22$ m.t

Resisting moment $= M_R = 0.9(P_s + P_o)(a/2) = (0.9)(2.52 + 17.64)(2.8)/2 = 25.4$ m.t

Stability ratio $= M_R/M_{OT} = (25.4)/(18.22) = 1.39 > 1.0$

3. *Check operating and longitudinal earthquake and eccentric (channel/sliding end)*:

Longitudinal operating earthquake load on exchangers:

$0.7\ V_{LEo} = 3.84$ applied at the center of each exchanger

Vertical load at top of piers from longitudinal operating earthquake load on exchangers

(owing to overturning moment):

$0.7\ P_{Eo} = (3.84)(0.85 + 2.5)/(3.4) = \pm 3.78$ tons

Axial loads at bottom of footing:

$P_{max} = P_s + P_o + 0.7 \quad P_{Eo} = (2.52) + (26.46) + (3.78) = 32.76$ tons

(Chapter 2, Table 2.25, Load Combination 3)

$P_{min} = 0.9\ (P_s + P_o) - 0.7\ P_{Eo} = (0.9)(2.52 + 26.46) - (3.78) = 22.3$ tons

(Chapter 2, Table 2.25, Load Combination 5a)

Moments at bottom of footing:

$M_L = (0.3$ at sliding end$)\ (3.84)\ (2$ exchangers$)\ (2.45)$

$+ (0.175)(2.15)(1.95/2 + 0.5) = 3.4$ m.t

M_{To} (from pipe eccentricity) = 1.18 m.t

$q = \frac{22.3}{(2.8)(2.1)} + \frac{6(3.4)}{2.1(2.8)^2} + \frac{6(1.18)}{2.8(2.1)^2} = 5.59\ t/m^2$

$q < q_{all}$

Stability ratio check using minimum axial load:

Overturning moment $= (M_{TO}) = [M_L + (0.7P_{Eo})(a/2)] = [(3.4) + (3.84)(2.45)/2] = 8.1$ m.t

Resisting moment $= (M_R) = 0.9(P_s + P_o)(a/2) = (0.9)(2.52 + 26.46)(2.8)/2 = 36.51$ m.t

Stability ratio $= M_R/M_{TO} = (36.51)/(8.1) = 4.5 > 1.0$

4. *Check stability for case: empty and longitudinal earthquake and eccentric (channel/sliding end) loads*:

Longitudinal empty earthquake load on exchangers:

$0.7\ V_{LEe} = 2.8$ tons applied at the center of each exchanger

Vertical load at top of piers from longitudinal empty earthquake load on exchangers (owing to overturning moment):

$0.7\ P_{Ee} = (2.8)(0.85 + 2.5)/(3.4) = \pm 2.8$ tons

Minimum axial load at bottom of footing:

$P_{min} = 0.9\ (P_s + P_e) - 0.7\ P_{Ee} = (0.9)(2.15 + 18.27) - (2.8) = 15.6$ tons

(Chapter 2, Table 2.25, Load Combination 5b)

Longitudinal moment at bottom of footing:

$M_L = (0.3$ at sliding end$)(2.8)(2$ exchangers$)(2.45)$
$+ (0.175)(2.15$ wt$)(1.95/2 + 0.45) = 4.65$ m.t

Stability ratio check using minimum axial load:

Overturning moment $(M_{TO}) = [M_L + (0.7P_{Ee})(a/2)] = [(4.65) + (2.8)(2.45)/2] = 8.08$ m.t

Resisting moment$(M_R) = 0.9(P_s + P_e)(a/2) = (0.9)(2.15 + 18.27)(2.45)/2 = 22.5$ m.t

Stability ratio $= M_R/M_{TO} = (22.5)/(8.08) = 2.78 > 1.0$

5. *Check stability for case: empty and longitudinal earthquake and eccentric (shell/fixed end)*:

Longitudinal operating earthquake load on exchangers:

$0.7\ V_{LEe} = 2.8$ tons applied at the center of each exchanger

Vertical load at top of piers from longitudinal empty earthquake load on exchangers (owing to overturning moment):

$0.7\ P_{Ee} = (2.8)(0.85 + 2.5)/(3.4) = \pm 2.8$ tons

Minimum axial load at bottom of footing:

$P_{min} = 0.9\ (P_s + P_e) - 0.7\ P_{Ee} = (0.9)(2.15 + 12.18) - (2.8) = 10.1$ tons

(Chapter 2, Table 2.25, Load Combination 5b)
Longitudinal moment at bottom of footing:
$M_L = (0.7 \text{ at fixed end})(2.8)(2 \text{ exchangers})(2.45) + (0.175)(2.15)(1.95/2 + 0.45 \text{ ft}) = 10.14$ m.t
Stability ratio check using minimum axial load:
Overturning moment $(M_{TO})=[M_L+(0.7P_{Ee})(a/2)]=[(10.14)+(2.8)(2.45)/2] = 13.57$ m.t
Resisting moment $(M_R) = 0.9(P_s + P_e)(a/2) = (0.9)(2.15 + 12.18)(2.45)/2 = 15.8$ tons
Stability ratio $= M_R/M_{TO} = (15.8)/(13.57) = 1.16 > 1.0$

6. *Check operating and transverse earthquake and eccentric (channel/sliding end)*:
Transverse operating earthquake load on exchangers:
$0.7\ V_{TEo} = 2.37$ tons applied at the center of each exchanger
Axial loads at bottom of footing:
$P_{max} = P_s + P_o = (2.15) + (25.2) = 27.35$ tons
(Chapter 3, Table 3.19, Load Combination 3)
$P_{min} = 0.9\ (P_s + P_o) = (0.9)(2.15 + 25.2) = 24.62$ tons
(Chapter 2, Table 2.26, Load Combination 5a)
Transverse moment at bottom of footing:
$M_T = (2.37)\ [(0.85\ +2.45) + (1.75 + 0.85\ +2.45)]\ (0.6 \text{ channel end}) + (0.108)(2.15)(1.95/2 + 0.45) + (1.17 \text{ m.t pipe eccentricity}) = 13.23$ m.t
Soil-bearing check using maximum axial load:
$e = M_T/P_{max} = (13.23)/(27.3) = 0.48 \text{ m} > b/6 = (5.5)/6 = 0.283$ m
$q_{max} = 2\ P_{max}/[3\ a\ (b/2 - e)] = 2\ (27.35)/[3\ (2.45)\ (1.7/2 - 0.48)] = 20.11\ \text{t/m}^2$
Stability ratio check using minimum axial load:
$M_R = (P_{min})\ (b/2) = (24.62)(1.7/2) = 20.9$ tons
Stability ratio $= M_R/M_{TO} = 20.9/13.23 = 1.58 > 1.0$

7. *Check operating and transverse earthquake and eccentric (shell/fixed end)*:
Transverse operating earthquake load on exchangers:
$0.7\ V_{TEo} = 2.37$ tons applied at the center of each exchanger
Axial loads at bottom of footing:
$P_{max} = P_s + P_o = (2.15) + (16.8) = 18.95$ tons
(Chapter 2, Table 2.26, Load Combination 3)
$P_{min} = 0.9\ (P_s + P_o) = (0.9)(18.95) = 17.1$ tons

The load will be 0.4 at the shell so the overturning moment M_T will be less than the above case, so this case will be safe.

$M_R = (P_{min})\ (b/2) = (17.1)(1.7/2) = 14.54$ m.t

8. *Check stability for case: empty and transverse earthquake and eccentric (channel/sliding end)*:

Transverse empty earthquake load on exchangers:

$0.7\ V_{TEe} = 1.72$ tons applied at the center of each exchanger

Minimum axial load at bottom of footing:

$P_{min} = 0.9\ (P_s + P_e) = (0.9)(2.15 + 18.27) = 18.4$ tons

(Chapter 2, Table 2.25, Load Combination 5a)

Transverse moment at bottom of footing:

$M_T = (3.79)[(0.85\ + 2.45) + (1.7 + 0.85 + 2.45)]$ (0.6 channel end)

$+ (0.108)(2.15)(1.95 + 0.45) + (1.41$ pipe eccentricity$) = 10.31$ m.t

Stability ratio check using minimum axial load:

$M_R = (P_{min})\ (b/2) = (18.4)(1.65/2) = 15.2$ tons

Stability ratio $= M_R/M_{TO} = 15.2/10.31 = 1.47 > 1.0$

9. *Check stability for case: empty and transverse earthquake and eccentric (shell/fixed end)*:

The load will be 0.4 at the shell so the previous case is covered in this case.

10. *Check empty and bundle pull and eccentric (channel/sliding end; pulling top bundle out)*:

Vertical load from bundle pull on top exchanger:

$P_B = (8.62)(0.80 + 1.7)/(3.4) = 6.34$ tons

Vertical load on sliding end at top of pier:

$P_{SL} = P_e + P_B = (18.9) + (6.34) = 25.24$ tons

Horizontal load at sliding end:

$V_{SL} = \mu\ (P_{SL}) = (0.4)(25.24) = 10.1$ tons

Note that the horizontal load on the sliding end computed on the basis of friction is greater than half of the total bundle pull (8.62 tons). Therefore, because the two pedestals and footings are equal in size and thus even in stiffness, the actual horizontal load will be the same on both pedestals.

$V_{SL} = V_{FX} = 8.62/2$ piers $= 4.31$ tons

Maximum axial load at bottom of footing:

$P_{max} = P_s + P_{SL} = (2.52) + (25.24) = 27.76$ tons

(Load Combination 8)

Moments at bottom of footing:
$M_L = (V_{SL})(8.0) = (4.31)(2.45) = 10.6$ m.t
M_{Te} (from pipe eccentricity) $= 1.41$ m.t
Soil-bearing check using maximum axial load:

$$q_{max} = \frac{26.9}{2.8 \times 2.1} + \frac{6(10.6)}{2.1(2.8)^2} + \frac{6(1.41)}{2.8(2.1)^2} = 9.1 \ \ t/m^2$$

$q_{max} < q_{all} = 15$ t/m^2

11. *Check stability for case: empty and bundle pull and eccentric (shell/fixed end; pulling top bundle out)*:

Vertical load from bundle pull on top exchanger:
$P_B = (8.6)(0.8 + 1.7)/(3.4) = 6.32$ tons
Vertical load on fixed end at top of pier:
$P_{FX} = P_e - P_B = (12.6) - (6.32) = 6.28$ tons
Horizontal load at fixed end:
$V_{FX} = V_{SL} = 8.62/2$ piers $= 4.31$ tons

Minimum axial load at bottom of footing, in this case as a critical in overturning we add the weight of the foundation to the pedestal weight:

$P_{min} = P_s + P_{FX} = (12.56) + (6.28) = 18.84$ tons
(Chapter 2, Table 2.25) Load Combination 8)
Moments at bottom of footing:
$M_L = (V_{FX})(2.45) = (4.31)(2.45) = 10.6$ m.t
M_{Te} (from pipe eccentricity) $= 0.575$ m.t
Stability ratio check using minimum axial load:
$M_R = P_{min}\ (a/2) = (18.84)(2.8/2) = 26.38$ m.t
Stability ratio $= M_R/M_{TO} = 26.38/11.17 = 2.36 > 1.0$

The footing dimensions will be 2.45 m by 1.7 m with thickness 0.5 m

The footing dimensions will be 2.8 m by 2.1 m with thickness 0.5 m

5. *Footing Design*

Use the load combinations and strength design load factors from Chapter 2, Table 2.26.

From the previous analysis for foot sizing the maximum stresses on the soil are from the case of operating and thermal and eccentric (channel/sliding end):

Load factors are from Chapter 2, Table 2.26, Load Combination 1.
Thermal force at top of pier:
$V_{uThermal} = 1.4\ (V_T) = 1.4\ (10.1) = 14.14$ tons

Axial load at bottom of footing:

$P_u = 1.4\ (P_s + P_o) = 1.4\ (2.52 + 26.46) = 40.57$ tons

Moments at bottom of footing:

$M_{uL} = (V_T)(2.45) = (14.14)(2.45) = 34.64$ m.t

M_{uTo} (from pipe eccentricity) $= 1.4\ (M_{To}) = 1.66$ m.t It is small with respect to the moment at the bottom.

Maximum factored soil bearing:

$e_{u1} = M_{uL}/P_u = (34.64)/(40.57) = 0.85\ \text{m} > a/6 = (2.45)/6 = 0.4\ \text{m}$

Because the transverse eccentricity is very small, it can be ignored in calculations of factored soil bearing for the design of footing reinforcing.

$S_{max} = 2\ (P_u)/(3b)(a/2 - e_{u1}) = (2)(40.57)/(3)(2.1)[(2.8)/2 - (0.85)] = 23.4\ \text{tons/m}^2$

Bearing length (longitudinal direction) $= 3\ (a/2 - e_{u1}) = 3\ [(2.8)/2 - (0.85)] = 1.65$ m

Factored soil bearing at face of pier (for checking moment) is equal to zero (Fig. 3.6).

The force effecting moment:

$F = ½(1.65)(23.4) = 19.31$ tons

$M = ½(1.65)(23.4).\ 2/3(1.65) = 21.24$ m.t/m'

Check for two-way action (punching shear) according to *ACI 318-08M*, Section 11.12 (at distance "*d*/2" from face of pier):

By inspection, operating weight at channel end will govern as shown in Eq. (3.9).

$P_u = 1.4\ (P_s + P_o) = 1.4\ (27.3) = 38.22$ tons

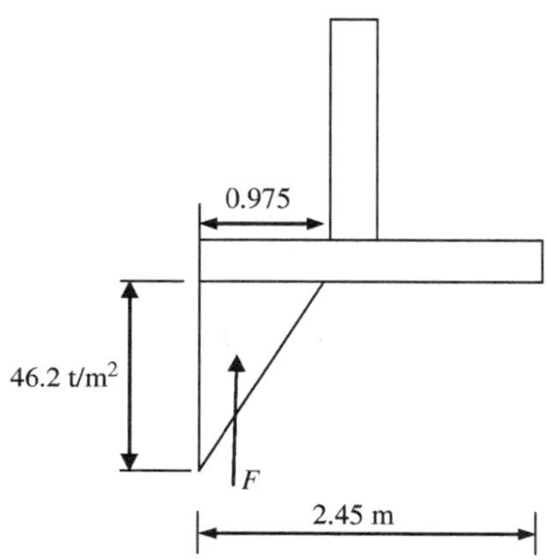

Figure 3.6 Sketch indicating stresses on foundation.

$q = P_u/(ab) = (38.22)/(2.8)\ (1.1) = 6.5\ t/m^2$
$V_u = P_u - (q)\ (1.1 + d)\ (0.4 + d)$
$= 38.22 - (9.18)(1.1 + 0.4)\ (0.4 + 0.4) = 27.2$ tons
$= 38.22 - (6.5)\ (1.1 + 0.4)\ (0.4 + 0.4) = 27.2$ tons

Start designing the foundations as normal procedure for the project code.

3.8 SEPARATOR DESIGN EXAMPLE

Design Data

In this example it is required to design the foundation underneath a horizontal vessel as in Fig. 3.7 with dimensions in Fig. 3.8 and the final foundation drawing sketch is shown in Fig. 3.9.

Figure 3.7 Horizontal vessel (Separator).

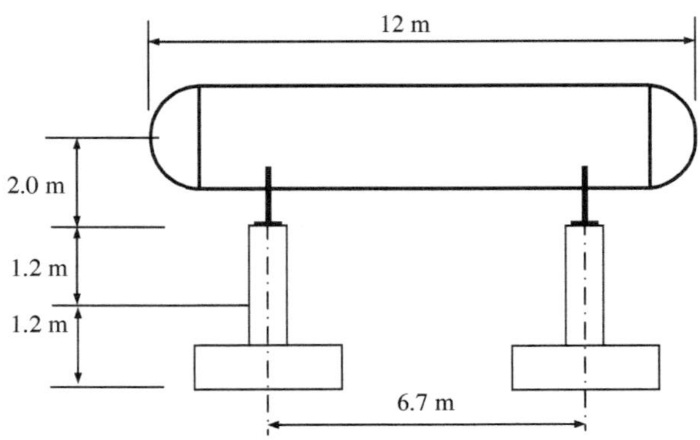

Figure 3.8 Sketch of horizontal separator.

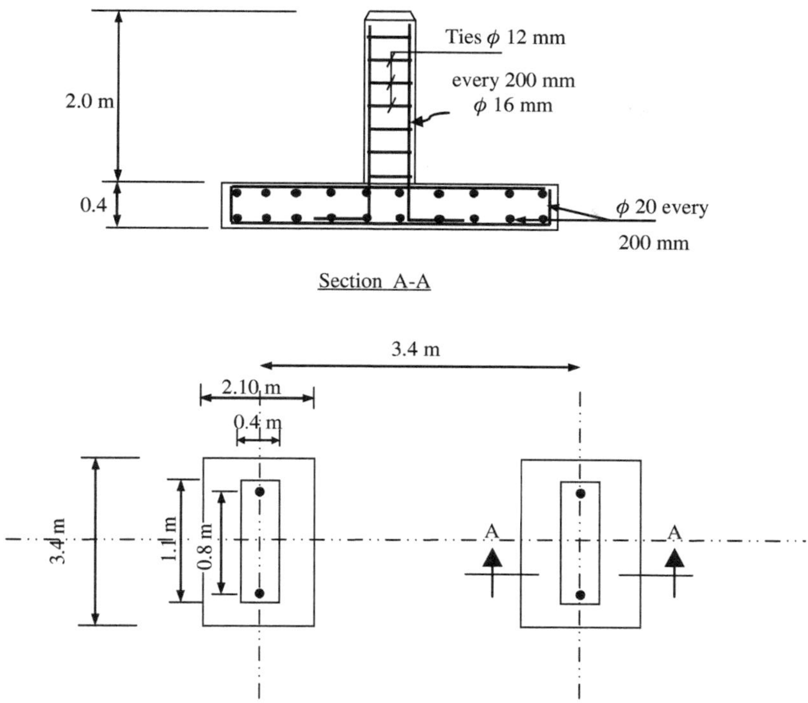

Figure 3.9 Foundation sketch.

Vessel Data:
Empty weight = 50 tons
Operating weight = 140 tons
Test weight = 170 tons
Basic diameter = 4.0 m
Maximum design temperature = 500°F
Vessel material: carbon steel
Bolts: two 25 mm diameter, *ASTM F1554*, Grade 36 (galvanized) per pier
Bolt spacing: 800 mm and the Saddle dimensions are 3.50 m by 0.25 m
Design Criteria:
Concrete: $f_c' = 27.5\ \text{N/mm}^2$
Reinforcing: $f_y = 414\ \text{N/mm}^2$
Soil unit weight: $\gamma = 2.0\ \text{t/m}^3$
At 1.5 m depth, the allowable net soil-bearing capacity: $q_{\text{all}} =$ 0.15 MPa

Wind loads: *ASCE/SEI 7-05*
Earthquake loads: *ASCE/SEI 7-05*
Use a 20% increase in soil allowable pressure for test load combinations.

Solution:

(A) Load calculation

- ***Empty, Operating, and Test Loads***

  Include an additional 10% of the applicable weight (empty, operating, or test) to account for piping supported on the horizontal vessel (refer to this practice, Section 4.2, "Vertical Loads"):

  Total empty load, $D_e = (1.10)(44.4)\ (50) = 55$ tons
  Total operating load, $D_o = (1.10)(152)\ (140) = 154$ tons
  Total test load, $D_t = (1.10)(178.7)\ (170) = 187$ tons

- ***Transverse Moment from Pipe Eccentricity***

  Eccentricity = (basic diameter)/2 + (0.45) = (4.0)/2 + (0.45) = 2.45 m
  Empty transverse moment per pier, $M_{Te} = (50)(0.1)(0.5$ per pier$)(2.45) = 6.13$ m.t
  Operating transverse moment per pier, $M_{To} = (140)(0.1)(0.5$ per pier$)(2.45) = 17.15$ m.t
  Test transverse moment per pier, $M_{Tt} = (170)(0.1)(0.5$ per pier$)(2.45) = 20.83$ m.t

Wind Loads:
Wind load calculations are beyond the scope of this practice.
Vessel wind is applied at the center of the vessel.
Transverse wind: $H_w = 6.5$ tons
Longitudinal wind: $H_w = 1.5$ tons
Transverse or longitudinal wind on each pier: $H_w = 0.45\ t/m^2$
Thermal Force:

1. Compute sliding force (assume that a low-friction manufactured slide plate assembly is used):

   $\mu = 0.10$ (maximum based on the manufacturer's literature)
   Operating load at one pier:
   $P_o = (140)/(2\ \text{piers}) = 70$ tons
   $F_f = \mu\ (P_o) = (0.10)(70) = 7.0$ tons

2. Compute force required to deflect pier:

   Assume pier is 3.7 m long by 0.35 m wide by 2.0 m high:
   Moment of inertia: $I = b\ (h)^3/12 = (3.7)(0.35)^3/12 = 0.01322\ m^4$

Based on *ACI 318-08*, Section 8.5.1

Modulus of elasticity: $E_c = 57{,}000\ (f_c')^{0.5} = 24518.4$ MPa

Thermal expansion coefficient for carbon steel at 500°F: $\varepsilon =$ 1.96 mm/m as obtained from Table 6.1

Thermal growth between saddles: $\Delta = (\varepsilon)(L) = (2.94\ \text{mm/m})(6.7) =$ 19.88 $T = 3\ \Delta\ E\ I\ /\ 2\ H^3 = 3\ (0.0202)(2500000)(0.01322)/2\ (2.0)^3 =$ 125 tons

Because $F_f < T$, and in order to reduce high-friction forces, use a low-friction manufactured slide plate assembly.

(B) Design Element

Size Low-Friction Manufactured Slide Plate Elements

Upper element width = (saddle width) + 25 mm = 250 + 25 = 275 mm = 300 mm

Maximum load on sliding end (from test weight): $P_t = (187)/(2$ piers) = 93.5 tons

Operating load on sliding end: $P_o = (167.2)/(2$ piers) = 83.6 tons

Operating load on sliding end: $P_o = (154)/(2$ piers) = 77 tons

According to the manufacturer's recommendations, we will provide the required number of side plates and the components required for each assembly, as in this example, with seven layers with the lower element length equal to 220 mm with width 65 mm and the upper element with length equal to 280 mm and width 90 mm.

From the vendor specifications, we can obtain bearing around ($\mu = 0.055$) based on the bearing load.

Revised operating frictional force = $F_f = (0.055)(77) = 4.24$ tons

Steel Bearing Plate Dimensions

Steel bearing plate dimensions:

Width = (lower slide plate element width) + (25) = (200) + (25) = 225 mm = 250 mm

Length = (saddle length) + (25 mm) = (3.50) + (0.025) = 3.525 m = 3.55 m

Check bearing stress (test load case):

$P_u = 1.4\ P_t = 1.4\ (187/2$ piers) = 130.9 tons

Based on *ACI 318-09*, Section 10.17

$P_n = \varphi \quad 0.85 f_c' \quad A_1 = (0.65)(0.85)(27.5)(225)(3600)/1000 = 12307$ kN = 1255 tons

- Choose a bearing plate 3.6 m by 0.25 m by 10 mm.

Pier Size

Choose the pier length from the maximum of the following two limits:

1. (c/c anchor bolts) + (2)(125 minimum anchor bolt edge distance) = (3350) + 2 (125) = 3600 mm
2. (bearing plate length) + (100) = (3550) + (100) = 3650 mm
   So choose pier length 3.65 m.
   Choose pier width from the maximum of the following four limits:
1. 250 mm
2. 10% of pier height = (0.10)(2.05) = 0.205 m (based on assumed pier height)
3. (2)(150 mm minimum anchor bolt edge distance) = 2 (150) = 300 mm
4. (bearing plate width) + (100) = (250) + (100) = 350 mm

- Choose a pier size of 0.35 m by 3.65 m.

3.8.3.4 Anchor Bolt Design, Pier Design, and Footing Design

Anchor bolt design, pier design, and footing design are very similar to the previous example.

3.9 VERTICAL VESSEL FOUNDATION DESIGN

The vertical vessel is traditionally used in gas plant as shown in Fig. 3.10.

3.9.1 Dead Loads

The following nominal loads shall be considered as dead loads when applying load factors used in strength design.

A. Structure dead load (D_s)—Vessel's foundation weight, which is defined as the combined weight of footing, pedestal dead load (D_p), and the overburden soil.

B. Erection dead load (D_f)—Fabricated weight of vessel, generally taken from the certified vessel drawing.

C. Empty dead load (D_e)—Empty weight of the vessel, including all attachments, trays, internals, insulation, fireproofing, agitators, piping, ladders, and platforms, generally taken from the certified vessel drawing.

D. Operating dead load (D_o)—Empty dead load of the vessel plus the maximum weight of contents (including packing/catalyst) during normal operation. Operating dead load shall be taken from the certified vessel drawing and data sheet.

Figure 3.10 Photo for vertical vessel.

E. Test dead load (D_t)—Empty dead load of the vessel plus the weight of test medium contained in the system. The test medium shall be as specified in the contract documents.

Unless otherwise specified, a minimum specific gravity of 1.0 shall be used for the test medium. Cleaning load shall be used for the test dead load if cleaning fluid is heavier than the test medium. Whether to test or clean in the field should be determined. In design, the test dead load is generally desirable because unforeseen circumstances may occur.

Eccentric vessel loads caused by large pipes or reboilers shall be considered for the applicable load cases.

3.9.1.1 Live Loads

Live loads (L) shall be calculated as per Chapter 2 to be considered as a guideline if the data from the client is not available. In general, the live load has a minor effect with respect to the dead load of the vessel so it will have less influence in the design of the vessel horizontally or vertically.

3.9.1.2 Wind Loads

Wind loads (W) shall be calculated in accordance with the requirements of the client based on the site location, "Meteorological and Seismic Design Data," "Structural Design Criteria for Non-Building Structures," and the guidelines of ASCE "Wind Load and Anchor Bolt Design for Buildings and Other Structures."

The engineer is responsible for determining the wind loads used for the foundation design.

Partial wind load (W_p) shall be based on the requirements of ASCE 37-02, Section 6.2.1, for the specified test or erection duration. The design wind speed shall be *75%* of the actual wind speed.

When calculating or checking wind loads, due consideration shall be given to factors which may significantly affect total wind loads, such as the application of dynamic gust factors or the presence of spoilers, platforms, ladders, piping, etc., on the vessel.

If detailed information (number of platforms, platform size, etc.) is unknown at the time of foundation design, the following simplified method may be used:

a. For the projected width, add 5 ft (1.52 m) to the diameter of the vessel, or add 3 ft (0.91 m) plus the diameter of the largest pipe to the diameter of the vessel, whichever is greater. This will account for platforms, ladders, nozzles, and piping below the top tangent line.

b. The vessel height should be increased one (1) vessel diameter to account for a large diameter and platform attached above the top tangent, as it is common with most tower arrangements.

c. The increases in vessel height or diameter to account for wind on appurtenances should not be used in calculating the h/D ratio for force coefficients or flexibility.

d. The force coefficient (C_f) should be determined as discussed in Chapter 2.

e. If most design detail items such as platforms, piping, ladders, and other accessories of the vessel are known, the detailed method of the guidelines of the ASCE standard "Wind Load and Anchor Bolt Design for Buildings and Other Structures" shall be used.

3.9.1.3 Earthquake Loads(E)

Seismic forces shall be calculated in accordance with ASCE or UBC according to the basis of design and the seismic zone will be defined by the owner based on the site location or from the meteorological and seismic design data.

The seismic load is usually calculated by the vessel vendor. This data should be transferred to the structural engineer to include in designing the foundation.

For skirt-supported vertical vessels classified as SUG III in accordance with ASCE 7-02, Section 9, use the critical earthquake provisions and implied load combinations of ASCE 7-02, as illustrated in Chapter 2.

3.9.1.4 Other Loading

Thrust forces caused by thermal expansion of piping shall be included in the operating load combinations, if deemed advisable. *Dead load* factors shall be applied to the resultants of piping thermal loadings. The piping stress engineer shall be consulted for any thermal loads that are to be considered.

Consideration shall be given to operation mode conditions that could occur and could increase loading on the foundation.

3.9.1.5 Load Combinations

General

The structure, equipment, and foundations shall be designed for the appropriate load combinations from ASCE 7 and any other probable realistic combination of loads. The load combination for vertical vessels is presented in Chapter 2.

Table 2.17 is a load combination guide in the case of designs by the allowable stress design, and Table 2.18 is a load combination guide to be used in the case of designs by the strength design method or load resistance factor designs.

3.9.2 Pedestal Design

The concrete pedestal dimensions shall be sized on the available wooden or steel form on site so it is usually taken octagon pedestal dimensions shall be sized with pedestal faces in an incremental to 50 mm.

The following criteria shall be used to determine the size and shape for the pedestal.

Face-to-face pedestal size shall not be less than the largest of the following:

If you will not use a sleeve choose the largest from:

$$D_B + 250\ \text{mm} \tag{3.10}$$

$$D_B + 12(d)(\text{for high-strength anchor bolts}) \tag{3.11}$$

$$D_B + 8(d)(\text{for anchor bolts grade 36 or A307}) \tag{3.12}$$

$$D_B + d_s + 250 - d \tag{3.13}$$

In the case of using a sleeve:

$$D_B + d_s + 11(d)(\text{for high-strength anchor bolts}) \tag{3.14}$$

$$D_B + d_s + 7(d)(\text{for anchor bolts Grade 36 or A307}) \tag{3.15}$$

where:

D_B = diameter for the circle of bolts (mm)

d = bolt diameter (mm)

d_s = sleeve diameter (mm)

Pedestals 1.8 m and larger shall be in an octagonal shape. Dimensions for octagonal pedestals are provided in Table 3.6. The octagon faces have 50 mm increments to be matched with the regular wooden or steel form dimensions.

Table 3.6 Various dimensions for octagonal pedestal

D (m)	B (m)	C (m)	E (m)	Area (m^2)	S_d (m^3)
1.81	0.53	0.75	1.96	2.72	0.60
1.93	0.57	0.80	2.09	3.09	0.73
2.05	0.60	0.85	2.22	3.49	0.87
2.17	0.64	0.90	2.35	3.91	1.04
2.29	0.67	0.95	2.48	4.36	1.22
2.41	0.71	1.00	2.61	4.83	1.42
2.53	0.74	1.05	2.74	5.32	1.65
2.66	0.78	1.10	2.87	5.84	1.89
2.78	0.81	1.15	3.01	6.39	2.16
2.90	0.85	1.20	3.14	6.95	2.46

(*Continued*)

Table 3.6 (Continued)

D (m)	B (m)	C (m)	E (m)	Area (m^2)	S_d (m^3)
3.02	0.88	1.25	3.27	7.54	2.78
3.14	0.92	1.30	3.40	8.16	3.13
3.26	0.95	1.35	3.53	8.80	3.50
3.38	0.99	1.40	3.66	9.46	3.91
3.50	1.03	1.45	3.79	10.15	4.34
3.62	1.06	1.50	3.92	10.86	4.80
3.74	1.10	1.55	4.05	11.60	5.30
3.86	1.13	1.60	4.18	12.36	5.83
3.98	1.17	1.65	4.31	13.15	6.39
4.10	1.20	1.70	4.44	13.95	6.99
4.22	1.24	1.75	4.57	14.79	7.63
4.35	1.27	1.80	4.70	15.64	8.30
4.47	1.31	1.85	4.83	16.53	9.01
4.59	1.34	1.90	4.96	17.43	9.76
4.71	1.38	1.95	5.10	18.36	10.55
4.83	1.41	2.00	5.23	19.31	11.39
4.95	1.45	2.05	5.36	20.29	12.26
5.07	1.48	2.10	5.49	21.29	13.18
5.19	1.52	2.15	5.62	22.32	14.14
5.31	1.56	2.20	5.75	23.37	15.15
5.43	1.59	2.25	5.88	24.44	16.21
5.55	1.63	2.30	6.01	25.54	17.32
5.67	1.66	2.35	6.14	26.66	18.47
5.79	1.70	2.40	6.27	27.81	19.67
5.91	1.73	2.45	6.40	28.98	20.93
6.04	1.77	2.50	6.53	30.18	22.24
6.16	1.80	2.55	6.66	31.40	23.60
6.28	1.84	2.60	6.79	32.64	25.01
6.40	1.87	2.65	6.92	33.91	26.48
6.52	1.91	2.70	7.06	35.20	28.01
6.64	1.94	2.75	7.19	36.51	29.60
6.76	1.98	2.80	7.32	37.85	31.24
6.88	2.02	2.85	7.45	39.22	32.95
7.00	2.05	2.90	7.58	40.61	34.71
7.12	2.09	2.95	7.71	42.02	36.54
7.24	2.12	3.00	7.84	43.46	38.43
7.36	2.16	3.05	7.97	44.92	40.38
7.48	2.19	3.10	8.10	46.40	42.40
7.60	2.23	3.15	8.23	47.91	44.48

(Continued)

Table 3.6 (Continued)

D (m)	B (m)	C (m)	E (m)	Area (m^2)	S_d (m^3)
7.73	2.26	3.20	8.36	49.44	46.63
7.85	2.30	3.25	8.49	51.00	48.86
7.97	2.33	3.30	8.62	52.58	51.14
8.09	2.37	3.35	8.75	54.19	53.50
8.21	2.40	3.40	8.88	55.82	55.94
8.33	2.44	3.45	9.02	57.47	58.44
8.45	2.47	3.50	9.15	59.15	61.02
8.57	2.51	3.55	9.28	60.85	63.67
8.69	2.55	3.60	9.41	62.58	66.40
8.81	2.58	3.65	9.54	64.33	69.21
8.93	2.62	3.70	9.67	66.10	72.09
9.05	2.65	3.75	9.80	67.90	75.05
9.17	2.69	3.80	9.93	69.72	78.09
9.29	2.72	3.85	10.06	71.57	81.22
9.42	2.76	3.90	10.19	73.44	84.42
9.54	2.79	3.95	10.32	75.34	87.71
9.66	2.83	4.00	10.45	77.25	91.08
9.78	2.86	4.05	10.58	79.20	94.54
9.90	2.90	4.10	10.71	81.17	98.09
10.02	2.93	4.15	10.84	83.16	101.72
10.14	2.97	4.20	10.98	85.17	105.44
10.26	3.01	4.25	11.11	87.21	109.25
10.38	3.04	4.30	11.24	89.28	113.15
10.50	3.08	4.35	11.37	91.37	117.15
10.62	3.11	4.40	11.50	93.48	121.23
10.74	3.15	4.45	11.63	95.61	125.41
10.86	3.18	4.50	11.76	97.78	129.69
10.98	3.22	4.55	11.89	99.96	134.06
11.11	3.25	4.60	12.02	102.17	138.53
11.23	3.29	4.65	12.15	104.40	143.09
11.35	3.32	4.70	12.28	106.66	147.76
11.47	3.36	4.75	12.41	108.94	152.52
11.59	3.39	4.80	12.54	111.25	157.39
11.71	3.43	4.85	12.67	113.58	162.36
11.83	3.46	4.90	12.80	115.93	167.44
11.95	3.50	4.95	12.93	118.31	172.61
12.07	3.54	5.00	13.07	120.71	177.90

Pedestals smaller than 1.8 mm shall be square, or round if forms are available.

It is normally desirable to make the pedestal deep enough to contain the anchor bolts and keep them out of the footing. Consideration shall be given to anchor bolt development and foundation depth requirements. Pedestal size may need to be increased to provide adequate projected concrete area for anchor bolts when additional reinforcement for anchor bolts is not used.

Pedestal reinforcement shall be tied to the footing with sufficient dowels around the pedestal perimeter to prevent separation of the pedestal and footing. Development length of reinforcing steel shall be checked.

Dowels shall be sized by computing the maximum tension existing at the pedestal perimeter due to overturning moments.

Conservatively, the following formula may be used to calculate the tension in anchor bolts. More exact tension loads may be obtained by using *ACI 318* strength design or any other standard such as BS 8110; based on project specifications and standards in this book, we will take *ACI 318* equations into consideration.

$$\text{Tension, } F_u = 4(M_{up})/[(N_d)(D_C)] \; - 0.9[(D_e \text{ or } D_o) + W_p]/N_d \quad (3.16)$$

In the case of wind loads, use empty weight, but use empty or operating for earthquake loads depending on which condition is used to calculate M_{uped}.

$$\text{As required} = \text{Tension/design stress} = F_u/\varphi f_y \quad (3.17)$$

where:

F_u = maximum ultimate tension in reinforcing bar

M_{up} = maximum factored overturning moment at base of pedestal, calculated by using load factors in load combinations for uplift cases in Table 2.20

N_d = number of dowels (assumed); shall be a multiple of 8

D_C = dowel circle diameter (assume pedestal size minus 150 mm)

$D_e + D_p$ = empty weight of vessel and pedestal weight

$D_o + D_p$ = operating weight of vessel and pedestal weight

W_p = the pedestal weight

ϕ = strength reduction factor = 0.90

f_y = yield strength of reinforcing steel

The minimum pedestal reinforcing shall be as follows, noting that the ties steel bars will be 10 mm diameter as a minimum at 300 mm maximum spacing:

For top reinforcement, a mat of reinforcing steel at the top of the pedestal shall be provided. It shall be minimum steel with 12 mm diameter at 300 mm maximum spacing across the flats in only two directions.

1. For octagons 1.8–2.5 m: A minimum of 16 bars with diameter 12 mm verticals
2. For octagons larger than 2.5–3.5 m: A minimum of 24 bars with bar diameter 16 mm vertical
3. For octagons larger than 3.5 mm: The minimum bar diameter is equal to 16 mm at 450 mm maximum spacing

3.9.2.1 Anchor Bolts

Conservatively, the maximum tension on an anchor bolt may be determined using the following formula. More exact tension loads may be obtained by using *ACI 318* strength design methodology.

$$N_u = 4M_u/[(N_b)(D_B)] - 0.9(D_e \text{ or } D_o)/N_b \tag{3.18}$$

where

N_u = factored maximum tensile load on an anchor bolt

M_u = factored moment at the base of vessel, calculated using load factors in load combinations for uplift cases as presented in Chapter 2.

N_b = number of anchor bolts

D_B = bolt circle diameter

(D_e or D_o) = nominal empty or operating vessel weight. Use empty weight for wind loads. Use empty or operating for earthquake loads depending on which condition is used to calculate M_u.

For most cases, there is no shear on the anchor bolts because the load is resisted by friction caused primarily by the overturning moment. If friction cannot resist the load, the bolts shall be designed to resist the entire shear load, or other methods may be used to resist the shear load. The friction resistance can be calculated using the following formulas:

$$P_u = M_u/LA + 0.9(D_e \text{ or } D_o) \tag{3.19}$$

$$V_f = \mu P_u \tag{3.20}$$

where

P_u = factored compression force at top of pedestal

LA = lever arm between centroid of tension loads on bolts and centroid of the compression load on the pedestal. This is a complicated distance to precisely determine. A conservative approximation is to use two-third of the bolt circle diameter as the lever arm.

μ = coefficient of friction. For the normal case of grout at the surface of the pedestal, $\mu = 0.55$.

V_f = frictional resisting force (factored)

To have no shear load on the bolts:

$$V_u \leq \varphi V_f \tag{3.21}$$

where

V_u = factored shear load at the base of the vessel, calculated using load factors in load combinations for uplift cases in Table 2.18 for load combinations for strength design.

ϕ = strength reduction factor as per ACI = 0.75

3.9.3 Footing Design

The size of spread footings may be governed by stability requirements, sliding, soil bearing pressure, or settlement.

Footings for vertical vessels shall be octagonal or square and sized based on standard available form sizes. When form information is not available, footing dimensions shall be sized with footing faces in 50 mm increments to allow use of standard manufactured wooden forms. Octagons listed in Table 3.6 are those having faces in 2-in. increments. If extended to the recommended depth specified in the geotechnical report, the pedestal may be adequate without a footing.

Footings smaller than 2.0 m in diameter shall be square.

Where a footing is required, the footing thickness shall be a minimum of 300 mm. The footing thickness shall be adequate to develop pedestal reinforcement and satisfy the shear requirements of *ACI 318*.

The footing thickness shall also be checked for top tension without top reinforcement in accordance with *ACI 318*. If the thickness is not adequate, either a thicker footing or top reinforcing steel is required. Note that increasing the footing thickness is typically more cost effective for construction than adding a top mat of reinforcing steel except where seismic effects create tensile stresses requiring top reinforcement.

For the *first trial*, the diameter of an octagonal footing may be approximated by the following formula:

$$\text{Diameter, } D = 2.6\left(M_{\text{ftg}}/q\right)^{1/3} \tag{3.22}$$

where

M_{ftg} = overturning moment at base of footing, m.t

q_{all} = allowable gross soil bearing, t/m^2

A common assumption in the design of soil-bearing footings is that the footing behaves as a rigid unit. Hence, the soil pressure beneath a footing is assumed to vary linearly when the footing is subjected to axial load and moment. The following footing formulas are based on the linear pressure assumption.

The footings shall be designed so that under operating loads the total settlement and the differential settlement between footings do not exceed the established limits. The maximum allowable amount of total settlement and differential settlement is typically set by the project Structural Engineer based on the sensitivity of the equipment or structure being supported and the connected piping to the equipment. This will be agreed upon by the structure engineer and the piping stress analysis engineer.

3.9.4 Soil Bearing on the Octagon Footing

The octagon shape has its own characteristic as shown in Fig. 3.11, which should be known to calculate the soil stresses beneath this footing.

A = Area (m^2) = 0.8284272D^2

C = Length of side = 0.4142136D

$B = C \times \sin 45^\circ$ = 0.2928932D

E = Length of diameter = 1.0823922D

S_d = Sec. mod. diameter (m^3) = 0.1011422D^3

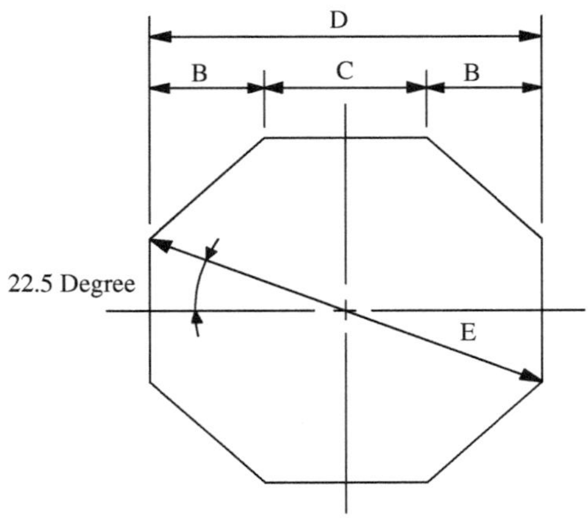

Figure 3.11 Octagon properties and dimensions.

S_f = Sec. mod. flat = S_d (E/D)

I = Moment of inertia = S_d $(E/2)$

Soil-bearing pressure shall be checked for maximum allowable on the diagonal, and footing design shall be computed on the flat.

Where the total octagonal footing area is in compression ($e/D \leq 0.122$ on the diagonal and $e/D \leq 0.132$ on the flat), the soil-bearing pressure shall be computed using the following formulas:

$$q = (P/A) \pm M_{ftg}/S \tag{3.23}$$

$$q(\text{diagonal}) = P/A \pm M_{ftg}/S_{\text{Diagonal}} \tag{3.24}$$

$$q(\text{diagonal}) = P/A[1 \pm (8.19e/D)] \tag{3.25}$$

$$q(\text{flat}) = P/A \pm M_{ftg}/S_{\text{Flat}} \tag{3.26}$$

$$q(\text{flat}) = P/A[1 \pm (7.57e/D)] \tag{3.27}$$

where

D = distance between parallel sides (m)

q = toe pressure (t/m^2)

P = total vertical load including soil and foundation (tons)

A = bearing area of octagonal footing $(0.828\ D^2)$ (m^2)

M_{ftg} = overturning moment at base of footing (m.t)

S = section modulus (m^3)

The section modulus at diagonal; $S_D = 0.1011422D^3$

The section modulus at flat; $S_F = 0.10948\ D^3$

e = eccentricity (M_{ftg}/P) (m)

The total octagonal footing area is not in compression if the following condition are verified:

$e/D > 0.122$ on the diagonal

and

$e/D > 0.132$ on the flat,

The soil-bearing pressure shall be computed using Fig. 3.11 and the following formula:

$$q = L(P/A) \tag{3.28}$$

Where the values of (L) and (K) will be obtained from Table 6.8 corresponding to The e/D ratios for octagon footings (Fig. 3.12; Table 3.7).

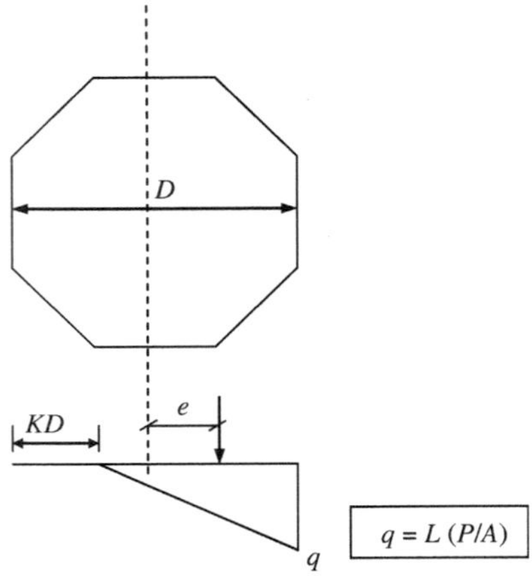

Figure 3.12 Pressure distribution under foundation.

Table 3.7 Foundation pressures for octagon bases for large eccentricities at the Face

e/D	*K*	*L*	*e/D*	*K*	*L*	*e/D*	*K*	*L*
0.15	0.05	2.2	0.28	0.45	4.0	0.41	0.76	12.2
0.16	0.09	2.2	0.29	0.48	4.2	0.42	0.79	14.0
0.17	0.13	2.3	0.30	0.49	4.5	0.43	0.81	16.3
0.18	0.16	2.4	0.31	0.52	4.8	0.44	0.84	19.4
0.19	0.19	2.5	0.32	0.55	5.2	0.45	0.86	23.8
0.20	0.22	2.6	0.33	0.57	5.6	0.46	0.89	30.4
0.21	0.25	2.8	0.34	0.60	6.0	0.47	0.92	41.5
0.22	0.28	2.9	0.35	0.62	6.6	0.48	0.94	63.6
0.23	0.3	3.0	0.36	0.64	7.2	0.49	0.97	130.2
0.24	0.34	3.2	0.37	0.67	7.8	0.50	See Note	See Note
0.25	0.37	0.40	0.38	0.69	8.7	>0.50	See Note	See Note
0.26	0.40	3.6	0.39	0.71	9.6			
0.27	0.42	3.8	0.40	0.74	10.8			

$e/D \geq 0.500$, assume soil bearing as a line load with a length equal to the face dimension of the octagon. Footing applied at a distance of (*e*) from the centerline of the footing.

3.9.5 Check Stability/Sliding

All foundations subject to buoyant forces shall be designed to resist a uniformly distributed uplift equal to the full hydrostatic pressure. The minimum safety factor against flotation shall be *1.20*, considering the highest anticipated water level.

The stability ratio is "The ratio of the resisting moment to overturning moment about the edge of rotation." The minimum stability safety factor against overturning for service loads other than earthquake shall be *1.5*.

Compute the stability factor of safety ratio (*FS*) using the following formula:

$$FS = M_R / M_O \tag{3.29}$$

where

M_R is the resistance moment

M_o is the overturning moment

The moment and loads shall be factored in accordance with load combinations in Table 2.19 for the allowable Stress Design (Service Loads).

Eccentricity $e = \Sigma M_o / \Sigma P$

Resisting Moment $M_R = \Sigma P \times D/2$

D = dimension of footing in the direction of the overturning moment, m

e = eccentricity (m) = overturning moment at the base of the footing divided by the total vertical load, m

So the stability factor of safety can be calculated from the following equation:

$$FS = D/2e$$

3.9.6 Check for Foundation Sliding

The minimum safety factor against sliding for service loads other than earthquakes shall be *1.5*. The coefficient of friction used in computing the safety factor against sliding for cast-in-place foundations shall be *0.40*, unless otherwise specified in the soil investigation report.

The passive earth pressure from backfill shall not be considered in computing these safety factors.

The minimum overturning "stability ratio" and the minimum factor of safety against sliding for earthquake service loads shall be 1.5.

In addition, the minimum overturning stability factor of safety for the anchorage and foundations of skirt-supported vertical vessels shall be 1.2 as per ASCE 7-02 as it classified as SUG III.

3.9.7 Reinforced Concrete Design Guide

For this foundation, the reinforced concrete design shall be done using factored strength design loads as per the project design code and standard. The critical section for moment shall be taken with respect to the face of a square with an area equivalent to that of the pedestal.

The bending moment shall be calculated at the face of the equivalent square. Moment shall be calculated for a 1 m-wide strip as a simple cantilever from the edge of the equivalent square.

The punching shear may need to be checked in some circumstances. The resulting reinforcing steel shall be placed continuously across the entire footing in two layers, with minimum bottom reinforcement 12-mm bars at 200-mm spacing in each direction. The minimum bottom steel cross-sectional area shall be according to the project code and standard as presented in ACI Code Section 10.5.4. This percentage depends on the grade of steel bars.

3.9.7.1 Top Reinforcement

Except where seismic effects create tensile stresses, top reinforcement in the footing is not necessary if the factored tensile stress at the upper face of the footing does not exceed the flexural strength of structural plain concrete.

The effective thickness of the footing for tensile stress calculations should be 50 mm less than the actual thickness for footings cast against soil.

For footings cast against a seal slab, the actual thickness of the footing may be used for the effective thickness. If the factored tensile stress exceeds the flexural strength of structural plain concrete, top reinforcement should be used if an increase in the footing thickness is not feasible.

See the following formulas for footing thicknesses that do not require top reinforcing steel:

For footings cast against soil:

$$t = t_c + 50 \text{ mm} \tag{3.30}$$

For footings cast against a seal slab:

$$t = t_c \tag{3.31}$$

where t_c is calculated as follows:

$$t_c = \left(6M_u/f_t'\right)^{1/2} \tag{3.32}$$

where

t = required footing thickness with no top reinforcing steel, m

t_c = calculated footing thickness, m

M_u = factored moment caused by the weight of soil and concrete acting on 1 m strip in the footing at the face of the equivalent square pedestal, m.t/m', calculated using a load factor of 1.4

f_c' = flexural strength of structural plain concrete, t/m^2

3.9.7.2 Shear Consideration

Both *wide-beam* action and *two-way* action (punching shear) must be checked to determine the required footing depth. Shear, as a measure of diagonal tension, shall be checked at the critical section specified in *ACI 318*, Section 11.1.3.1 (at a distance *d* from the face of the equivalent square).

The shear shall be calculated for a 1-ft-wide strip as a simple cantilever from the edge of the equivalent square. Beam action assumes that the footing acts as a wide beam with a critical section across its entire width.

Punching shear may need to be checked in some situations in accordance with *ACI 318-08*. Two-way action (at a distance *d*/2 from the face of the equivalent square) for the footing checks "punching" shear strength. The critical section for punching shear is a perimeter around the supported member with the shear strength computed in accordance with ACI Code.

For footing design, the depth must be selected, so that shear reinforcement is not required. If either permissible shear is exceeded, the thickness of the footing must be increased. The shear strength equations may be summarized as follows based on *ACI 318-08*:

- For beam action:

$$\begin{aligned} V_u &\le \phi V_n \\ &\le \phi(0.17\sqrt{f'_c}\, b_w d) \end{aligned}$$

  where (b_w) is the critical width and (d) is the beam depth
- Two-way action:

  $V_u \le$ minimum of the following:
- $0.17(1+(2/\beta))\sqrt{f'_c}\, b_o d$
- $0.083(\alpha_s d/b_o + 2)\sqrt{f'_c}\, b_o d$
- $1/3\sqrt{f'_c}\, b_o d$

where

β = ratio of long side to short side of the column, concentrated load or reaction area

$\alpha_s = 40$ for interior columns

$\alpha_s = 30$ for edge columns

$\alpha_s = 20$ for corner columns

B_o = perimeter of critical section

3.10 EXAMPLE OF VERTICAL VESSEL

Design Data

Vessel Data

Empty wt. (D_e) = 80 tons

Oper. wt. (D_o) = 160 tons

Test wt. (D_t) = 280 tons

Structural Data

Allowable soil bearing = 2.0 kg/cm^2 = 20.0 t/m^2

$\gamma = 2.0$ t/m^3

$f_c' = 30$ N/mm^2

$f_y = 420$ N/mm^2

Wind Load

In accordance with *SEI/ASCE 7-02*,

$V = 115$ mph

$V = 25$ tons

$M = 262.94$ m.t = 260 m.t (at top of grout)

The vertical vessel dimensions and the assumed foundation dimensions are presented in Figs. 3.13 and 3.14, respectively.

Anchor Bolts

The anchor bolt diameter is 65 mm type *ASTM F1554*, Grade 36. There is a long sleeve with diameter 100 mm × 400 mm length and 350 mm projection on 4.53-m diameter of bolt circle (nonpretension).

Solution:

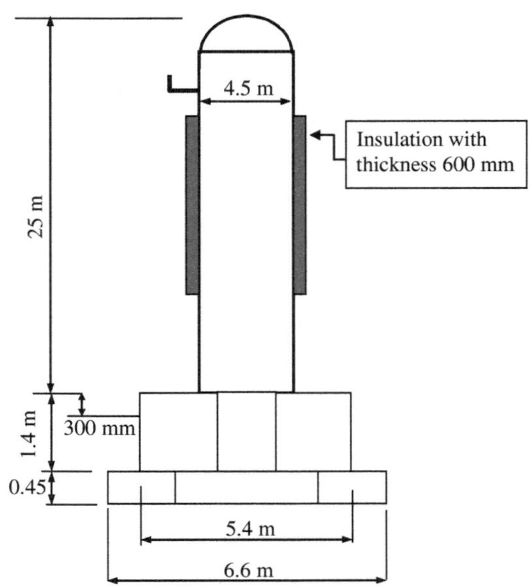

Figure 3.13 Sketch of vertical vessel.

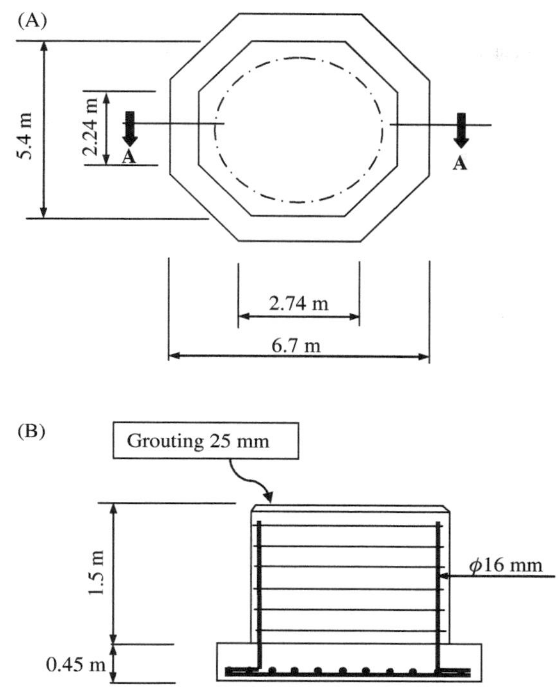

Figure 3.14 (A) Plan view. (B) Section A-A.

(A) Pedestal Design

Pedestal Dimensions and Weight

D_B + 250 mm = 4530 + 250 mm = 4780 mm

$D_B + 8(d)$ = 4530 + 8(40) = 4850 mm

$D_B + d_s$ + 250 mm − d = 4530 + 100 + 250 − 40 = 4840 mm

$D_B + d_s + 7(d)$ = 4530 + 100 + 7(40) = 4910 mm controls

- Use 4910 mm octagon.

Note: Pedestal "diameter" had to be increased to 5.40 m to provide a sufficiently large projected concrete failure area to resist the tensile load in the anchor bolts. Alternatively, additional reinforcing steel may be used to transfer anchor bolt forces to concrete.

Pedestal Reinforcement

Pedestal area = 24.13 m^2 as per Table 3.6

Pedestal weight (W_p) = (24.13 m^2)(1.4 m)(2.5 t/m^3) = 84.5 tons

M_p = O.T.M. at pedestal base = (260 m.t) + (1.5 m)(25 tons) = 297.5 m.t

$M_{up} = 1.6\ M_p = 1.6(297.5\ \text{m.t}) = 476\ \text{m.t}$

Load factors shall be in accordance with in Chapter 2, Table 2.24. Load Combination 4.

Assume the number of dowels, N_d, is equal to 40:

$D_C = (5.4) - (0.15) = 5.25\ \text{m}$

$D_e + W_p =$ empty weight of vessel + pedestal weight

$= 80 + 84.5 = 164.5$ tons

$F_u = 4(M_{uped})/[(N_d)(D_C)] - 0.9(D_e + D_p)/N_d$

$= 4(466.176)/[(40)(5.25)] - 0.9(156.75)/40 = 5.35$ tons

$A_s = F_u/\varphi f_y = (5.35)/(0.9)(0.042) = 141.5\ \text{mm}^2$

- Choose 40 with 16 mm diameter ($A_s = 201\ \text{mm}^2$) with 12 mm diameter at 350 mm c/c (minimum reinforcement controls)

(B) Anchor Bolt Check

Maximum Tension on Anchor Bolt

$N_u = 4\ M_u/[(N_b)(D_B)] - 0.9(D_e)/N_b$

$N_u = 4[(1.6)(260)]/[(24)(4.53)] - 0.9(80)/24 = 12.31$ tons

Load factors shall be in accordance with Table 2.24, Load Combination 4.

Maximum Shear on Anchor Bolt

V_u at base top of grout $= 1.6(25) = 40.0$ tons

Check whether shear load can be taken by friction between base of vessel and top of grout.

$P_u = M_u/LA + 0.9(D_e)$

Choose $LA = 2/3\ D_B$ as a conservative.

$LA = 2/3 \times 4.53 = 3.0\ \text{m}$

$P_u = 1.6 \times (260/3) + 0.9 \times (80) = 210.67$ tons

$V_f = \mu P_u = (0.55)(210.67) = 115.87$ tons

$\varphi V_f = (0.75)(115.87) = 86.90 > 32.48$ tons

- Therefore, the anchor bolts are not required to resist shear.

Projected Concrete Failure Area

Often several iterations were required to determine that $D = 4.9$ m would not provide enough projected concrete failure area to resist the maximum tensile load on the anchor (N_u) equal to 12.58 tons, regardless of what embedment depth (l_{dh}) was used. To save space, these trial calculations are not shown here. Reinforcing steel should be either added to transfer the tensile load from the anchor bolts to the pedestal, or the pedestal diameter should be increased. This second alternative is shown here.

Try increasing D to 5.4 mm.

(C) Footing Design

Select a Trial Octagon Size

M_{of} = O.T.M. at footing base = (260) + (1.95)(25 tons) = 308.75 m.t

$q_{all} = 20.0\ t/m^2$

Trial diameter = $(2.6)(M_{ftg}/q)^{1/3} = 6.6$ m

Try a 6.7 m octagon. Area = 37.1 m^2

Check Required Thickness for Pedestal Reinforcing Embedment

For 16 mm hooked bar based on *ACI 318M-08*, the development length will be calculated from the following equation:

$l_{dh} = [(0.24\ \Psi f_y)/\ \lambda\ (f_c')^{0.5}](d_b)$ (Section 12.5.2)

where $\Psi = 1.2$ in the case of epoxy coating

= 1.0 in the case of no coating

$\lambda = 1$ for normal concrete

$l_{dh} = 308$ mm

$t_f = 308 + 2$ (20 mm bars) + (75 mm cover) = 423 mm

Assume the foundation thickness, t_f = 450 mm

Footing Weights

Weight of pedestal $D_p = 24.13 \times 1.4 \times 2.5 = 84.5$ tons

The maximum normal force is usually due to hydrostatic test load if it is with water.

Total (D_s) = 84.5 tons

$P_t = D_t + D_s = 280 + 84.5 = 364.5$ tons

Check Soil Bearing and Stability

Empty + Wind (see Table 2.23, *Load Combination 3):*

$P = P_e = 80 + 84.5 = 164.5$ tons

$M_{of} = 308.75$ m.t

$e = M_{of}/P = (308.75)/(164.5) = 1.88$ m

Stability ratio = $D/2e = (6.6)/[2(1.8)] = 1.76 > 1.5$, therefore O.K

$e/D = (1.88)/(6.6) = 0.28 > 0.122$, therefore $L_{diag} = 2.85$

(From Table 3.7)

$q = LP/A = (2.85)(164.5)/(37.1) = 12.64\ t/m^2 < 20.0\ t/m^2$

Operating + Wind (see Table 2.23, *Load Combination 2):*

$P = P_o = D_s + D_o = 84.5 + 160 = 244.5$ tons
$M_{of} = 308.75$ m.t
$e = M_{of}/P = (308.75)/(244.5) = 1.26$ m
$e/D = (1.26)/(6.6) = 0.19 > 0.122$, therefore $L_{diag} = 2.25$
From Table 3.7
$q = LP/A = (2.25)(244.5)/(36.33) = 15.14\ t/m^2 < 20\ t/m^2$
Test + Partial Wind as Load Combination 6 from Table 2.23:
$P = P_t = D_s + D_t = 84.5 + 280 = 364.5$ tons
Partial wind velocity = 68 mph
$M_{of} = (68\ mph/115\ mph)^2(308.75) = 107.95$ m.t
$e = M_{of}/P = 107.95/364.5 = 0.30$ m
$e/D = 0.3/6.7 = 0.044 < 0.122$
$q = P/A\ [1 + (8.19)(e/D)]$
$= [(364.5)/(35.2)][1 + (8.19)(0.044)] = 14.1\ t/m^2 < 20\ t/m^2$

- Choose 6.6-m octagon.

Bottom Reinforcement

1. *Check Operating + Wind:*
   $[1.2(D_s + D_o) + 1.6\ W]$
   $P_u = 1.2(84.5 + 160) = 293.4$ m.t
   $M_u = 1.6(308.75) = 494$ m.t
   $e = M_u/P_u = 494/293.4 = 1.68$ m
   $e/D = 1.68/6.6 = 0.25 > 0.132$
   $L = 2.70$ (flat) $K = 0.225$ (flat)
   $KD = (0.225)(6.6) = 1.5$
   $q = LP/A = (2.7)293.4/36.33 = 21.81\ t/m^2$
   Find the equivalent square for pedestal:
   side of square = $(24.13\ m^2)^{0.5} = 4.9$ m
   projection = (6.6-4.9)/2 = 0.85 m
   q at the face of an equivalent square:
   $= 21.81(5.13 - 0.85)/(5.13) = 18.2\ t/m^2 < 20\ t/m^2$
   $M = \frac{18.2(0.85)^2}{2} + \frac{(21.81 - 18.2)(0.85)^2}{3} = 7.37$ m.t
2. *Check Empty + Wind:*
   $[0.9(D_{e+}D_s) + 1.6\ W]$
   $P_u = 0.9(80 + 84.5) = 148.05$ tons
   $M_u = 1.6(308.75) = 394$ m.t
   $e = M_u/P_u = (494)/(148.1) = 3.3$ m
   $e/D = (3.3)/(6.6) = 0.5 > 0.132$ (flat)
   $L = 7.63$ (flat) $K = 0.660$ (flat) (see Table 3.7)
   $KD = (0.660)(6.6) = 4.4$ m

$q = LP_u/A = (7.63)(148.05)/(36.33) = 31.1\ t/m^2$
Find the equivalent square for pedestal:
q at the face of an equivalent square:
$= (31.1)(2.2 - 0.85)/(2.2) = 19.1\ t/m^2$
$M = \frac{19.1(0.85)^2}{2} + \frac{(31.1 - 19.1)(0.85)^2}{3} = 9.78\ m.t$
By design the concrete section based on ACI
The concrete section is subjected to $M = 9.78$ m.t

- Choose 20 mm diameter every 200 mm, and the footing effective depth is 0.4 m.

For check shear

- *Beam Shear*

Apply the case of Empty + Wind:
$q = 31.1\ t/m^2$ at the edge
$q = 19.1\ t/m^2$ at the face of pedestal
$V_u = 0.85\ [(31.1 + 19.1)/2] = 21.34\ t/m^2$

$$\phi V_n = \varphi(0.17\sqrt{f_c'}b_w d$$
$$= 0.75(0.17)(30)0.5(1000)(400)/10,000 = 27.9\ t/m^2 > V_u$$

- *Punching Shear*

The punching shear will be calculated in the case of test load, and the load combination as shown in Table 2.18, Load Combination 7.
$P_u = 1.4(280) = 392$ tons
The length of the new octagon with $D = (4.9 + 0.4) = 5.3$ m
$b_o = (4 \times 4.9 + 0.4) = 20$ m
$V_u = 396.2/(17.6)(0.4) = 56.3\ t/m^2 = 0.55\ N/mm^2$
$V_u \leq$ minimum of the following:

- $0.17(1 + (2/\beta))\sqrt{f_c'}$
  $\phi V_n = 0.75\ (0.17\ (1 + (2/1))\ (30)^{0.5} = 2.1$
- $0.083(\alpha_s d/b_o + 2)\sqrt{f_c'}$
  $\phi V_n = (0.75)(0.083)\ (40\ (400/21200) + 2)(30)0.5 = 0.94\ MPa \leq V_u$
- $1/3\sqrt{f_c'}b_o d$
  $\phi V_n = (0.75)\ (1/3) \times (30)0.5 = 1.35$ MPa

3.11 PIPE SUPPORT

The pipe is usually obtained by the pipeline stress analysis engineer, but the following equations and table provide a guideline for choosing the pipe span. The maximum allowable span depends on the bending stress, vertical deflection, and natural frequency.

The formulation and equation obtained depend on the end conditions. By assuming a straight pipeline simply support at both ends; this end condition gives a higher stress and sag and therefore results in a conservative span.

$$L = \sqrt{\frac{0.33ZS}{w}} \quad \text{(stress limitation)} \tag{3.33}$$

$$L = \sqrt[4]{\frac{\Delta EI}{22.5w}} \quad \text{(deflection limitation)} \tag{3.34}$$

For less conservative the end condition may be assumed between simply supported at both ends and fixed condition at both ends.

$$L = \sqrt{\frac{0.4ZS}{w}} \quad \text{(stress limitation)} \tag{3.35}$$

$$L = \sqrt[4]{\frac{\Delta EI}{13.5w}} \quad \text{(deflection limitation)} \tag{3.36}$$

where

L = the allowable pipe span, m

Z = the modulus of pipe sections

S = the allowable tensile stress for the pipe materials at design temperature (known as allowable hot stress)

W = the total weight of the pipe which is equal to the summation of pipe self weight, content weight, and insulation weight

Δ = the deflection

I = the moment of inertia

E = the modulus of elasticity of the pipe material at design temperature

3.11.1 Natural Frequency

For most refinery piping a natural frequency is about 4 cycles per second (cps), which is sufficient to avoid resonance in nonpulsating pipe lines.

However, the natural frequency in cps is related to the maximum deflection, as in the following equation:

$$f_n = \frac{1}{2\pi}\sqrt{\frac{g}{\Delta}} \tag{3.37}$$

where g is the gravity acceleration and f_n is the natural frequency.

The deflection is limited to 25 mm; the reason for limiting the deflection is to provide a stiff enough pipe with high enough natural frequency to avoid large amplitude under any small disturbing force. From a practical point of view, it seems too low as the natural frequency will be higher because of the following reasons:

- End moments are neglected in the previous equation, which will raise the natural frequency by more than 15%.
- The critical span is usually limited by stress and is rarely reached.
- The piping weight assumption is often larger than the actual load.

Table 3.8 provides quick reference values of span based on the following assumption:

- Pipe material is carbon steel A53 grade, and this table applies conservatively to all other steels.
- Temperature ranges from zero to 650°F. At 650°F, $S = 12000$ psi. the modulus of elasticity $E = 25.2 \times 10^6$ from piping code.
- Assume the fluid is water.
- Assume density of insulationis 11 Ib/ft^3.
- Thickness of insulation is 1.5 in. for pipe sizes from 1 to 4 in.
  - 2.0 in. for pipe sizes from 6 to 14 in.
  - 2.5 in. for pipe sizes from 16 to 24 in.
- The pipe was treated as a horizontal beam, supported at both ends, carrying a uniform load equal to the summation of metal weight, water, and insulation.
- The maximum static deflection was 25 mm.
- The maximum bending stress was equated to allowable weight stress equal to half the allowable hot stress, S.

Example:

Piping 18″ diameter with schedule standard carrying oil, the recommended distance between support is:

Solution:

Based on Table 3.9 the minimum spacing between support is 13.4 m.

Table 3.8 Maximum spans of horizontal pipe Lines (ft)

Schedule	Limit	Pipe size (in.)													
		1	1.5	2	3	4	6	8	10	12	14	16	18	20	24
Schedule 10	*L*	13	15	17	20	22	25	29	30	32	37	38	39	39	41
	L′	13	16	18	21	24	28	31	34	37	41	42	44	46	48
Schedule 20	L							33	35	36	39	41	42	45	47
	L′							33	37	39	42	44	46	49	52
Schedule 30	*L*							34	37	39	42	43	46	49	52
	L′							34	38	41	43	45	48	51	55
Standard	*L*	13	16	18	23	26	31	35	38	41	42	43	44	45	47
	L′	13	16	18	23	26	31	35	38	41	43	45	47	49	52
Schedule 40	*L*	13	16	18	23	26	31	35	38	41	43	46	49	51	56
	L′	13	16	18	23	26	31	35	38	42	44	45	50	52	57
Schedule 60	*L*							36	40	43	46	49	52	55	60
	L′							35	39	43	45	48	51	54	59
Extra- strong	*L*	13	17	19	24	27	33	37	41	43	44	46	48	49	51
	L′	13	17	19	23	26	32	36	40	43	44	46	49	51	54
Schedule 80	*L*	13	17	19	24	27	33	37	42	46	48	52	55	58	63
	L′	13	17	19	23	26	32	36	40	44	46	50	52	55	61
Schedule 100	*L*							38	43	47	49	53	56	59	65
	L′							37	41	45	47	50	53	56	51
Schedule 120	*L*					28	34	39	44	48	51	54	57	61	67
	L′					27	32	37	41	45	47	51	54	57	62
Schedule 140	*L*					28	34	39	44	49	51	54	58	61	67
	L′					27	33	37	42	45	48	51	54	57	62
Schedule 160	*L*	13	17	20	25	29	35	40	45	49	51	55	58	62	68
	L′	13	17	19	23	27	33	37	42	45	48	51	54	57	63

Table 3.9 Maximum spans of horizontal pipe lines (m)

Schedule	Limit	Pipe size (in.)													
		1	1.5	2	3	4	6	8	10	12	14	16	18	20	24
Schedule 10	L	4.0	4.6	5.2	6.1	6.7	7.6	8.8	9.1	9.8	11.3	11.6	11.9	11.9	12.5
	L'	4.0	4.9	5.5	6.4	7.3	8.5	9.4	10.4	11.3	12.5	12.8	13.4	14.0	14.6
Schedule 20	L	0.0	0.0	0.0	0.0	0.0	0.0	10.1	10.7	11.0	11.9	12.5	12.8	13.7	14.3
	L'	0.0	0.0	0.0	0.0	0.0	0.0	10.1	11.3	11.9	12.8	13.4	14.0	14.9	15.8
Schedule 30	L	0.0	0.0	0.0	0.0	0.0	0.0	10.4	11.3	11.9	12.8	13.1	14.0	14.9	15.8
	L'	0.0	0.0	0.0	0.0	0.0	0.0	10.4	11.6	12.5	13.1	13.7	14.6	15.5	16.8
Standard	L	4.0	4.9	5.5	7.0	7.9	9.4	10.7	11.6	12.5	12.8	13.1	13.4	13.7	14.3
	L'	4.0	4.9	5.5	7.0	7.9	9.4	10.7	11.6	12.5	13.1	13.7	14.3	14.9	15.8
Schedule 40	L	4.0	4.9	5.5	7.0	7.9	9.4	10.7	11.6	12.5	13.1	14.0	14.9	15.5	17.1
	L'	4.0	4.9	5.5	7.0	7.9	9.4	10.7	11.6	12.8	13.4	13.7	15.2	15.8	17.4
Schedule 60	L	0.0	0.0	0.0	0.0	0.0	0.0	11.0	12.2	13.1	14.0	14.9	15.8	16.8	18.3
	L'	0.0	0.0	0.0	0.0	0.0	0.0	10.7	11.9	13.1	13.7	14.6	15.5	16.5	18.0
Extra- strong	L	4.0	5.2	5.8	7.3	8.2	10.1	11.3	12.5	13.1	13.4	14.0	14.6	14.9	15.5
	L'	4.0	5.2	5.8	7.0	7.9	9.8	11.0	12.2	13.1	13.4	14.0	14.9	15.5	16.5
Schedule 80	L	4.0	5.2	5.8	7.3	8.2	10.1	11.3	12.8	14.0	14.6	15.8	16.8	17.7	19.2
	L'	4.0	5.2	5.8	7.0	7.9	9.8	11.0	12.2	13.4	14.0	15.2	15.8	16.8	18.6
Schedule 100	L	0.0	0.0	0.0	0.0	0.0	0.0	11.6	13.1	14.3	14.9	16.2	17.1	18.0	19.8
	L'	0.0	0.0	0.0	0.0	0.0	0.0	11.3	12.5	13.7	14.3	15.2	16.2	17.1	15.5
Schedule 120	L	0.0	0.0	0.0	0.0	8.5	10.4	11.9	13.4	14.6	15.5	16.5	17.4	18.6	20.4
	L'	0.0	0.0	0.0	0.0	8.2	9.8	11.3	12.5	13.7	14.3	15.5	16.5	17.4	18.9
Schedule 140	L	0.0	0.0	0.0	0.0	8.5	10.4	11.9	13.4	14.9	15.5	16.5	17.7	18.6	20.4
	L'	0.0	0.0	0.0	0.0	8.2	10.1	11.3	12.8	13.7	14.6	15.5	16.5	17.4	18.9
Schedule 160	L	4.0	5.2	6.1	7.6	8.8	10.7	12.2	13.7	14.9	15.5	16.8	17.7	18.9	20.7
	L'	4.0	5.2	5.8	7.0	8.2	10.1	11.3	12.8	13.7	14.6	15.5	16.5	17.4	19.2

L is calculated with limiting bending stress of S divided by 2.
L' is calculated with limiting static deflection 25.4 mm.

REFERENCES

Process Industry Practices (PIP), PIP STC01015—Structural Design Criteria.
Process Industry Practices (PIP), PIP STE05121—Anchor Bolt Design Guide.
American Concrete Institute (ACI), ACI 318/318R-05—Building Code Requirements for Structural Concrete and Commentary.
American Society of Civil Engineers (ASCE), ASCE/SEI 7-05—Minimum Design Loads for Buildings and Other Structures.
ASTM International (ASTM), ASTM F1554—Standard Specification for Anchor Bolts, Steel, 36, 55, and 105-ksi Yield Strength.

CHAPTER 4

Steel and Concrete Storage Tank

4.1 INTRODUCTION

Tanks are very famous structures in industrial projects. There are different types of storage tanks, as they can be used for different purposes in industrial projects.

The dimensions of the tank and its material depends on the fluids that will be stored. The oil storage usually has a steel structure for a shell and reinforced concrete ring beams that transfer the load to the soil.

In some applications, a reinforced concrete tank is used for the storage of salinity water, which will be treated and used in the water flood system traditionally used in some process plants.

The process of treating the oil can be done by using a reinforced concrete tank with a weir to perform the separation between the oil and water in the treating process.

For the LNG project, which is needed to cool the gas into liquid in order to transport it through the harbor for export, prestressed tanks with steel shells inside are needed. In this chapter, the types of retaining walls and their design methodology will be presented in addition to the main features in the design and construction of reinforced concrete and steel storage tanks.

4.2 STEEL STORAGE TANKS

Steel storage tanks are usually used in petrochemical plants. Fig. 4.1 presents a construction of a steel storage tank.

The design, construction, and manufacture of the steel storage tanks are usually performed based on API codes. The following sections are the guidelines for best practices with interaction with international codes in designing steel tanks.

4.2.1 Tank Capacity

The client shall specify the maximum capacity and the overfill protection level or volume requirement based on API RP 2350.

Onshore Structural Design Calculations.
DOI: http://dx.doi.org/10.1016/B978-0-08-101944-3.00004-8

Figure 4.1 Construction of a storage tank.

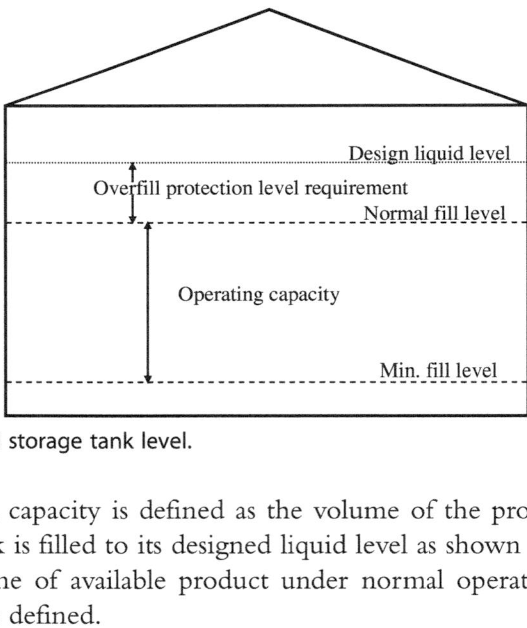

Figure 4.2 Steel storage tank level.

Maximum capacity is defined as the volume of the product in a tank when the tank is filled to its designed liquid level as shown in Fig. 4.2.

The volume of available product under normal operating conditions should also be defined.

By knowing of the design and operating level from the storage tanks mechanical design, the load in case of empty, operating, and test condition can be defined.

4.2.2 Bottom Plates

All bottom plates shall have a minimum nominal thickness of 6 mm (1/4 in.). The plate's weight is around 49.8 kg/m^2, and you should add any corrosion allowance specified by the owner for the bottom plates based on his Statement of Requirement (SOR). Unless otherwise agreed to by the owner, all rectangular sketch plates shall have a minimum nominal width of 1800 mm (72 in.).

Bottom plates of sufficient size shall be ordered so that, when trimmed, at least a 50 mm (2 in.) width will project outside the shell. The bottom plates shall be welded based on the welding procedure.

According to API 650, the tank bottoms shall have a minimum slope of 1:120 upwards toward the center of the tank, unless otherwise specified on the data sheet. The material shall be carbon steel with minimum thickness 3 mm (1/8 in.).

4.2.3 Annular Bottom Plates

Based on the API 650, if the bottom shell course is designed by using the allowable stress for materials as shown in Table 4.3, butt-welded annular bottom plates shall be used. When the stress is less than or equal to 172 MPa, lap-welded bottom plates may be used in lieu of butt-welded annular bottom plates.

Annular bottom plates shall have a radial width that provides at least 600 mm (24 in.) between the inside of the shell and any lap-welded joint in the remainder of the bottom.

A greater radial width for the annular plate is required when calculated as follows:

In SI units:

$$\frac{215\, t_b}{(HG)^{0.5}}$$

where:

t_b = thickness of the annular plate in mm,

H = maximum design liquid level in m,

G = design-specific gravity of the liquid to be stored.

The thickness of the annular bottom plates shall not be less than the greater thickness determined using Table 4.1 for product design, in addition to any specified corrosion allowance or for hydrostatic test design.

Table 4.1 is applicable for effective product height of $H \times G \leq 23$ m. Beyond this height, an elastic analysis must be made to determine the annular plate thickness.

Table 4.1 Plate thickness for annular bottom—(t_{br})

Plate thickness[a] of shell course (mm)	Stress in first shell course (MPa)			
	≤ 190	≤ 210	≤ 220	≤ 250
$t \leq 19$	6	6	7	9
$19 < t \leq 25$	6	7	10	11
$25 < t \leq 32$	6	9	12	14
$32 < t \leq 38$	8	11	14	17
$38 < t \leq 45$	9	13	16	19

[a]Plate thickness refers to shell plate thickness exclusive of corrosion allowance for product design and thickness as constructed for hydrostatic test design.

Table 4.2 Relation between tank diameter and minimum plate thickness

Nominal tank diameter, m (ft)	Nominal plate thickness, mm (in.)
<15 (50)	5 (3/16)
15 (50)–36 (120)	6 (1/4)
36 (120)–60 (200)	8 (5/16)
>60 (200)	10 (3/8)

The stress to be used is the maximum stress (as in Table 4.1) in the first shell course (greater of product or hydrostatic test stress). The stress may be determined using the required thickness divided by the thickness from "a" then multiplied by the applicable allowable stress:

Product Stress = $(t_d - CA/\text{corroded } t)(S_d)$

Hydrostatic Test Stress = $(t_t/\text{nominal } t)(S_t)$.

It is worth to mention that the thickness specified in the table are based on the foundation providing uniform support under the full width of the annular plate.

The thicknesses specified in Table 4.2 are based on the foundation providing uniform support under the full width of the annular plate. As will be discussed later in this chapter, unless the foundation is properly compacted inside a concrete ring wall, settlement will produce additional stresses in the annular plate.

The ring of annular plates shall have a circular outside circumference, but may have a regular polygonal shape inside the tank shell, with the number of sides equal to the number of annular plates to which these pieces shall be welded.

In lieu of annular plates, the entire bottom may be butt-welded provided that the requirements for annular plate thickness, welding, materials, and inspection are met for the annular requirement.

4.2.4 Shell Design

The required shell thickness shall be the greater of the design shell thickness, including any corrosion allowance, or the hydrostatic test shell thickness, but in no case shall the shell thickness be less than the thicknesses specified in Table 4.2.

Unless otherwise specified by the owner specification, the nominal tank diameter shall be the centerline diameter of the bottom shell course plate. Nominal plate thickness refers to the tank shell as constructed. The thickness will be specified based on erection requirements. In some cases, the client specified a plate with a minimum nominal thickness of 6 mm. If the tank diameter is less than 15 m (50 ft) but greater than 3.2 m (10.5 ft), then for the lowest shell course only, the minimum thickness is 6 mm.

The shell plates shall have a minimum nominal width of 1800 mm (72 in.) unless specified by another value from the owner or the available factories. Plates that are to be butt-welded shall be properly squared.

The calculated stress for each shell course shall not be greater than the stress permitted for the particular material used for the course. For any case and under any circumstances, no shell course shall be thinner than the course above it.

The tank shell shall be checked against wind speed load for stability due to buckling. If required for stability, intermediate girders, increased shell plate thicknesses, or both shall be used.

It is essential to isolate the radial loads on the tank shell, such as those caused by heavy loads on platforms and elevated walkways between tanks; these shall be distributed by rolled structural sections, plate ribs, built-up members, or any other structural configuration to transfer the load to the ground.

4.2.4.1 Allowable Stress

The maximum allowable product design stress, S_d, shall be as shown in Table 4.3. The net plate thicknesses—the actual thicknesses less any corrosion allowance—shall be used in the calculation.

According to API 650, the design stress basis S_d shall be either two-thirds the yield strength (2/3 Fy) or two-fifths (2/5 Ft) the tensile strength, whichever is less.

The maximum allowable hydrostatic test stress S_t shall be as shown in Table 4.3. The gross plate thicknesses, including any corrosion allowance,

Table 4.3 Plate materials and allowable stresses

Plate specification	Grade	Minimum yield strength (MPa)	Minimum tensile strength (MPa)	Product design stress S_d (MPa)	Hydrostatic test stress S_t (MPa)
A 283M (A 283)	C	205	380	137	154
A 285M (A 285)	C	205	380	137	154
A 131M (A 131)	A,B,CS	235	400	157	171
A 36M (A 36)		250	400	160	171
A 131M (A 131)	EH36	360	490a	196	210
A 573M (A 573)	400	220	400	147	165
A 573M (A 573)	450	240	450	160	180
A 573M (A 573)	485	290	485a	193	208
A 516M (A 516)	380	205	380	137	154
A 516M (A 516)	415	220	415	147	165
A 516M (A 516)	450	240	450	160	180
A 516M (A 516)	485	260	485	173	195
A 662M (A 662)	B	275	450	180	193
A 662M (A 662)	C	295	485a	194	208
A 537M (A 537)	1	345	485a	194	208
A 537M (A 537)	2	415	550a	220	236
A 633M (A 633)	C,D	345	485a	194	208
A 678M (A 678)	A	345	485a	220	208
A 678M (A 678)	B	415	550a	194	236
A 737M (A 737)	B	345	485a	194	208
A 841M (A 841)	CLASS 1	345	485a	194	208
A 841M (A 841)	CLASS2	415	550a	220	236

shall be used in the calculation. The hydrostatic test basis shall be either three-fourths the yield strength (3/4 Fy) or three-sevenths the tensile strength (3/7 Fy), whichever is less.

According to API 650, in cases of shell thickness less than or equal to one-half inch (1/2 in.), an alternative shell design with a fixed allowable stress of 145 MPa (21,000 psi^2) and a joint efficiency factor of 0.85 or 0.70 is permitted.

4.2.4.2 Calculation of Thickness by the 1-Foot Method

The 1-foot method is required to calculate the thicknesses at design points 0.3 m (1 ft) above the bottom of each shell course. It is important to know that this method shall not be used for tanks larger than 60 m (200 ft) in diameter.

The required minimum thickness of shell plates shall be the greater of the values computed by the following formulas:

$$t_d = \frac{4.9D(H - 0.3)G}{S_d} + CA \tag{4.1}$$

$$t_t = \frac{4.9D(h - 0.3)}{S_t} \tag{4.2}$$

where:

t_d = design shell thickness in millimeters;

t_t = hydrostatic test shell thickness in millimeters;

D = nominal tank diameter in meters;

H = design liquid level in meters;

h = height from the bottom of the course under consideration to the top of the shell including the top angle, if any, or to the bottom of any overflow that limits the tank filling height; or to restricted by an internal floating roof;

G = design-specific gravity of the liquid to be stored as defined by the owner;

CA = corrosion allowance in mm as specified by the owner requirement;

S_d = allowable stress for the design condition as specified in Table 4.3 in milli-Pascals;

S_t = allowable stress for the hydrostatic test condition as specified in Table 4.3 in milli-Pascals.

4.2.4.3 Calculation of Thickness by the Variable-Design-Point Method

Before using this method we must know that this procedure normally provides a reduction in shell-course thicknesses and total material weight, but more important is its potential to permit construction of larger diameter tanks within the maximum plate thickness limitation.

This calculation is based on the study performed by Zick and McGrath (1968) for "Design of Large Diameter Cylindrical Shells." Design by the variable-design-point method gives shell thicknesses at design points that result in the calculated stresses being relatively close to the actual circumferential shell stresses. This method may only be used when the client has not specified that the 1-foot method be used and when the following equation is verified.

In SI units:

$$\frac{L}{H} \leq \frac{1000}{6} \tag{4.3}$$

where:

$L = (500Dt)^{0.5}$;

D = tank diameter in m;

t = bottom-course shell thickness, excluding any corrosion allowance, in mm;

H = maximum design liquid level in m.

The minimum plate thicknesses for both the design condition and the hydrostatic test condition shall be determined as outlined.

The complete calculations shall be made for all of the courses for the design condition, exclusive of any corrosion allowance, and for the hydrostatic test condition. The required shell thickness for each course shall be the greater of the design shell thickness plus any corrosion allowance or the hydrostatic test shell thickness.

To calculate the bottom-course thicknesses, preliminary values t_{pd} and t_{pt} for the design and hydrostatic test conditions shall first be calculated.

The bottom-course thicknesses t_{d1} and t_{1t} for the design and hydrostatic test conditions shall be calculated using the Eqs. (4.4) and (4.2), respectively.

$$t_{d1} = \left(1.06 - \frac{0.0696D}{H}\sqrt{\frac{HG}{S_d}}\right)\left(\frac{4.9HGD}{S_d}\right) + CA \tag{4.4}$$

$$t_{1t} = \left(1.06 - \frac{0.0696D}{H}\sqrt{\frac{H}{S_t}}\right)\left(\frac{4.9HD}{S_t}\right) \tag{4.5}$$

Note: For the design condition, t_{d1} need not be greater than t_{pd}, and for the hydrostatic test condition, t_{1t} need not be greater than t_{pt}.

To calculate the second-course thicknesses for both the design condition and the hydrostatic test condition, the value of the following ratio shall be calculated for the bottom course:

$$R = \frac{h_1}{(rt_1)^{0.5}} \quad (4.6)$$

where:

h_1 = height of the bottom shell course in millimeters (inches);
r = nominal tank radius in millimeters (inches);
t_1 = calculated thickness of the bottom shell course, less any thickness added for corrosion allowance, in mm (in.), used to calculate t_2 (design). The calculated hydrostatic thickness of the bottom shell course shall be used to calculate t_2 (hydrostatic test).
If $R \leq 1.375$ $\quad t_2 = t_1$,
$R \geq 2.625$ $\quad t_2 = t_{2a}$,
$1.375 \leq R < 2.625$,

$$t_2 = t_{2a} + (t_1 - t_{2a})\left[2.1 - \frac{h_1}{1.25(rt_1)^{0.5}}\right] \quad (4.7)$$

where:

t_2 = minimum design thickness of the second shell course excluding any corrosion allowance, in millimeters (inches);
t_{2a} = thickness of the second shell course, in millimeters (inches), as calculated for an upper shell course as described in Eqs. (4.8) and (4.9), without any corrosion allowance. For second shell calculation the course thickness (t_2) for design case and hydrostatic test case, applicable values of t_2a and t_1 shall be used.

The preceding formula for t_2 is based on the same allowable stress being used for the design of the bottom and second courses.

For tanks where the value of the ratio is greater than or equal to 2.625, the allowable stress for the second course may be lower than the allowable stress for the bottom course.

To calculate the upper-course thicknesses for both the operating condition and the hydrostatic test condition, a preliminary value t_u for the upper-course thickness shall be calculated using Eqs. (4.1) and (4.2), excluding any corrosion allowance. The distance x of the variable design point from the bottom of the course shall be calculated using the lowest value obtained from the following:

$$X_1 = 0.61(rt_u)^{0.5} + 320\ CH$$
$$X_2 = 1000CH$$
$$X_3 = 1.22(rt_u)^{0.5}$$

where:

t_u = thickness of the upper course at the girth joint calculation from Eqs. (4.1) and (4.2), exclusive of any corrosion allowance in millimeters,

$C = [K^{0.5}(K-1)]/(1+K^{1.5})$,

$K = t_L/t_u$,

t_L = thickness of the lower course at the girth joint exclusive of any corrosion allowance in millimeters,

H = design liquid level in meters.

The minimum thickness t_x for the upper shell courses shall be calculated for both the design condition (t_{d_x}) and the hydrostatic test condition (t_{t_x}) using the minimum value of x.

$$t_{d_x} = \frac{4.9D\left(H - \frac{x}{1000}\right)G}{S_d} + CA \tag{4.8}$$

$$t_{t_x} = \frac{4.9D\left(H - \frac{x}{1000}\right)}{S_t} \tag{4.9}$$

4.2.5 Roof Systems

The designer will decide the type of roof system based on the tank diameter and the use of the tank. In some tanks there will be a floating roof, but in this section, the discussion about the structure system will be for a fixed roof.

The fixed roof types will be as follows:

1. A supported cone roof is a roof formed to approximately the surface of a right cone that is supported principally either by rafters on girders and columns or by rafters on trusses with or without columns.
2. A self-supporting cone roof is a roof formed to approximately the surface of a right cone that is supported only at its periphery.
3. A self-supporting dome roof is a roof formed to approximately a spherical surface that is supported only at its periphery.

4. A self-supporting umbrella roof is a modified dome roof formed so that any horizontal section is a regular polygon with as many sides as there are roof plates, and is supported only at its periphery.

All roofs and supporting structures shall be designed for load combinations (1), (2), (3), and (4).

Roof plates shall have a minimum nominal thickness of 5 mm (3/16 in.) or 7-gauge sheet. Thicker roof plates may be required for self-supporting roofs. The corrosion allowance should be included in the design of the roof thickness.

In general, the roof plates of supported cone roofs shall not be attached to the supporting members unless otherwise approved by the client. Continuously attaching the roof to cone supporting members may be beneficial when interior coating systems are required, however, the tank roof cannot be considered frangible.

All internal and external structural members shall have a minimum nominal thickness of 4 mm in any component. The method of providing a corrosion allowance, if any, for the structural members shall be a matter of agreement between the client and the contractor.

Roof plates shall be attached to the top angle of the tank with a continuous fillet weld on the top side only.

A roof is considered frangible, as for emergency venting requirements, if the roof-to-shell joint will fail prior to the shell-to-bottom joint in the event of excessive internal pressure. When a client specifies a tank with a frangible roof, the tank design shall comply with all of the following:

1. The tank shall have a 15-m diameter or greater.
2. The slope of the roof at the top-angle attachment does not exceed 1:6.
3. The roof is attached to the top angle with a single continuous fillet weld that does not exceed 5 mm.
4. The roof support members shall not be attached to the roof plate.
5. All members in the region of the roof-to-shell junction, including insulation rings, are considered as contributing to the cross-sectional area (A).
6. The cross-sectional area (A) of the roof-to-shell junction is less than the limit shown here in the following equation in SI units.

$$A = \frac{D_{ls}}{1390 \tan \theta} \tag{4.10}$$

For all types of roofs, the plates may be stiffened by sections welded to the plates but may not be stiffened by sections welded to the supporting rafters or girders.

The contractor shall provide a roof designed and constructed to be as safe as otherwise provided for in this standard. In the roof design, particular attention should be given to preventing failure through instability.

4.2.5.1 Allowable Stresses

All parts of the roof structure shall be proportioned so that the sum of the maximum static and dynamic stresses shall not exceed the limitations specified in the AISC Specification for Structural Steel Buildings, or, with the agreement of the client, an equivalent structural design code recognized by the government of the country where the tank is located.

The portion of the specification, "Allowable Stress Design," shall be used in determining allowable unit stresses; based on API 650, the plastic design is not allowed to be used in tank design.

The minimum thickness of any structural member, including any corrosion allowance on the exposed side or sides, shall not be less than 6 mm for columns, knee braces, beams, or stiffeners, which by design normally resist axial compressive forces, or 4 mm for any other structural member.

For columns, the value L/r shall not exceed 180. For other compression members, the value L/r shall not exceed 200. For all other members, except tie rods whose design is based on tensile force, the value L/r shall not exceed 300.

where:

L = unbraced length in millimeters,

r = governing radius of gyration in millimeters.

4.2.5.2 Supported Cone Roofs

The slope of the roof shall be 1:16 or greater if specified by the client. If the rafters are set directly on chord girders, producing slightly varying rafter slopes, the slope of the flattest rafter shall conform to the specified or ordered roof slope.

Main supporting members, including those supporting the rafters, may be rolled or fabricated sections or trusses.

Although these members may be in contact with the roof plates, the compression flange of a member or the top chord of a truss shall be designed with no lateral support from the roof plates; it shall be laterally braced, if necessary, by other acceptable methods. The structural members which is using as rafter will be rolled hot or built up section. Rafters shall be designed for the dead load of the rafters and roof plates with the compression flange of the rafter has no lateral support from the roof plates and shall be laterally braced.

When considering additional dead loads or live loads, the rafters in direct contact with the roof plates applying the loading to the rafters may be considered as receiving adequate lateral support from the friction between the roof plates and the compression flanges of the rafters, with the following exceptions:

1. Trusses and open-web joints used as rafters.
2. Rafters with a nominal depth greater than 375 mm (15 in.).
3. Rafters with a slope greater than 1:6.

Rafters shall be spaced so that in the outer ring, their centers are not more than 0.6π m [2π ft (6.28 ft)] apart measured along the circumference of the tank. Spacing on inner rings shall not be greater than 1.7 m.

Roof columns shall be made as pipe or structural shapes sections. Pipe columns shall either be sealed or have openings on both the top and bottom of the column.

Rafter clips for the outer row of rafters shall be welded to the tank shell.

Roof support columns shall be provided at their bases with details that provide for the following:

1. Column loads shall be distributed over a bearing area based on the specified soil-bearing capacity or foundation design. Where an unstiffened horizontal plate is designed to distribute the load, it shall be a minimum thickness of 12 mm (1/2 in.). Alternatively, the column load may be distributed by an assembly of structural beams. The plate or members shall be designed to distribute the load without exceeding the allowable stresses.
2. At each column a wear plate with a minimum 6 mm thickness shall be welded to the tank bottom with a 6 mm (1/4 in.) minimum fillet weld. A single adequate thickness plate may be designed for the dual functions of load distribution and corrosion and abrasion protection.
3. The design shall allow the columns to move vertically relative to the tank bottom without restraint in the event of tank overpressure or bottom settlement.
4. The columns shall be effectively guided at their bases to prevent lateral movement. The guides shall remain effective in the event of vertical movement of columns relative to tank bottom of up to 75 mm. The guides shall be located such that they are not welded directly to the tank bottom plates.

When supporting members are attached to the roof plate, consideration shall be given to the design of the supporting members and their attachment details when considering internal pressure.

4.2.5.3 Self-Supporting Cone Roofs

Self-supporting roofs whose roof plates are stiffened by sections welded to the plates need not conform to the minimum thickness requirements, but the thickness of the roof plates shall not be less than 5 mm when so designed by the contractor, subject to the approval of the engineer.

Self-supporting cone roofs shall conform to the following requirements:

$\theta \leq 37^\circ$ (slope = 3:4);

$\theta \geq 9.5^\circ$ (slope = 1:6).

In SI units:

$$t_{min} = \frac{D}{4.8 \sin \theta}\sqrt{\frac{T}{2.2}} \geq 5 \text{ mm}$$

$$t_{max} = 12.5 \text{ mm, exclusive of corrosion allowance}$$

where:

D = nominal diameter of the tank shell (m),

T = greater of load combinations (KPa),

θ = angle of cone elements to the horizontal (°).

4.2.5.4 Self-Supporting Dome and Umbrella Roofs

Self-supporting roofs whose roof plates are stiffened by sections welded to the plates need not conform to the minimum thickness requirements.

The thickness of the roof plates shall not be less than 5 mm.

Self-supporting dome and umbrella roofs shall conform to the following requirements:

minimum radius = $0.8D$,

maximum radius = $1.2D$.

$$t_{min} = \frac{r}{2.4}\sqrt{\frac{T}{2.2}} + CA \geq 5 \text{ mm}$$

$$t_{max} = 12.5 \text{ mm, exclusive of corrosion allowance}$$

The following are certain restrictions on types of top-angle joints:

1. Roof plates shall at a minimum be welded on the top side with a continuous full-fillet weld on all seams; butt-welds are also permitted.

Table 4.4 Tank diameter and top-angle size

Tank diameter (*D*)	**Minimum top-angle size[a] (mm)**
$D \leq 11$ m ($D \leq 35$ ft)	$51 \times 51 \times 4.8$
11 m $< D \leq$ 18 m (35 ft $< D \leq$ 60 ft)	$51 \times 51 \times 6.4$
$D > 18$ m ($D > 60$ ft)	$76 \times 76 \times 76$

[a]You can choose an equivalent section size to accommodate local materials availability.

2. For frangible roofs, roof plates shall be attached to the top angle of a tank with a continuous fillet weld on the top side only. For nonfrangible roofs, alternate details are permitted.
3. The top-angle sections, tension rings, and compression rings shall be joined by butt-welds having complete penetration and fusion.
4. For self-supporting roofs of the cone, dome, or umbrella type, the edges of the roof plates may be flanged horizontally to rest flat against the top angle to improve welding conditions.
5. Except as specified for open-top tanks, for tanks with frangible joints for self-supporting roofs, and for tanks with the flanged roof-to-shell detail described later, tank shells shall be supplied with top angles of not less than the sizes given in Table 4.4.

4.2.6 Tank Design Loads

The loads that shall be taken into design are important to be defined by the client in his SOR. For example in some countries there is snow or earthquakes; however, other countries are far away from the effect of these conditions.

Loads are defined as follows:

1. *Dead load (D_L)*: The weight of the tank or tank component, including any corrosion allowance, unless otherwise noted.
2. *Design external pressure (P_e)*: It shall not be less than 0.25 kPa (1 in. of water). This standard does not contain provisions for external pressures greater than 0.25 kPa (1 in. of water). Design requirements for vacuum exceeding this value and design requirements to resist flotation and external fluid pressure shall be a matter of agreement between the client and the manufacturer.
3. *Design internal pressure (P_i)*: It shall not exceed 18 kPa (2.5 lbf/in.^{2}).
4. *Hydrostatic test (H_t)*: The load due to filling the tank with water to the design liquid level.

5. *Minimum roof live load (L_r)*: 1.0 kPa (20 lb/ft^2) on the horizontal projected area of the roof.
6. *Seismic (E)*: Seismic loads determined based on the seismic zone of the tank location.
7. *Snow (S)*: The ground snow load shall be determined from ASCE 7, unless the ground snow load that equals or exceeds the value based on a 2% annual probability of being exceeded (50-year mean recurrence interval) is specified by the client. The design snow load shall be 0.84 times the ground snow load. Alternately, the design snow load shall be determined from the ground snow load in accordance with ASCE 7. The design snow load shall be reported to the client.
8. *Stored liquid (F)*: The load due to filling the tank to the design liquid level with liquid with the design-specific gravity specified by the client.
9. *Test pressure (P_t)*: The pressure due to filling the tank with water for hydro states test purposes until the design liquid level.
10. *Wind (W)*: The design wind speed (*V*) shall be 190 km/h (120 mph), the 3-s gust design wind speed determined from ASCE 7, Fig. 6.1, or specified by the client [this specified wind speed shall be for a 3-s gust based on a 2% annual probability of being exceeded (50-year mean recurrence interval)].

    The design wind pressure shall be 0.86 kPa $(V/190)^2$, [(18 lbf/ft^2) $(V/120)^2$] on vertical projected areas of cylindrical surfaces and 1.44 kPa $(V/190)^2$, [(30 lbf/ft^2)$(V/120)^2$] uplift on horizontal projected areas of conical or doubly curved surfaces, where *V* is the 3-s gust wind speed. The 3-s gust wind speed used shall be reported to the client.

    1. These design wind pressures are in accordance with ASCE 7 for wind exposure Category C. As an alternative, pressures may be determined in accordance with ASCE 7 (exposure category and importance factor provided by the client) or a national standard for the specific conditions for the tank being designed.
    2. The design uplift pressure on the roof (wind plus internal pressure) does not need to exceed 1.6 times the design pressure.
    3. Windward and leeward horizontal wind loads on the roof will not include in the above pressures, as they are conservatively equal and opposite.
    4. Fastest mile wind speed multiplied by 1.2 is approximately equal to the 3-s gust wind speed.

4.2.7 Load Combination

For the purposes of this standard, loads are combined in the following manner; design rules account for these load combinations, including the absence of any load other than D_L in the combinations:

1. Fluid and internal pressure:

$$D_L + F + P_i$$

2. Hydrostatic test:

$$D_L + (Ht + P_t)$$

3. Wind and internal pressure:

$$D_L + W + 0.4P_i$$

4. Wind and external pressure:

$$D_L + W + 0.4P_e$$

5. Gravity loads:

a. $D_L + (L_r \text{ or } S) + 0.4\ P_e$

b. $D_L + P_e + 0.4(Lr \text{ or } S)$

6. Seismic:

$$D_L + F + E + 0.1S + 0.4P_i$$

If the ratio of operating pressure to design pressure exceeds 0.4, the client should consider specifying a higher factor on design pressure in (3), (4), (5)(a), and (6).

4.2.8 Design Basis for Small Tanks

A small tank is a tank in which the stressed components have a maximum nominal thickness of 12 mm, including any corrosion allowance specified by the client. The stressed components include the shell and reinforcing plates, shell reinforcing plates for flush-type cleanout fittings and flush-type shell connections, and bottom plates that are welded to the shell.

For small tanks, API 650 provides Tables 4.5 and 4.6 to present the tank volume capacity based on the height and the diameter of the tank and the corresponding tank shell thickness, respectively.

In addition, Tables 4.5 and 4.6 are applicable only when specified by the client and are limited to design metal temperatures above −30°C (−20°F) [above −40°C (−40°F) when killed, fine-grain material is used].

Table 4.5 Typical sizes and corresponding nominal capacities (m^3) for tanks with 1800-mm courses

Tank diameter, m	Capacity per meter of height (m^3)	Tank height (in meters)/number of courses in completed tank								
		3.6/2	**5.4/3**	**7.2/4**	**9/5**	**10.8/6**	**12.6/7**	**14.4/8**	**16.2/9**	**18/10**
3	7.07	25	38	51	64	76				
4.5	15.9	57	86	115	143	172				
6	28.3	102	153	204	254	305	356	407		
7.5	44.2	159	239	318	398	477	557	636	716	795
9	63.6	229	344	458	573	687	802	916	1031	1145
10.5	86.6	312	468	623	779	935	1091	1247	1403	1559
12	113	407	611	814	1018	1221	1425	1629	1832	2036
13.5	143	515	773	1031	1288	1546	1804	2061	2319	2576
15	177	636	954	1272	1590	1909	2227	2545	2863	3818
18	254	916	1374	11,832	2290	2748	3206	3664	4122	4580
21	346	1247	1870	2494	3117	3741	4364	4988	5089	$D = 18$
24	452	1629	2443	3257	4072	4886	5700	5474	$D = 20$	
27	573	2061	3092	4122	5153	6184	6690	$D = 22$		
30	707	2545	3817	5089	6362	7634	$D = 26$			
36	1018	3664	5497	7329	9161	$D = 30$				
42	1385	4988	7481	9975	$D = 36$					
48	1810	6514	9772	11,966						
54	2290	8245	12,367	$D = 46$						
60	2827	10,179	15,268							
66	3421	12,316	16,303							
			$D = 62$							

Table 4.6 Typical sizes and corresponding nominal capacities (barrels) for tanks with 72-in. courses

Tank diameter, ft	Capacity per meter of height barrels (m^3)	Tank height (ft)/number of courses in completed tank								
		12/2	**18/3**	**24/4**	**30/5**	**36/6**	**42/7**	**48/8**	**54/9**	**60/10**
10	14.0	170	250	335	420	505				
15	31.5	380	565	755	945	1130				
20	560	670	1010	1340	1680	2010	2350	2690		
25	87.4	1050	1570	2100	2620	3150	3670	4200	4720	5250
30	126	1510	2270	3020	3780	4530	5290	6040	6800	7550
35	171	2060	3080	4110	5140	6170	7200	8230	9250	10,280
40	224	2690	4030	5370	6710	8060	9400	10,740	12,100	13,430
45	283	3400	5100	6800	8500	10,200	11,900	13,600	15,300	17,000
50	350	4200	6300	8400	10,500	12,600	14,700	16,800	18,900	21,000
60	504	6040	9060	12,100	15,110	18,130	21,150	24,190	37,220	28,260
70	685	8230	12,340	16,450	20,580	24,700	28,800	32,930	30,970	$D = 58$
80	895	10,740	16,120	21,500	26,880	32,260	37,600	35,810	$D = 64$	
90	1133	13,600	20,400	27,220	43,030	40,820	40,510	$D = 73$		
100	1399	16,800	25,200	33,600	42,000	48,400	$D = 83$			
120	2014	42,190	36,290	48,380	58,480	$D = 98$				
140	2742	32,930	49,350	65,860	$D = 118$					
160	3581	43,000	64,510	74,600						
180	4532	54,430	81,650	$D = 149$						
200	5595	67,200	100,800							
220	6770	81,310	102,830							
			$D = 202,335$							

The maximum tensile stress before the joint efficiency factor is applied shall be 145 MPa.

The stresses shall be computed on the assumption that the tank is filled with water (specific gravity = 1.0) or with liquid with a higher specific gravity than water.

The tension in each ring shall be computed 300 mm (12 in.) above the centerline of the lower horizontal joint of the course in question. When these stresses are computed, the tank diameter shall be taken as the nominal diameter of the bottom course.

The joint efficiency factor shall be 0.85 with the spot radiography. By agreement between the client and the contractor, the spot radiography may be omitted, and a joint efficiency factor of 0.70 shall be used.

4.2.9 Ring Beam Design Consideration

The load used to design the tank shell will be the same as that for the ring beam, except the static equipment expert or manufacturer which designs and builds the steel tank will deliver the wind load and seismic load that will affect the ring beam (Table 4.7).

The forces on the ring beam will be as shown in Fig. 4.3.

The design and specifications for construction of storage tank ring beam foundations shall be adequate for the structure-intended use, in accordance with API STD 650.

The types of storage tanks normally encountered in gas and oil plants have cylindrical shells, essentially flat bottoms, and either cone or floating roofs. The height of these tanks generally does not exceed 18 m (60 ft), and the diameter generally ranges from 3 to 110 m (10–360 ft).

Concrete ring walls are used to help distribute heavy shell loads, such as those from large dome roof tanks where the roof is supported entirely by the shell. Concrete ring walls are also used when the tank must be anchored for internal pressure, earthquake, or wind loads. Ring wall foundation design shall conform to this engineering guideline and to applicable requirements of the soil report.

Table 4.7 Minimum inside edge distance

Tank diameter, D (m)	Minimum inside edge distance edge, L (mm)
$D \leq 24$	100
$24 < D \leq 45$	125
$D \geq 45$	150

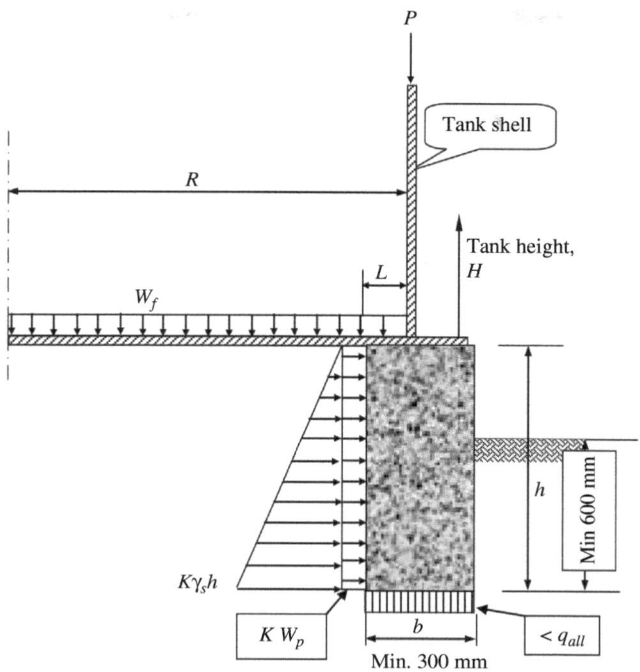

Figure 4.3 Loads that affect the ring beam.

A geotechnical investigation is required before the decision of constructing the tanks, and the soil investigation should be obtained from a geotechnical report according to Chapter 8, Soil Investigation and Pile Design.

The allowable soil-bearing pressure shall be based on the results of the geotechnical investigation and a consideration of permissible total and differential settlements.

Soil pressures shall be calculated under the action of vertical and lateral loads using load combinations that result in the maximum soil pressures.

Foundations shall be founded on either undisturbed soil or compacted fill. In the case of foundations supported on compacted fill, the geotechnical investigation shall govern the type of fill material and degree of compaction required.

In case of sites that have firm soils, recommended steps for site preparation are as follows:

1. Excavate all topsoil and organic material from the site.
2. Ring wall should be a minimum of 15 m above finished grade elevation to ensure adequate drainage is about 150 mm.

3. Slope interior tank pad surface to match tank bottom. A minimum of 1 in. to 10 ft is recommended, taking into account settlement and differential settlement between the tank center and the ring beam to ensure adequate tank drainage.
4. Provide a concrete ring wall to reduce differential settlement along the circumference of the tank shell.

In areas of weak compressible soils, special foundations and design procedures or soil improvement may be required to prevent failures or excessive settlement as recommended in the project soil investigation report.

Ring beam depth below grade should be determined from allowable soil-bearing pressures.

For tanks utilizing interior roof supports, the tank bottom shall be reinforced as required to distribute column loads and reduce local settlement.

In cases where the tank diameter is less than 4.5 m, the engineer may use a solid concrete octagon as opposed to a ring wall foundation.

For the ring beam, concrete with a compressive strength of at least 25 N/mm^2 at 28 days is used.

In summary, the loads effect on the beam in the two directions are as follows:

- *Vertical loading*: Vertical loads are considered in ring wall foundation design. Storage tank foundations shall be designed to support the tank weight plus the weight of stored product or water which is directly over the ring wall.
- *Horizontal loading*: Normally, horizontal loads are not considered in foundation design for storage tanks. However, for tanks having a shell height greater than the diameter $ht > D$, the effect of high winds shall be considered, and the foundations should be checked for necessary anchor bolts.

Wind loads shall be calculated in accordance with the "Structural Design Criteria for Non-Building Structures" and ASCE 7, as discussed in Chapter 3, Static Equipment Foundation Design.

4.2.9.1 Wind and Earthquake Stability and Pressure

Generally, wind stability is not a problem with storage tanks; however, if $ht/D > 1$, the stability should be checked.

For an unanchored tank, the overturning moment from wind pressure shall not exceed two-thirds of the dead load resisting moment, excluding any tank content, and shall be calculated as follows:

$$M_{\mathrm{O.T.}} \leq \frac{2}{3}\left(\frac{W \cdot D}{2}\right)$$

where:

$M_{\mathrm{O.T.}}$ = overturning moment from wind pressure in m.t;

W = shell weight available to resist uplift in tons, less any corrosion allowance, plus dead weight of any portion of roof supported by the shell;

D = tank diameter in meters.

For empty, small, high tanks, the resistance to sliding and overturning under the force of wind should be checked.

If $M_{\mathrm{O.T.}} > 2/3\ (W \cdot D/2)$, anchor bolts are required.

If the anchors are required, the design tension load per anchor shall be calculated as follows:

$$t_B = \frac{(4\ M_{\mathrm{O.T.}})}{dN} - \frac{W}{N}$$

where:

t_B = design tension load per anchor in tons,

d = diameter of anchor circle in meters,

N = number of anchors.

The anchors shall be spaced a maximum of 3 m (10 ft) apart.

4.2.9.2 Earthquake Stability

The seismic overturning moment in tank and shear forces shall be provided by the tank's vendor. The minimum safety factor against overturning and sliding shall be 1.5. For seismic forces computations, refer to ASCE 7 and API STD 650.

4.2.9.3 Soil Bearing

In selecting the proper type of foundation, the bearing capacity of the soil is the primary factor. A thorough knowledge of the soil properties is necessary to avoid excessive differential settlement and possible failure.

Foundations on sandy soil shall be checked using the weight of water in cases of tank hydrostatic tests, since settlement under this condition occurs very rapidly and should be essentially complete after the hydro test weight (which is the first loading).

Foundations on clay soils shall be checked using the weight of the commodity (i.e., as the operation load) because clay settles more evenly and slowly due to consolidation. In some instances, it is advisable to use the water test to consolidate clay layers by slowly filling the tank and waiting for settlement to occur.

In considering soil-bearing pressure due to an empty tank plus wind conditions, there are two possible cases. In the first case, the foundation has no uplift, which means that all of the ring beam is subject to compression.

If there is no uplift, obviously maximum soil pressure for an empty tank plus wind will be less than the maximum soil pressure for an operating tank plus wind. Thus, when there is no uplift, soil pressure for an empty tank, plus wind does not need to be checked.

The e/D_o ratio can be used to determine whether uplift is present. If uplift is *not* present, then:

$$q_{\min} = P/A - M_{\text{O.T.}}/S_{xx} \geq 0$$
$$\text{or } e/D_o \leq S_{xx}/A\ D_o$$

where:

$$S_{xx} = \pi(D_o^4 - D_i^4)/32\ D_o$$
$$A = \pi(D_o^2 - D_i^2)/4$$

Thus:

$$e/D_o \leq 0.125\{1 + (D_i/D_o)^2\}$$
$$\text{or } e/D_o \leq 0.20 \text{ for } D_i/D_o \geq 0.8$$

where:

D_o = ring beam outside diameter in meters,
D_i = ring beam inside diameter in meters,
e = eccentricity in meters.

4.2.9.4 Soil Pressure (With Uplift Present)

If uplift is present ($e/D_o > 0.20$), soil pressure should be calculated since the maximum pressure for an empty tank plus wind could conceivably be higher than an operating tank plus wind. In computing maximum pressure for a particular (e/D_o) ratio, two transcendental simultaneous equilibrium equations must be solved.

Since uplift will seldom be of major consequence with storage tanks, no design curves are developed to solve the two equilibrium equations. If $e/D_o > 0.33$, the engineer may either increase h_b or use another foundation type. However, he should not increase t_b to meet wind requirements.

Example: A steel storage tank has a diameter of 15.2 m and height 16.2 with conical closed roof. The wind velocity is 130 km/h and the seismic zone is 0.15 g, which is required to calculate the wind and seismic load. If you need the excel sheet for calculation, it is available here: www.elreedyma.comli.com. or www.elreedyman.tk.

Solution:

Overturning moment due to seismic forces (according to API Standard)

M = overturning moment due to seismic force applied to the bottom of the shell

$M = ZI\,(C_1\ Ws\ Xs + C_1\ Wr\ H_t + C_1\ W_1\ X_1 + C_2\ W_2\ X_2)$

$Z =$	0.150	Zone coefficient
$I =$	1.00	Essential facilities factor
H_t (m) =	16.20	Total height of the tank shell
D (m) =	13.81	Nominal tank diameter
H (m) =	16.00	Maximum design liquid level
Xs (ft)	8.10	Height from the bottom of the tank to the shell's center of gravity
$D/H =$	0.86	Ratio
Wt (ton) =	8147.10	Total weight of tank contents
Wr (ton) =	37.60	Total weight of tank roof
Ws (ton) =	47.90	Total weight of tank shell (wall)

W_1 = the weight of the effective mass of the tank contents that move in unison with the tank shell

W_2 = the weight of the effective mass of the tank contents that move in the first sloshing mode

By using* Fig. E-2 *from API 650

We can get W_1 and W_2

$W_1/Wt =$	0.82
$W_2/Wt =$	0.2
W_1 (ton) =	6680.619
W_2 (ton) =	1629.419
S (m) =	6

***By using* Fig. E-4**

$T = k(D^{0.5})$

$K =$	0.59
$T =$	2.19
$C_1 =$	0.6
$C_2 =$	0.68408

***By using* Fig. E-3**

We can get X_1 and X_2

$X_1/H =$	0.78
$X_2/H =$	0.42
X_1 (ft) =	12.48
X_2 (ft) =	6.72

Where S = site amplification factor

Where T = natural period of first sloshing mode

M (m.t) = 8716.978

Overturning moment due wind load (according to API 650 Standard)
This load based on $V \leq 190$ km/h (120 mph)
Projected area subjected to this wind pressure

$= Ht \times D$	assume we neglect the shell thickness
Area (m^2) =	223.76
Wind press (kN/m^2) =	0.86
Wind force (kN) =	192.433
Wind moment (kN m) =	1558.71

4.2.10 Concrete Ring Beam Design

The ring beam depth shall be determined from the allowable soil-bearing pressure requirements.

The thickness of ring wall t_b is determined from a consideration of bearing pressure. However, in order to minimize differential settlements, the bearing pressure under the ring wall must equal the bearing pressure beneath the tank at the same elevation as the ring wall bottom. The minimum ring wall width shall be 300 mm (12 in.).

Soil pressure under main portion of tank (t/m^2):

$$P_o = \gamma_o h_L + \gamma_S h \tag{4.11}$$

Soil pressure beneath ring beam (t/m^2):

$$P_o = \frac{(P)}{b} + \gamma_o \left(\frac{L \cdot h_L}{b} \right) + h\,\gamma_C \tag{4.12}$$

Equating Eqs. (4.11) and (4.12) and solving for b yields:

$$b = \frac{P + L \cdot h_L \cdot \gamma_o}{\gamma_o h_L - h(\gamma_C - \gamma_S)} \tag{4.13}$$

where:

γ_o = unit weight of tank product in t/m^3,
γ_S = unit weight of soil in t/m^3,
γ_C = unit weight of concrete in t/m^3 = 2.5 t/m^3,
h_L = height of tank product in meters,
h = height of ring beam in meters,
P = weight of tank wall and roof weight per meter of circumference in t/m',
b = width of ring wall in meters.

The beam width should satisfy the following equation:

$$q = \frac{P}{b} + \gamma_o \left(\frac{L \cdot h_L}{b} \right) + h\,\gamma_C \leq q_{all}$$

where q_{all} is the allowable bearing pressure.

4.2.10.1 Ring Wall Reinforcement

Horizontal reinforcing of the ring beam foundations shall be designed to resist horizontal active earth pressure in hoop tension, with horizontal reinforcing sized to take all of the tension.

Passive soil pressure is neglected and shall not be used to reduce the horizontal design loads. The normal horizontal (radial) forces acting at every section along the ring wall are shown later.

F_o = horizontal force due to products in tanks = $K_a\,\gamma_o\,(h)\,h_L$,

F_s = horizontal active soil = $K_a\,\gamma_S\,(h)^2/2$.

Total force: $F = F_o + F_s$

Thus:

$$F = K_a\,h\left(\gamma_o h_L + \frac{\gamma_S h}{2} \right) \tag{4.14}$$

where:

F = radial outward force on ring beam, t/m;

K_a = active earth pressure coefficient = $\tan^2(45° - \Phi/2)$;

Φ = angle of internal soil friction.

If Φ is not known, K_a should be taken as 0.3 for sand and 0.7 for clay.

The radial force F causes a hoop tension in the ring, which is presented in Fig. 4.4:

For working stress design:

$$T = \frac{1}{2}\,F\,D \tag{4.15}$$

For limit state design based on ACI:

$$T_U = 1.7\,T \tag{4.16}$$

where:

T = hoop tension in ring beam,

T_U = ultimate hoop tension in ring beam,

D = tank diameter.

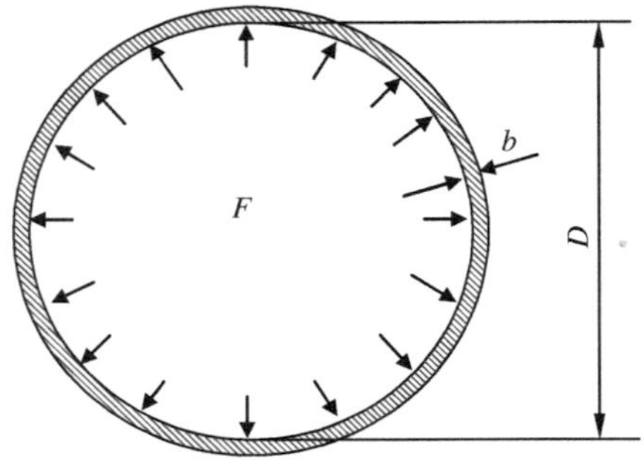

Figure 4.4 Hoop stresses.

The horizontal reinforcing shall be designed to carry all the tension.

By using ultimate stress design:

$$As = \frac{T_U}{\Phi fy}$$

The minimum horizontal reinforcing shall equal 0.0025 times the ring cross-section area.

$$As(\text{min.}) = 0.0025 \times h \times b$$

All circumferential tension steel shall be continuous with splices staggered. All splices shall be class B type (1.3 l_d), as given in ACI 318-99.

The minimum for vertical reinforcing shall equal 0.0015 times the ring cross-section area but not less than 5 bars with a diameter of 8 mm every meter ($5\phi 8$/m').

$$As(\text{min.}) = 0.0015 \times h \times b$$

In special cases for areas of weak compressible soils, special foundations and design procedures or soil improvement may be required to prevent excessive settlements. As a general guideline, if the differential settlement approaches 12–20 mm in a distance of 9–12 m of the shell circumference of an ordinary flat bottom tank, a detailed evaluation of the soil pressure shall be made. Under no circumstances shall the soil pressure exceed the allowable soil pressure.

Consideration shall also be given to the location of drain out nozzles at the bottom of the tank and appropriate block out details provided in the concrete ring wall foundation.

This will prevent any interference with the drain out nozzle flanges.

Example: Floating roof tank:

A—TANK DATA (from vendor drawings or tank data sheet):

Floating roof tank as shown in Fig. 4.5

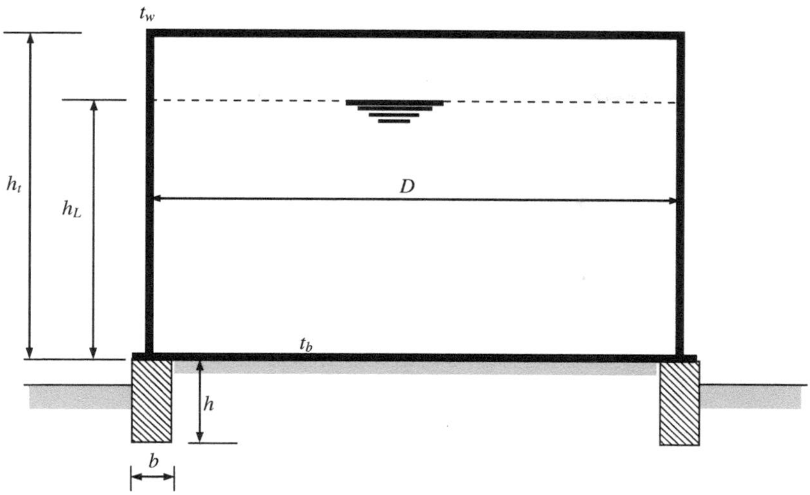

Figure 4.5 Sketch for example ring beam design.

Diameter $D = 110$ m
Tank height $h_T = 18$ m
Avg. wall plate thickness, $t_w = 22$ mm
Bottom plate thickness = 10 mm
Height of liquid $h_L = 17.7$ m
Unit weight of product $\gamma_o = 9$ t/m^3
Unit weight of water to hydrostatic test $\gamma_o = 1.0$ t/m^3

Assume that the weight of stairs and platform is 5% from the weight of the tank shell.

B—SOIL DATA (from the Geotechnical Soil Investigation Report as discussed in Chapter 8):

Unit weight of soil $\gamma_S = 2$ t/m^3
$K_a = 0.30$
Allowable bearing pressure = 0.2 MPa

Solution:

C—FOUNDATION MATERIAL:

$f'_c = 30$ MPa
$\gamma_C = 2.5$ t/m^3
Reinforcement $f_y = 360$ N/mm^2

D—Wind Load:

Basic wind speed = 93 mph per the Metocean data delivered from the owner

E—Seismic Load:

Seismic zone 0

F—Stability Analysis:

In this case, the tank diameter $D > h_L$, so wind analysis is not required. No uplift will occur due to wind load.

G—Tank Ring beam Design and Soil-Bearing Pressure:

Assume ring beam depth $h_b = 1250$ mm

Assume ring beam thickness $t_b = 450$ mm

Weight of tank wall per meter of circumference, $W_w = 0.022 \times 7.85 \times 18 \times 1.05 = 3.3$ t/m'

I—Forces Acting on Ring Beam:

1—Soil-Bearing Pressure Under Ring Beam:

$W_R = 0$ for floating root

$L = b/2$

$P_o = (W_W + W_R)/b + \gamma_o\, h_L/2 + h\,\gamma_C$

$P_o = (3.3 + 0)/0.45 + (0.9 \times 17.7)/2 + 1.25 \times 2.5$

$P_o = 7.33 + 7.97 + 3.125 = 18.43\ \text{t/m}^2 < 19.2\ \text{t/m}^2$ Ok

2—Soil-Bearing Pressure Under Tank:

$q_o = \gamma_o\, h_L + \gamma_S\, h$

$q_o = (0.9 \times 17.7) + (1.8 \times 1.25) = 18.18\ \text{t/m}^2 < 20\ \text{t/m}^2$ OK

Soil-bearing pressure under the ring wall and tank bottom is almost equal, therefore no settlement problem will occur.

3—Ring Wall Thickness:

$b = (W_W + W_R)/\{\gamma_P\, h_L - 2\,h\,(\gamma_C - \gamma_S)\}$

b (product) $= (3.3 + 0)/\{0.9 \times 17.7 - 2 \times 1.25\ (2.5 - 1.8)\} = 0.23$ m

b (hydro test) $= 3.3/(1.0 \times 17.7 - 2 \times 1.25\ (2.5 - 1.81)) = 0.21 < 0.45$ OK

J—Ringwall Reinforcement:

1—Horizontal Reinforcement:

Radial outward force F on ring beam (t/m)

For Hydro test, as a maximum loading case:

$F_H = 0.3 \times 1.25(1.0 \times 17.7 + 2X1.25/2) = 7.11$ t/m'

Hoop tension T in ring beam = ½ FD

$T_H = ½ \times 7.11 \times 109.78 = 390$ ton

Using ultimate stress design

$TU = 1.7 \times 390 = 663$ ton

As (required) $= TU/\Phi f_y$

As (required) $= 663\ (9800)/0.9 \times 360 =$ **20,054 mm²**

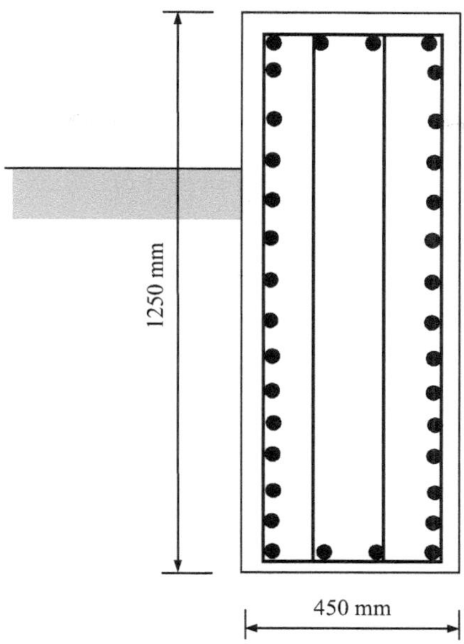

Figure 4.6 Reinforced detail for ring beam.

As (provided) = 22ϕ 25 mm (2faces) = 21 598 mm^2 > **20,054 mm^2** OK

As (minimum) = 0.0025 × 1250 × 450 = 1406 mm^2 < 20,054 mm^2 provided OK

2—Vertical Reinforcement:

As (minimum) = $0.0015 \times h_R \times t_R$

As (minimum) = 0.0015 × 1.25 × 0.45 × 1,000,000 = 843.75 mm^2

Choose 6ϕ 10 mm/m'

$As = 2 \times 78 \times 6 = 936$ mm^2 > As (min) OK

Detailing of the steel bars in the ring beam is illustrated as shown in Fig. 4.6.

4.2.11 Piping Flexibility

Piping systems connected to tanks shall consider the potential movement of the connection points during earthquakes and provide sufficient flexibility to avoid release of the product by failure of the piping system.

The piping system and supports shall be designed so as not to impart significant mechanical loading on the attachment to the tank shell. Local

Table 4.8 Design displacement for piping connecting to the tank

Condition	ASD design displacement (mm)
Mechanically anchored tanks	
Upward vertical displacement relative to support or foundation	25
Downward vertical displacement relative to support or foundation	13
Range of horizontal displacement (radial and tangential) relative to support or foundation	13
Self-anchored tanks	
Upward vertical displacement relative to support or foundation	
Anchorage ratio less than or equal to 0.785	25
Anchorage ratio greater than 0.785	100
Downward vertical displacement relative to support or foundation	
For tanks with a ring wall/mat foundation	13
For tanks with a berm foundation	25
Range of horizontal displacement	50

loads at piping connections shall be considered in the design of the tank shell. Mechanical devices which add flexibility such as bellows, expansion joints, and other flexible apparatus may be used when they are designed for seismic loads and displacements.

Unless otherwise calculated, piping systems shall provide for the minimum displacements in Table 4.8 at working stress levels (with the 33% increase for seismic loads) in the piping, supports, and tank connection.

For attachment points located above the support or foundation elevation, the displacements in Table 4.8 shall be increased to account for drift of the tank or vessel.

The values given in Table 4.8 do not include the influence of relative movements of the foundation and piping anchorage points due to foundation movements (such as settlement or seismic displacements).

The effects of foundation movements shall be included in the design of the piping system design, including the determination of the mechanical loading on the tank or vessel consideration of the total displacement capacity of the mechanical devices intended to add flexibility (Fig. 4.7).

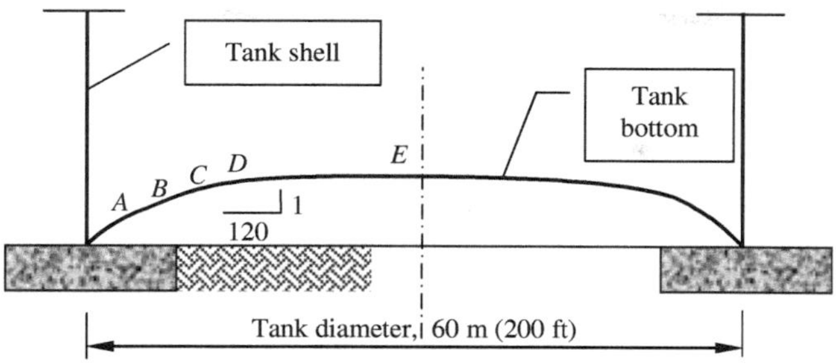

Figure 4.7 Tank bottom fabrication.

Table 4.9 Tank bottom coordinates based on EMUEA

Point	Distance from tank shell, m (ft)	Height above elevation (0,0) m, (ft)
A	3 (10)	0.22 (0.71)
B	6 (20)	0.39 (1.28)
C	9 (30)	0.53 (1.73)
D	12 (40)	0.61 (2.0)
E	30.5	0.76 (2.5)

4.3 DIFFERENTIAL SETTLEMENT TANK BOTTOM DESIGNS

Based on the degree and type of settlement expected as determined from similar installations in the area or from soil surveys, there are different methods to provide increasing effectiveness, although these come with increasing prices (Table 4.9). They are shown here in order from lowest to highest price:

1. standard lap-welded bottom,
2. annular plates with lap-welded bottom,
3. butt-welded bottoms.

Unless needed for a large differential settlement, the butt-welded tank bottom design is usually rejected on a cost/benefit basis.

Additional design features such as the following can provide further effectiveness:

4. deeper levels of soil compaction,
5. crushed stone ring walls,
6. reinforced concrete ring walls,
7. slabs on ring wall foundations.

Tank size

- For large tanks (15 m or greater), a concrete ring beam (preferred) or crushed stone ring wall should be used.
- For small tanks (6 m or less), a concrete slab foundation (preferred) or compacted granular fill foundation should be used (Fig. 4.8).

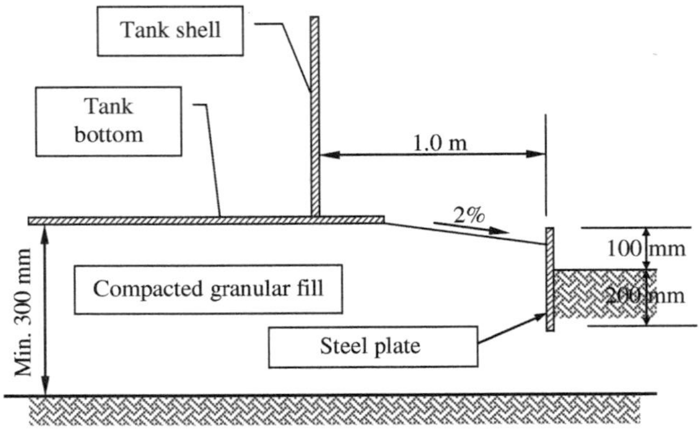

Figure 4.8 Compacted granular fill foundation.

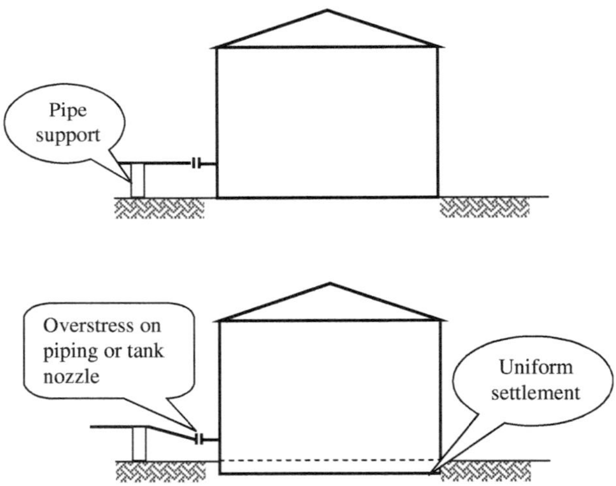

Figure 4.9 Uniform settlement.

- For medium tanks (6–15 m), the type of foundation should be at the discretion of the foundation design engineer.
- The effect of differential settlement on the tanks is presented in Figs. 4.9 and 4.10.

4.4 CONCRETE STORAGE TANKS

There are two classifications of tanks based on the load effects. If they are elevated tanks, they will be under hydrostatic pressure; if they are underground tanks, they will be affected by hydrostatic load and earth load.

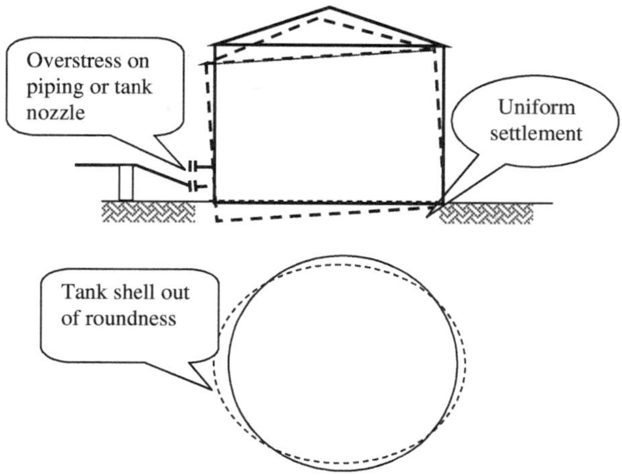

Figure 4.10 Effect of differential settlement in tank.

The tank shape is classified into one of the following categories (Fig. 4.11).

4.4.1 Rectangular Wall Concrete

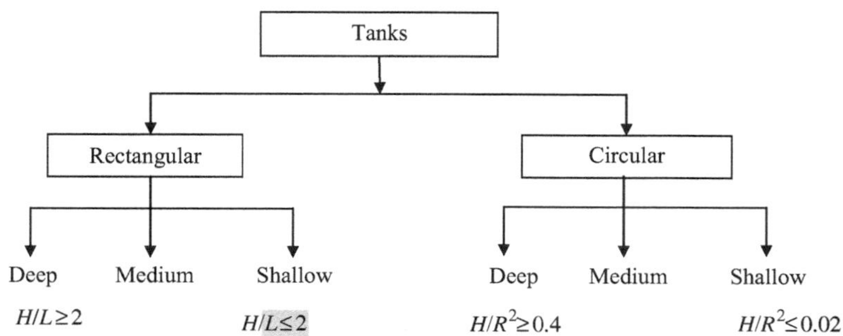

Figure 4.11 Types of reinforced concrete tanks.

Rectangular tanks are usually used in industrial projects for special applications. The wall of the rectangular tanks has a different statically system as shown in Fig. 4.12. Types of rectangular walls for the tanks will be as follow

1. A cantilever wall (up to 3 m).
2. A type of wall that is simply supported at the top and fixed at the bottom, $H \geq 3$ m.
3. A wall fixed at top and bottom.

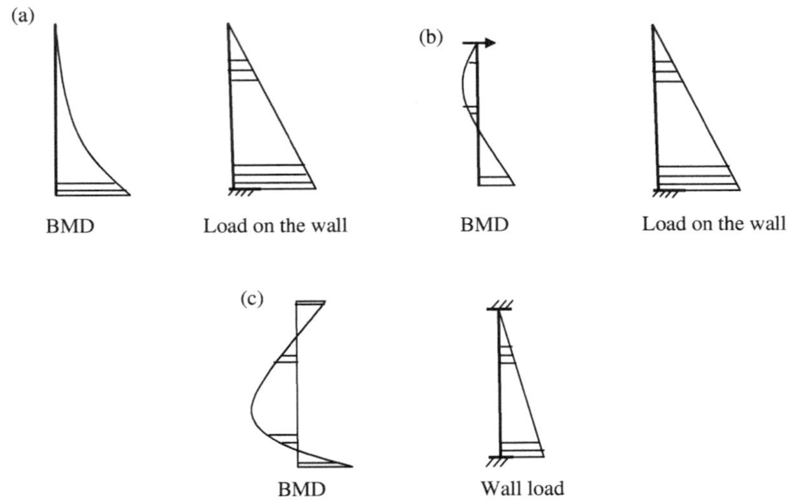

Figure 4.12 (a) Cantilever wall, (b) wall fixed support at the bottom and simply supported at the top and (c) different type of tank joints.

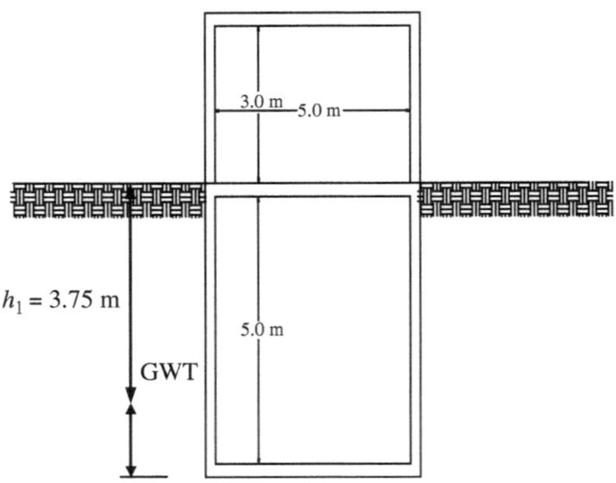

Figure 4.13 Sketch of a pump room.

Example of a Pump Room:

This example features a pump room embedded in the earth with volume $5 \times 5 \times 5$ m with a control room above ground with dimensions $5 \times 5 \times 3$ m height, as shown in Fig. 4.13.

A—Soil parameter:

Soil density, $\gamma_s = 1.8\ t/m^3$

Water density, $\gamma_w = 1.0\ t/m^3$

$K_a = 0.333$

$\Phi = 30^\circ$

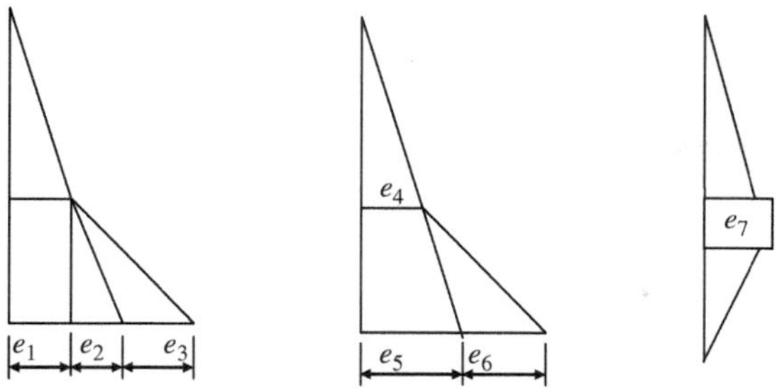

Figure 4.14 Distributed load on the walls.

Solution:

Earth pressure diagram will be as follows:
$r = 5/5 = 1.0$, $\alpha = \beta = 0.5$, according to Grashoff Table 4.10, will be used to distribute the load on the rectangular wall (Fig. 4.14, Table 4.10).

$$e_1 = \gamma_s\, h_1\, k_a = 1.8 \times 3.75 \times 0.3 = 2.25\ \text{t/m}^2$$
$$e_2 = (\gamma_s - \gamma_w)h^2 k_a = (1.8 - 1) \times 1.25 \times 0.3 = 0.33\ \text{t/m}^2$$
under ground water level (GWT)
$$e_3 = \gamma_w h_2 = 1 \times 1.25 = 1.25\ \text{t/m}^2$$
$$e_4 = \alpha e_1 = 0.5 \times 2.25 = 1.125\ \text{t/m}^2$$
$$e_4/e_5 = 3/4$$
$$e_5 = 1.5\ \text{t/m}^2$$
$$e_6 = (2.25 + 0.33 + 1.25) - 1.5 = 2.33\ \text{t/m}^2$$
$$e_7 = \beta e_1 = 0.5 \times 2.25 = 1.125\ \text{t/m}^2$$

B—Loads on the Roof:
Assume the thickness of the roof and the pump room are 100 and 120 mm, respectively.
Weight of on-roof slab = 250 kg/m^2—own weight
150 kg/m^2—plastering
200 kg/m^2—live load
Total = 600 kg/m^2
Weight on pump room slab = 300 kg/m^2—own weight
150 kg/m^2—plastering
200 kg/m^2—live load
Total = 650 kg/m^2
Total load from slab = (600 + 650) × 5 × 5 = 31,250 kg = 31.25 tons
Weight of brick walls = (4 × 3 × 5 × 0.25) × 1.7 = 25.5 tons
Weight of reinforced concrete walls = (4 × 5 × 5 × 0.2) × 2.5 = 50 tons
Total load = 106.75 tons

Table 4.10 Grashoff values for distributing the load in both directions (α,β)

r	1.0	1.1	1.2	1.3	1.4	1.5	1.6	1.7	1.8	1.9	2.0
α	0.5	0.595	0.672	0.742	0.797	0.834	0.867	0.893	0.914	0.928	0.914
β	0.5	0.405	0.328	0.258	0.203	0.166	0.107	0.107	0.086	0.072	0.059

r, is the ratio between the slab or wall dimensions, length and width.
where, α,β are the load distribution in short and long direction respectively.

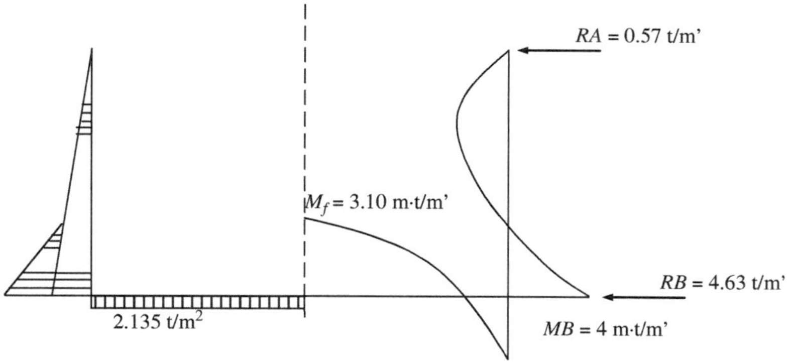

Figure 4.15 Longitudinal section bending moment and loads.

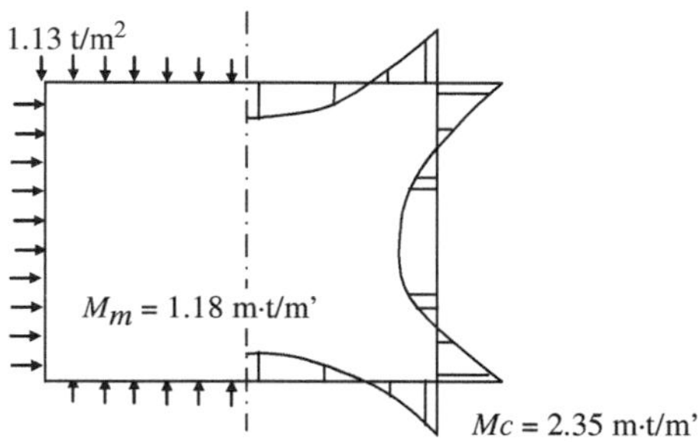

Figure 4.16 Shape of bending moment and load in cross section.

Load on floor/m^2 = 106.75/25 = 4.27 t/m^2

$r = 5/5 = 1.0$, $\alpha = \beta = 0.5$

Load in each direction = 0.5 × 4.27 = 2.135 t/m^2

C—Straining Action:

In the vertical direction, Fig. 4.15 presents a diagram of a longitudinal section with a load and bending moment.

In a horizontal direction, considering a 1.0 m strip at $h/4$ from the room floor, the structure analysis can provide us the bending moment diagram as shown in Fig. 4.16.

D—Design of Sections:

Assume a thickness of the floor (t_f) is equal to the wall thickness (t_w) equal to 200 mm for the bottom corner due to maximum moment will do a haunches with 350 mm with slope 3:1.

The steel detailing for the basement will be as shown in Fig. 4.17 considering the concrete section under the water will be design with no cracks as their tension zone adjacent to the water.

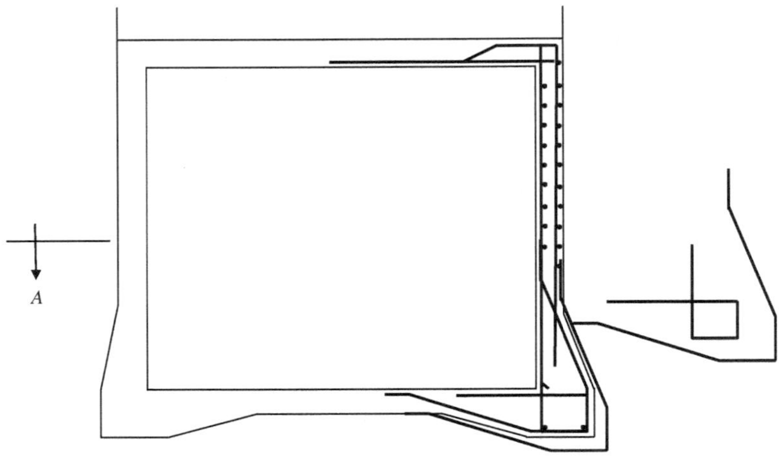

Figure 4.17 Reinforcement detail for the basement of the pump room.

4.4.2 Circular Tank

Based on Hilal (1988), The American Portland Cement Association has published the following series of tables giving the ring tension and cantilever moments in rectangular walls of cylindrical tanks free at the top and fixed or hinged at the bottom for different cases of loading. The tables include a lot of data very useful in the design of circular tanks.

Fig. 4.18 presents the construction of concrete tanks for a petrochemical project.

Tables 4.11 and 4.12 present the tension force and bending moment, respectively, on the wall for a tank fixed at the base and free at the top.

where: $T = \chi \times \gamma_w \times H \times (D/2)$

The positive sign indicates tension

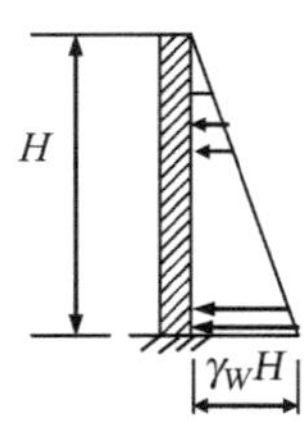

where:

H is the tank height,

D is the tank diameter.

t is the wall thickness.

Figure 4.18 Concrete tanks.

To calculate moment at the fixed end:

$$M = \mu \gamma_w H^3$$

Example:

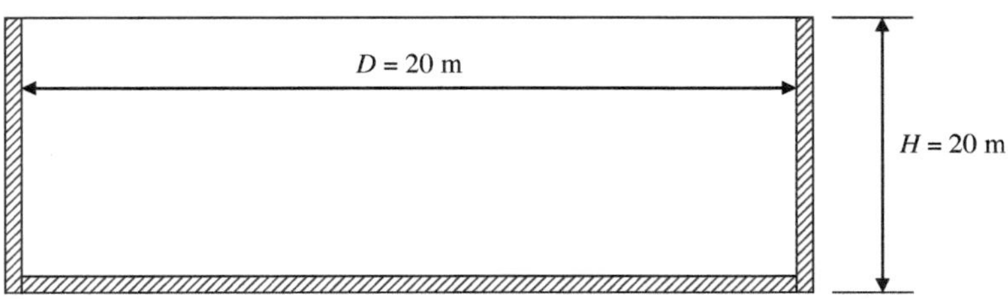

The circular tank is full of water, the tank wall upper end is free, and the bottom end is fixed.

Solution:

In this case, the water pressure will be resisted by ring tension in the horizontal direction and cantilever action in the vertical direction. We assume as a first trial that the wall thickness is 250 mm.

Table 4.11 Coefficient, χ, tension in circular rings with a triangular load, fixed base, and free top

H^2/Dt	0.0*H*	0.1*H*	0.2*H*	0.3*H*	0.4*H*	0.5*H*	0.6*H*	0.7*H*	0.8*H*	0.9*H*
0.4	+0.149	+0.134	+0.120	+0.101	+0.082	+0.066	+0.049	+0.029	+0.014	+0.004
0.8	+0.263	0.239	0.215	+0.190	0.160	+0.130	+0.096	+0.063	+0.034	+0.010
1.2	+0.283	+0.271	+0.254	+0.234	+0.209	+0.180	+0.142	+0.099	+0.054	+0.016
1.6	+0.265	+0.268	+0.268	+0.266	+0.250	+0.226	+0.185	+0.134	+0.075	+0.023
2.0	+0.234	+0.251	+0.273	+0.285	+0.285	+0.274	0.232	0.172	+0.104	+0.031
3.0	+0.134	+0.203	+0.267	+0.322	+0.357	+0.362	+0.330	+0.262	0.157	+0.052
4.0	+0.067	+0.164	+0.256	+0.339	+0.403	+0.429	+0.409	+0.334	+0.210	+0.073
5.0	+0.025	+0.137	+0.245	+0.346	+0.428	+0.477	+0.469	+0.398	+0.259	+0.092
6.0	+0.018	+0.119	+0.234	+0.344	+0.441	+0.504	+0.514	+0.447	+0.301	+0.112
8.0	−0.011	+0.104	+0.218	+0.335	+0.443	+0.534	+0.575	+0.530	+0.381	+0.151
10.0	−0.011	+0.098	+0.208	+0.323	+0.437	+0.542	+0.608	+0.589	+0.440	+0.179
12.0	−0.005	+0.097	+0.202	+0.312	+0.429	+0.543	+0.628	+0.633	+0.494	+0.211
14.0	−0.002	+0.098	+0.200	+0.306	+0.420	+0.539	+0.639	+0.666	+0.541	+0.241
16.0	0.000	+0.099	+0.199	+0.304	+0.412	+0.531	+0.641	+0.687	+0.582	+0.265

Table 4.12 Coefficient, μ, moment in cylindrical wall with a triangular load, fixed base and free top

H^2/Dt	0.1*H*	0.2*H*	0.3*H*	0.4*H*	0.5*H*	0.6*H*	0.7*H*	0.8*H*	0.9*H*	1.0*H*
0.4	+0.0005	+0.0014	+0.0021	+0.0007	−0.0042	−0.0150	−0.0302	−0.0529	−0.0816	−0.1205
0.8	+0.0011	+0.0037	+0.0063	+0.0080	+0.0070	+0.0023	−0.0068	−0.0224	−0.0465	−0.0795
1.2	+0.0012	+0.0042	+0.0077	+0.0103	+0.0112	+0.0090	+0.0022	−0.0108	−0.0311	−0.0602
1.6	+0.0011	+0.0041	+0.0075	+0.0107	+0.0121	+0.0111	+0.0058	−0.0051	−0.0232	−0.0505
2.0	+0.0010	+0.0035	+0.0068	+0.0099	+0.0120	+0.0115	+0.0075	−0.0021	−0.0185	−0.0436
3.0	+0.0006	+0.0024	+0.0047	+0.0071	+0.0090	+0.0097	+0.0077	+0.0012	−0.0189	−0.0333
4.0	+0.0003	+0.0015	+0.0028	+0.0047	+0.0066	+0.0077	+0.0069	+0.0023	−0.0080	−0.0268
5.0	+0.0002	+0.0008	+0.0016	+0.0029	+0.0046	+0.0059	+0.0059	+0.0028	−0.0058	−0.0222
6.0	+0.0001	+0.0003	+0.0008	+0.0019	+0.0032	+0.0046	+0.0051	+0.0029	−0.0041	−0.0187
8.0	0.0000	+0.0001	+0.0002	+0.0008	+0.0016	+0.0028	+0.0038	+0.0029	−0.0022	−0.0146
10.0	0.0000	0.0000	+0.0001	+0.0004	+0.0007	+0.0019	+0.0029	+0.0028	−0.0012	−0.0122
12.0	0.0000	−0.0001	+0.0001	+0.0002	+0.0003	+0.0013	+0.0023	+0.0026	−0.0005	−0.0104
14.0	0.0000	0.0000	0.0000	0.0000	+0.0001	+0.0008	+0.0019	+0.0023	−0.0001	−0.0090
16.0	0.0000	0.0000	−0.0001	−0.0002	+0.0001	+0.0004	+0.0013	+0.0019	+0.0001	−0.0079

From the table $H2/Dt = 5$:

Point	H^2/Dt	0.0*H*	0.1*H*	0.2*H*	0.3*H*	0.4*H*	0.5*H*	0.6*H*	0.7*H*	0.8*H*	0.9*H*
Ring tension	χ	+0.025	+0.137	+0.245	+0.346	+0.428	+0.477	+0.469	+0.398	+0.259	+0.092
	T (t/m')	1.25	6.85	12.25	17.3	21.4	23.85	23.45	19.9	12.9	4.6
Bending	μ	0.0	+0.0002	+0.0008	+0.0016	+0.0029	+0.0046	+0.006	+0.006	+0.003	−0.006
moment	*M* (m.t/m')	0.0	0.025	0.10	0.20	0.36	0.58	0.74	0.74	0.35	−0.72

REFERENCES

API 650 Welded Steel Tanks for Oil Storage, 10th ed.
Hilal, M., 1988. Theory and Design of Reinforced Concrete Tanks. J & Marcou Co, Cairo, Egypt.
Zick, L.P., McGrath, R.V., 1968. Design of large diameter cylindrical shells, Proceedings—Division of Refining, vol. 48. American Petroleum Institute, New York, pp. 1114–1140.

CHAPTER 5

Steel Structures in Industry

5.1 INTRODUCTION

Steel is the most popular material used in industrial projects, as a steel has advantages which are matched with industrial projects' targets. It is easy to strengthen steel if required without using advanced technology, it is easier and faster to erect as it can be assembled in a short period of time, and it retains its value when the structure needs to be deconstructed.

The main disadvantage of steel is that it requires periodic painting to protect it from corrosion; however, it is well-matched with the function of the industrial field prospective, as there is always a large budget for maintenance and competent corrosion engineers specifically for oil and gas fields.

5.2 STRESS—STRAIN BEHAVIOR OF STRUCTURAL STEEL

Structural steel is an important construction material. It possesses attributes such as strength, stiffness, toughness, and ductility that are very desirable in modern construction.

Strength is the ability of a material to resist stresses. It is measured in terms of the material's yield strength, F_y, and ultimate or tensile strength, F_u. For steel, the ranges of F_y and F_u ordinarily used in constructions are 36—50 ksi (248—345 MPa) and 58—70 ksi (400—483 MPa), respectively, although higher strength steels are becoming more common. Stiffness is the ability of a material to resist deformation. It is measured as the slope of the material's stress—strain curve. Fig. 5.1, in which uniaxial engineering stress—strain curves obtained from tests for various grades of steels are shown, indicates that the modulus of elasticity, E, does not vary appreciably for different steel grades. Therefore, a value of 29,000 ksi (200 GPa) is often used for design purposes.

Toughness is the ability of a material to absorb energy before failure. It is measured as the area under the material's stress—strain curve. As shown in Fig. 5.1, most (especially lower grade) steels possess high toughness, which makes them suitable for both static and seismic applications.

Onshore Structural Design Calculations.
DOI: http://dx.doi.org/10.1016/B978-0-08-101944-3.00005-X

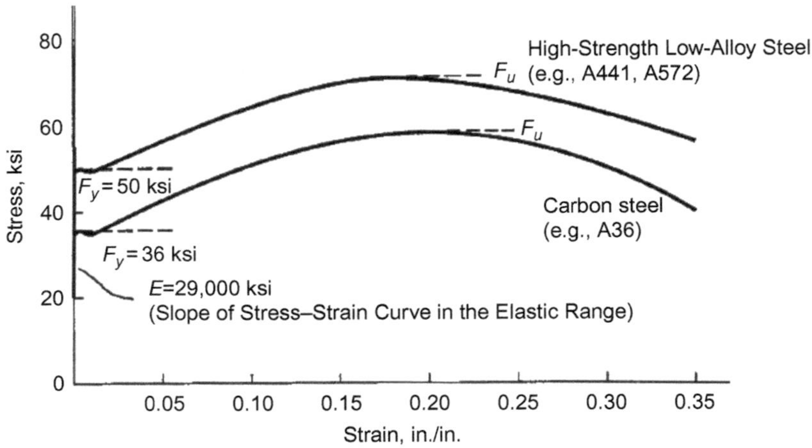

Figure 5.1 Stress-Strain beahvior of steel.

Ductility is the ability of a material to undergo large inelastic, or plastic, deformation before failure. It is measured in terms of percent elongation or percent reduction in area of the specimen tested in uniaxial tension.

Now we go through the principal of the design of the main component.

5.3 DESIGN PROCEDURE

Normally, we start by creating a preliminary design; this is a rapid approximate manual method of designing a structure, as opposed to carrying out rigorous analysis and detailed design. The overall aim for a given structure is to identify critical loads, estimate design actions, and select sections. The problem is bound up with conceptual design, alternative systems, idealization, identification of critical members, and rationalization. The process depends greatly on the designer's experience and use of appropriate design aids. The term is also often applied to manual design. Preliminary design is needed for the following reasons:

- to obtain sections and weights for cost estimation,
- to compare alternative proposals,
- to obtain initial sections for computer analysis,
- to check a completed design.

The need for approximate manual methods is more important than ever. It is necessary to know if an output is right, wrong, or complete nonsense.

Methods of preliminary analysis that are not dependent on member sizes are set out for both elastic and plastic theories.

After defining the structure system in the preliminary design, we work on the detailed engineering by building the computer model and then select the safe members. Then, you are ready to deliver the complete engineering tender package.

5.3.1 Tension Members

Tension members are structural elements that are subjected to pure tensile forces. The selection of their cross section is one of the simplest and most straightforward problems encountered in steel design. Since stability is of minor concern with tension members, the problem is reduced to selecting a section with sufficient area to carry the design load without exceeding the allowable tensile stress, as determined by the factor of safety.

A structural member is considered a compression member if it is under a compressive load, either alone or with other loadings. Columns, beam-columns, plates, component parts of frames, and the compression flange of beams or plate girders could be included in this category.

The two main differences between tension and compression members are:

- Tension members are held straight by means of tensile loads, while in compression members, the compressive loads tend to bend the member out of the plane of loading.
- For riveted and bolted connections, the net area will govern the strength of a tension member, while for compression members the rivets and holes are assumed to fill the holes.

For tension members, stability phenomena are not criteria in the design, but they are required to prevent sagging for tension members if they are too long or utilized to support vibrating equipment. In this case, the slenderness ratio ($\lambda = kL/r$), where L is the tension member length and r is the radius of gyration, should be not higher than 300 and the buckling length factor is equal to 1.0.

5.3.1.1 Slenderness Ratio

The slenderness ratio is obtained from the following equation:

$$\lambda = \frac{kL}{r}$$

where:

L = the slenderness ratio,
k = the buckling length factor.

For compression members, k depends on the rotational restraint at the member ends and the means available to resist lateral movements. For tension members, $k = 1$

L = The unsupported length for tension or compression members.

r = The radius of gyration corresponding to the member's effective buckling length (kL).

The buckling factor K is displayed in Table 5.1. The values in the table depend on ideal cases for pure fixation and hinges in the support. Practically, a hinge support will allow some restraint, and the fixed support will allow some rotation.

As indicated in Table 5.1, the condition of (a), (b), (c), and (e) the lower end of the column is assumed to be fixed; however the other remaining cases (d), (f) the end is considered to be hinged. The case of fixed support means that the column is adequately embedded in concrete footing or generally by moment resisting bolted connection.

For case (a), the top end is considered fixed when the column is in a frame with a heavy girder which has many times more rigidity than the column itself.

Table 5.1 *K*, value for different conditions

	(a)	(b)	(c)	(d)	(e)	(f)
	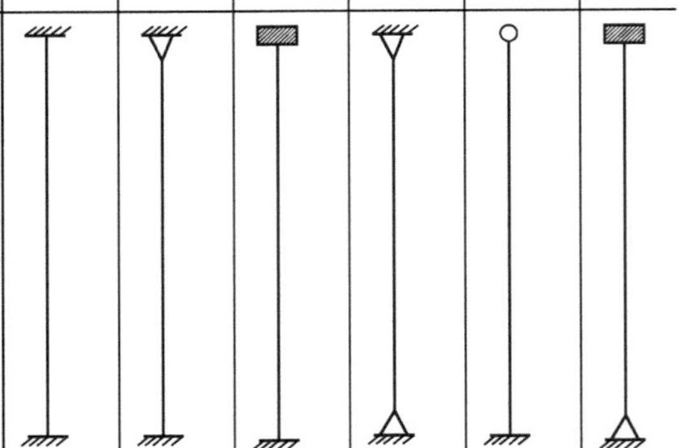					
Theoretical K value	0.5	0.7	1.0	1.0	2.0	2.0
Recommended design value when ideal condition are approximated	0.65	0.8	1.2	1.0	2.1	2.0
Case end conditions		Rotation fixed and translation fixed				
	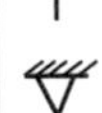	Rotation free and translation fixed				
	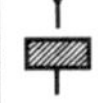	Rotation fixed and translation free				
		Rotation free and translation free				

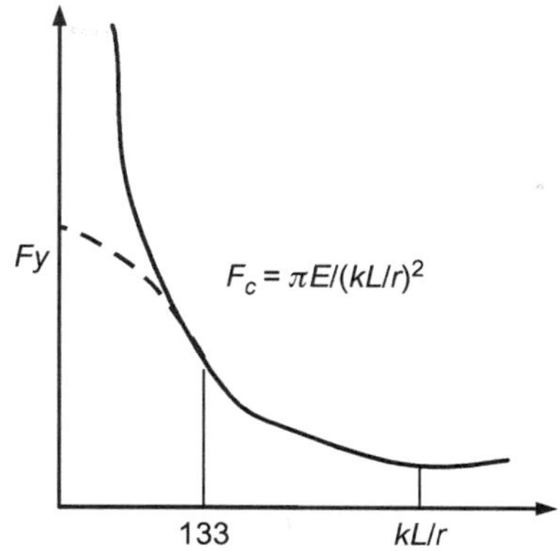

Figure 5.2 Relation between slenderness ratio and strength.

In Table 5.1 the cases (c) and (f) present the top of the column have a translation motion and this case exists in cases of column in unbraced frame with heavy connected beam.

The main method for calculating the strength based on the buckling is with the Euler formula; all other methods simply modify and enhance the Euler formula. The principal of the member strength and its slenderness ratio is shown in Fig. 5.2. In general, by decreasing ($\lambda = kL/r$), the strength will increase, but only if r has a higher value. Theoretically, the steel section can carry the load to infinity, but in reality, the yield strength is constrained by the section strength as shown in Fig. 5.2. So there are two formulas to calculate the member strength before and after $\lambda = 133$ which present the maximum load that can be carried by the member without it becoming buckling or deformed.

Example 1: Tension member for Allowable Stress Design (ASD):

Design the lower chord tension member. Design force = 30 t geometric length of the member $L_x = 400$ cm (using bolts 16 mm for bolted connection) with $F_y = 235$ Mpa.

Solution:

1. Construction guideline:
   $D = 16$ mm (bolts size)
   $\phi = 16 + 2$ mm $= 18$ hole size
   $a - t < 3\phi < 5.4$ cm
2. A_{req} $30/(2 \times 0.6 \times F_y \times 0.85) = 12.50\ \text{cm}^2$
3. Choose two equal angles $65 \times 65 \times 11$ mm ($A = 13.2\ \text{cm}^2$)
4. $A_{net} = 2(13.2 - \ (1.8 \times 1.1)) = 22.44\ \text{cm}^2$
5. $f_t = 30/22.44 = 1.34 < 0.6\ F_y$

6. Check $\lambda_{max} = kL/r = 300$

Noting that,

For two equal angles, $r = 0.3\ a$

For one angle, $r = 0.2\ a$

For star angles, $r = 0.385\ a$

$r = 0.3 \times 65 = 51.9$ mm

$kL/r = 4000/51.9 = 138.1 < 300$ OK

When you apply the Load Resistance Factor Design (LRFD), the following will be the strength in the case of a tension member.

LRFD for Tension Member

1. For yielding in the gross section

$\phi_t = 0.90,\ P_n = F_y\ A_g$

2. For fracture in the net section

$\phi_t = 0.75,\ P_n = F_u A_e$

where:

A_e = effective net area,

A_g = gross area of member,

F_y = specific min. yield stress,

F_u = specific min. tensile strength,

P_n = nominal axial strength.

Example 2: Previous example in the case of LRFD:

Design the lower chord tension member. Design force = 30 t for dead load. $L_x = 4000$ mm (using bolts 16 mm for bolted connection), using steel, $F_y = 235$ Mpa. Choose two equal angles $55 \times 55 \times 10$ mm ($A = 1010\ \text{mm}^2$).

Solution:

The slenderness constrains as in the previous example.

1. Load

$P_u = 1.4\ P_D = 1.4 \times 30 = 42\ \text{t} \leq \phi_t\ P_n$

2. Strength

$\phi_t\ P_n = 0.9 \times 2.35 \times 20.2 = 42.7 > P_u$ OK

$\phi_t\ P_n = 0.75 \times 3.6 \times 16.6 = 44.8 > P_u$ OK

5.3.2 Compression Members

Columns are structural elements subjected to a pure compression force. They are obviously in support for roof trusses or system girders. The main concern in compression members is buckling.

5.3.2.1 Steps of Preliminary Design

1. Select the convenient C.S.
2. In general, check if the section is noncompact which means it can achieve yield stress at the upper or lower fiber of the section without local plate buckling to the section, as shown in Tables 5.2 and 5.3 and Tables 5.4 and 5.5 for LRFD.
3. Obtain the value of the slenderness ratio (K)
4. Calculate the slenderness ratio (kL/r_{min}), must be less than l_{max}
5. Based on Table 5.2 as per (C-33) from (AISC)
6. Compute the actual stress $f_c = DF/Ag$
7. If f_c is less than F_c or not more than 3%, OK

Example 3: For column in the case of ASD:

Design a column to carry a total load of 950 kN. The column is hinged at both ends with buckling length:

$l_x = l_y = 6800$ mm

Solution:

$r = kL/\lambda = (7500/180) = 37.8$ mm

Try W10 × 60

$A = 11{,}600\ \text{mm}^2$, $r_x = 112$ mm, $r_y = 64.6$ mm, $t_f = 17.3$ mm,

$t_w = 10.7$ mm

$c = (256 - 10.7 - 2 \times 17.3)/2 = 210.7$

$c/t_f = 210.7/17.3 = 12.18 < 14.8$ OK

$d_w = (259.6 - 4 \times 17.3) = 190.4$ mm

$d_w/t_w = 17.8 < 41.3$ OK

$l_y/r_y = 105.2$, $F_c = 85\ \text{N/mm}^2$ (from tables)

$f_c = 950{,}000/11{,}600 = 81.8\ \text{N/mm}^2 < F_c$ OK

Example 4: LRFD for column:

Design a column to carry a dead load of 60 t and a live load of 40 t. The column is hinged at both ends with buckling length:

$$l_x = l_y = 800\ \text{cm}$$

Solution:

1. Load

$P_u = 1.2 \times 60 + 1.6 \times 40 = 136$ t

2. Strength

Try W12 × 65

$A = 1250\ \text{mm}^2$, $r_x = 134$ mm, $r_y = 76.4$ mm, $t_f = 15.4$ mm,

$t_w = 9.9$ mm

$\lambda = 8000/76.4 = 104.7$

$\phi_c F_c = 17.3 \times 0.07 = 119\ \text{N/mm}^2$

$\phi_c P_n = 119 \times 12{,}500/1000 = 1487.5 > 1360$ kN OK

Table 5.2 For compression members of 36 ksi (248 N/mm^2) specified yield stress steel

kL/r	F_a		*kL/r*	F_a		*kL/r*	F_a		*kL/r*	F_a		*kL/r*	F_a	
	ksi	MPa		ksi	MPa		ksi	MPa		ksi	MPa		ksi	MPa
1	21.56	148.65	41	19.11	131.76	81	15.24	105.08	121	10.14	69.91	161	5.76	39.71
2	21.52	148.38	42	19.03	131.21	82	15.13	104.32	122	9.99	68.88	162	5.69	39.23
3	21.48	148.1	43	18.95	130.66	83	15.02	103.56	123	9.85	67.91	163	5.62	38.75
4	21.44	147.82	44	18.86	130.04	84	14.90	102.73	124	9.70	66.88	164	5.55	38.27
5	21.39	147.48	45	18.78	129.48	85	14.79	101.97	125	9.55	65.84	165	5.49	37.85
6	21.35	147.20	46	18.70	128.93	86	14.67	101.15	126	9.41	64.88	166	5.42	37.37
7	21.30	146.86	47	18.61	128.31	87	14.56	100.39	127	9.26	63.85	167	5.35	36.89
8	21.25	146.51	48	18.53	127.76	88	14.44	99.56	128	9.11	62.81	168	5.29	36.47
9	21.21	146.24	49	18.44	127.14	89	14.32	98.73	129	8.97	61.85	169	5.23	36.06
10	21.16	145.89	50	18.35	126.52	90	14.20	97.91	130	8.84	60.95	170	5.17	35.65
11	21.10	145.48	51	18.26	125.90	91	14.09	97.15	131	8.70	59.98	171	5.11	35.23
12	21.05	145.13	52	18.17	125.28	92	13.97	96.32	132	8.57	59.09	172	5.05	34.82
13	21.00	144.79	53	18.08	124.66	93	13.84	95.42	133	8.44	58.19	173	4.99	34.40
14	20.95	144.45	54	17.99	124.04	94	13.72	94.60	134	8.32	57.36	174	4.93	33.99
15	20.89	144.04	55	17.90	123.42	95	13.60	93.77	135	8.19	56.47	175	4.88	33.65
16	20.83	143.62	56	17.81	122.80	96	13.48	92.94	136	8.07	55.64	176	4.82	33.23
17	20.78	143.27	57	17.71	122.11	97	13.35	92.04	137	7.96	54.88	177	4.77	32.89
18	20.72	142.86	58	17.62	121.49	98	13.23	91.22	138	7.84	54.05	178	4.71	32.47
19	20.66	142.45	59	17.53	120.87	99	13.10	90.32	139	7.73	53.30	179	4.66	32.13
20	20.60	142.03	60	17.43	120.18	100	12.98	89.49	140	7.62	52.54	180	4.61	31.78
21	20.54	141.62	61	17.33	119.49	101	12.85	88.60	141	7.51	51.78	181	4.56	31.44
22	20.48	141.20	62	17.24	118.87	102	12.72	87.70	142	7.41	51.09	182	4.51	31.10
23	20.41	140.72	63	17.14	118.18	103	12.59	86.80	143	7.30	50.33	183	4.46	30.75

24	20.35	140.31	64	17.04	117.49	104	12.47	85.98	144	7.20	49.64	184	4.41	30.41
25	20.28	139.83	65	16.94	116.80	105	12.33	85.01	145	7.10	48.95	185	4.36	30.06
26	20.22	139.41	66	16.84	116.11	106	12.20	84.12	146	7.01	48.33	186	4.32	29.79
27	20.15	138.93	67	16.74	115.42	107	12.07	83.22	147	6.91	47.64	187	4.27	29.44
28	20.08	138.45	68	16.64	114.73	108	11.94	82.32	148	6.82	47.02	188	4.23	29.16
29	20.01	137.96	69	16.53	113.97	109	11.81	81.43	149	6.73	46.40	189	4.18	28.82
30	19.94	137.48	70	16.43	113.28	110	11.67	80.46	150	6.64	45.78	190	4.14	28.54
31	19.87	137	71	16.33	112.59	111	11.54	79.57	151	6.55	45.16	191	4.09	28.20
32	19.80	136.52	72	16.22	111.83	112	11.40	78.60	152	6.46	44.54	192	4.05	27.92
33	19.73	136.03	73	16.12	111.14	113	11.26	77.63	153	6.38	43.99	193	4.01	27.65
34	19.65	135.48	74	16.01	110.39	114	11.13	76.74	154	6.30	43.44	194	3.97	27.37
35	19.58	135	75	15.90	109.63	115	10.99	75.77	155	6.22	42.89	195	3.93	27.10
36	19.50	134.45	76	15.79	108.87	116	10.85	74.81	156	6.14	42.33	196	3.89	26.82
37	19.42	133.9	77	15.69	108.18	117	10.71	73.84	157	6.06	41.78	197	3.85	26.54
38	19.35	133.41	78	15.58	107.42	118	10.57	72.88	158	5.98	41.23	198	3.81	26.27
39	19.27	132.86	79	15.47	106.66	119	10.43	71.91	159	5.91	40.75	199	3.77	25.99
40	19.19	132.31	80	15.36	105.90	120	10.28	70.88	160	5.83	40.20	200	3.73	25.72

Table 5.3 For compression members of 50 ksi (345 N/mm^2) specified yield stress steel

kL/r	F_a		kL/r	F_a		kL/r	F_a		kL/r	F_a		kL/r	F_a	
	ksi	MPa		ksi	MPa		ksi	MPa		ksi	MPa		ksi	MPa
1	29.94	206.43	41	25.69	177.13	81	18.81	129.69	121	10.2	70.33	161	5.76	39.71
2	29.87	205.95	42	25.55	176.16	82	18.61	128.31	122	10.03	69.15	162	5.69	39.23
3	29.8	205.47	43	25.4	175.13	83	18.41	126.93	123	9.87	68.05	163	5.62	38.75
4	29.73	204.98	44	25.26	174.16	84	18.2	125.48	124	9.71	66.95	164	5.55	38.27
5	29.66	204.50	45	25.11	173.13	85	17.99	124.04	125	9.56	65.91	165	5.49	37.85
6	29.58	203.95	46	24.96	172.09	86	17.79	122.66	126	9.41	64.88	166	5.42	37.37
7	29.5	203.40	47	24.81	171.06	87	17.58	121.21	127	9.26	63.85	167	5.35	36.89
8	29.42	202.84	48	24.66	170.02	88	17.37	119.76	128	9.11	62.81	168	5.29	36.47
9	29.34	202.29	49	24.51	168.99	89	17.15	118.25	129	8.97	61.85	169	5.23	36.06
10	29.26	201.74	50	24.35	167.89	90	16.94	116.80	130	8.84	60.95	170	5.17	35.65
11	29.17	201.12	51	24.19	166.78	91	16.72	115.28	131	8.7	59.98	171	5.11	35.23
12	29.08	200.50	52	24.04	165.75	92	16.5	113.76	132	8.57	59.09	172	5.05	34.82
13	28.99	199.88	53	23.88	164.65	93	16.29	112.32	133	8.44	58.19	173	4.99	34.40
14	28.9	199.26	54	23.72	163.54	94	16.06	110.73	134	8.32	57.36	174	4.93	33.99
15	28.8	198.57	55	23.55	162.37	95	15.84	109.21	135	8.19	56.47	175	4.88	33.65
16	28.71	197.95	56	23.39	161.27	96	15.62	107.70	136	8.07	55.64	176	4.82	33.23
17	28.61	197.26	57	23.22	160.10	97	15.39	106.11	137	7.96	54.88	177	4.77	32.89
18	28.51	196.57	58	23.06	158.99	98	15.17	104.59	138	7.84	54.05	178	4.71	32.47
19	28.4	195.81	59	22.89	157.82	99	14.94	103.01	139	7.73	53.30	179	4.66	32.13
20	28.3	195.12	60	22.72	156.65	100	14.71	101.42	140	7.62	52.54	180	4.61	31.78
21	28.19	194.36	61	22.55	155.48	101	14.47	99.77	141	7.51	51.78	181	4.56	31.44
22	28.08	193.61	62	22.37	154.24	102	14.24	98.18	142	7.41	51.09	182	4.51	31.10

23	27.97	192.85	63	22.2	153.06	103	14	96.53	143	7.3	50.33	183	4.46	30.75
24	27.86	192.09	64	22.02	151.82	104	13.77	94.94	144	7.2	49.64	184	4.41	30.41
25	27.75	191.33	65	21.85	150.65	105	13.53	93.29	145	7.1	48.95	185	4.36	30.06
26	27.63	190.51	66	21.67	149.41	106	13.29	91.63	146	7.01	48.33	186	4.32	29.79
27	27.52	189.74	67	21.49	148.17	107	13.04	89.91	147	6.91	47.64	187	4.27	29.44
28	27.4	188.92	68	21.31	146.93	108	12.8	88.25	148	6.82	47.02	188	4.23	29.16
29	27.28	188.09	69	21.12	145.62	109	12.57	86.67	149	6.73	46.40	189	4.18	28.82
30	27.15	187.20	70	20.94	144.38	110	12.34	85.08	150	6.64	45.78	190	4.14	28.54
31	27.03	186.37	71	20.75	143.07	111	12.12	83.56	151	6.55	45.16	191	4.09	28.20
32	26.9	185.47	72	20.56	141.76	112	11.9	82.05	152	6.46	44.54	192	4.05	27.92
33	26.77	184.58	73	20.38	140.52	113	11.69	80.60	153	6.38	43.99	193	4.01	27.65
34	26.64	183.68	74	20.1	138.58	114	11.49	79.22	154	6.3	43.44	194	3.97	27.37
35	26.51	182.78	75	19.99	137.83	115	11.29	77.84	155	6.22	42.89	195	3.93	27.10
36	26.38	181.88	76	19.8	136.52	116	11.1	76.53	156	6.14	42.33	196	3.89	26.82
37	26.25	180.99	77	19.61	135.21	117	10.91	75.22	157	6.06	41.78	197	3.85	26.54
38	26.11	180.02	78	19.41	133.83	118	10.72	73.91	158	5.98	41.23	198	3.81	26.27
39	25.97	179.06	79	19.21	132.45	119	10.55	72.74	159	5.91	40.75	199	3.77	25.99
40	25.83	178.10	80	19.01	131.07	120	10.37	71.50	160	5.83	40.20	200	3.73	25.72

Table 5.4 Design stress for compression members of 36 ksi (248 N/mm^2) specified yield stress steel, $\phi_c = 0.85$ based on AISC LRFD

kL/r	$\Phi_c F_{cr}$		***kL/r***	$\Phi_c F_{cr}$		***kL/r***	$\Phi_c F_{cr}$		***kL/r***	$\Phi_c F_{cr}$		***kL/r***	$\Phi_c F_{cr}$	
	ksi	**N/mm^2**		**ksi**	**N/mm^2**		**ksi**	**N/mm^2**		**ksi**	**N/mm^2**		**ksi**	**N/mm^2**
1	30.6	211.0	41	28	193.1	81	21.7	149.6	121	14.2	97.9	161	8.23	56.7
2	30.6	211.0	42	27.9	192.4	82	21.5	148.2	122	14	96.5	162	8.13	56.1
3	30.6	211.0	43	27.8	191.7	83	21.3	146.9	123	13.8	95.1	163	8.03	55.4
4	30.6	211.0	44	27.6	190.3	84	21.1	145.5	124	13.6	93.8	164	7.93	54.7
5	30.6	211.0	45	27.5	189.6	85	20.9	144.1	125	13.4	92.4	165	7.84	54.1
6	30.5	210.3	46	27.4	188.9	86	20.7	142.7	126	13.3	91.7	166	7.74	53.4
7	30.5	210.3	47	27.2	187.5	87	20.5	141.3	127	13.1	90.3	167	7.65	52.7
8	30.5	210.3	48	27.1	186.8	88	20.4	140.7	128	12.9	88.9	168	7.56	52.1
9	30.5	210.3	49	27	186.2	89	20.2	139.3	129	12.7	87.6	169	7.47	51.5
10	30.4	209.6	50	26.8	184.8	90	20	137.9	130	12.6	86.9	170	7.38	50.9
11	30.4	209.6	51	26.7	184.1	91	19.8	136.5	131	12.4	85.5	171	7.3	50.3
12	30.4	209.6	52	26.5	182.7	92	19.6	135.1	132	12.2	84.1	172	7.21	49.7
13	30.3	208.9	53	26.4	182.0	93	19.4	133.8	133	12.1	83.4	173	7.13	49.2
14	30.3	208.9	54	26.3	181.3	94	19.2	132.4	134	11.9	82.0	174	7.05	48.6
15	30.2	208.2	55	26.1	180.0	95	19	131.0	135	11.7	80.7	175	6.97	48.1
16	30.2	208.2	56	25.9	178.6	96	18.8	129.6	136	11.5	79.3	176	6.89	47.5
17	30.1	207.5	57	25.8	177.9	97	18.7	128.9	137	11.4	78.6	177	6.81	47.0
18	30.1	207.5	58	25.6	176.5	98	18.5	127.6	138	11.2	77.2	178	6.73	46.4
19	30	206.8	59	25.5	175.8	99	18.3	126.2	139	11	75.8	179	6.66	45.9
20	30	206.8	60	25.3	174.4	100	18.1	124.8	140	10.9	75.2	180	6.59	45.4
21	29.9	206.2	61	25.2	173.7	101	17.9	123.4	141	10.7	73.8	181	6.51	44.9

22	29.8	205.5	62	25	172.4	102	17.7	122.0	142	10.6	73.1	182	6.44	44.4
23	29.8	205.5	63	24.8	171.0	103	17.5	120.7	143	10.4	71.7	183	6.37	43.9
24	26.7	184.1	64	24.7	170.3	104	17.3	119.3	144	10.3	71.0	184	6.3	43.4
25	29.6	204.1	65	24.5	168.9	105	17.1	117.9	145	10.2	70.3	185	6.23	43.0
26	29.5	203.4	66	24.3	167.5	106	16.9	116.5	146	10	68.9	186	6.17	42.5
27	29.5	203.4	67	24.2	166.9	107	16.8	115.8	147	9.87	68.1	187	6.1	42.1
28	29.4	202.7	68	24	165.5	108	16.6	114.5	148	9.74	67.2	188	6.04	41.6
29	29.3	202.0	69	23.8	164.1	109	16.4	113.1	149	9.61	66.3	189	5.97	41.2
30	29.2	201.3	70	23.6	162.7	110	16.2	111.7	150	9.48	65.4	190	5.91	40.7
31	29.1	200.6	71	23.5	162.0	111	16	110.3	151	9.36	64.5	191	5.85	40.3
32	29	199.9	72	23.3	160.6	112	15.8	108.9	152	9.23	63.6	192	5.79	39.9
33	28.9	199.3	73	23.1	159.3	113	15.6	107.6	153	9.11	62.8	193	5.73	39.5
34	28.8	198.6	74	22.9	157.9	114	15.4	106.2	154	9	62.1	194	5.67	39.1
35	28.7	197.9	75	22.8	157.2	115	15.3	105.5	155	8.88	61.2	195	5.61	38.7
36	28.6	197.2	76	22.6	155.8	116	15.1	104.1	156	8.77	60.5	196	5.55	38.3
37	28.5	196.5	77	22.4	154.4	117	14.9	102.7	157	8.66	59.7	197	5.5	37.9
38	28.4	195.8	78	22.2	153.1	118	14.7	101.4	158	8.55	59.0	198	5.44	37.5
39	28.3	195.1	79	22	151.7	119	14.5	100.0	159	8.44	58.2	199	5.39	37.2
40	28.1	193.7	80	21.9	151.0	120	14.3	98.6	160	8.33	57.4	200	5.33	36.7

Table 5.5 Design stress for compression members of 50 ksi (345 N/mm^2) specified yield stress steel, $\phi_c = 0.85$ based on AISC LRFD

kL/r	$\Phi_c F_{cr}$		*kL/r*	$\Phi_c F_{cr}$		*kL/r*	$\Phi_c F_{cr}$		*kL/r*	$\Phi_c F_{cr}$		*kL/r*	$\Phi_c F_{cr}$	
	ksi	N/mm^2		ksi	N/mm^2		ksi	N/mm^2		ksi	N/mm^2		ksi	N/mm^2
1	42.5	293.0	41	28	193.1	81	18.81	129.7	121	14.2	97.9	161	8.23	56.7
2	42.5	293.0	42	27.9	192.4	82	18.61	128.3	122	14	96.5	162	8.13	56.1
3	42.5	293.0	43	27.8	191.7	83	18.41	126.9	123	13.8	95.1	163	8.03	55.4
4	42.5	293.0	44	27.6	190.3	84	18.2	125.5	124	13.6	93.8	164	7.93	54.7
5	42.4	292.3	45	27.5	189.6	85	17.99	124.0	125	13.4	92.4	165	7.84	54.1
6	42.4	292.3	46	27.4	188.9	86	20.7	142.7	126	13.3	91.7	166	7.74	53.4
7	42.4	292.3	47	27.2	187.5	87	20.5	141.3	127	13.1	90.3	167	7.65	52.7
8	42.3	291.6	48	27.1	186.8	88	20.4	140.7	128	12.9	88.9	168	7.56	52.1
9	42.3	291.6	49	27	186.2	89	20.2	139.3	129	12.7	87.6	169	7.47	51.5
10	42.2	291.0	50	26.8	184.8	90	20	137.9	130	12.6	86.9	170	7.38	50.9
11	42.1	290.3	51	26.7	184.1	91	19.8	136.5	131	12.4	85.5	171	7.3	50.3
12	42.1	290.3	52	26.5	182.7	92	19.6	135.1	132	12.2	84.1	172	7.21	49.7
13	42	289.6	53	26.4	182.0	93	19.4	133.8	133	12.1	83.4	173	7.13	49.2
14	41.9	288.9	54	26.3	181.3	94	19.2	132.4	134	11.9	82.0	174	7.05	48.6
15	41.8	288.2	55	26.1	180.0	95	19	131.0	135	11.7	80.7	175	6.97	48.1
16	41.7	287.5	56	25.9	178.6	96	18.8	129.6	136	11.5	79.3	176	6.89	47.5
17	41.6	286.8	57	25.8	177.9	97	18.7	128.9	137	11.4	78.6	177	6.81	47.0
18	41.5	286.1	58	25.6	176.5	98	18.5	127.6	138	11.2	77.2	178	6.73	46.4
19	41.4	285.4	59	25.5	175.8	99	18.3	126.2	139	11	75.8	179	6.66	45.9
20	41.3	284.75	60	25.3	174.4	100	18.1	124.8	140	10.9	75.2	180	6.59	45.4
21	41.2	284.1	61	25.2	173.7	101	17.9	123.4	141	10.7	73.8	181	6.51	44.9
22	41	282.7	62	25	172.4	102	17.7	122.0	142	10.6	73.1	182	6.44	44.4

23	40.9	282.0	63	24.8	171.0	103	17.5	120.7	143	10.4	71.7	183	6.37	43.9
24	40.8	281.3	64	24.7	170.3	104	17.3	119.3	144	10.3	71.0	184	6.3	43.4
25	40.6	279.9	65	24.5	168.9	105	17.1	117.9	145	10.2	70.3	185	6.23	43.0
26	40.5	279.2	66	24.3	167.5	106	16.9	116.5	146	10	68.9	186	6.17	42.5
27	40.3	277.9	67	24.2	166.9	107	16.8	115.8	147	9.87	68.1	187	6.1	42.1
28	40.1	276.5	68	24	165.5	108	16.6	114.5	148	9.74	67.2	188	6.04	41.6
29	40	275.8	69	23.8	164.1	109	16.4	113.1	149	9.61	66.3	189	5.97	41.2
30	39.8	274.4	70	23.6	162.7	110	16.2	111.7	150	9.48	65.4	190	5.91	40.7
31	29.1	200.6	71	23.5	162.0	111	16	110.3	151	9.36	64.5	191	5.85	40.3
32	29	199.9	72	23.3	160.6	112	15.8	108.9	152	9.23	63.6	192	5.79	39.9
33	28.9	199.3	73	23.1	159.3	113	15.6	107.6	153	9.11	62.8	193	5.73	39.5
34	28.8	198.6	74	22.9	157.9	114	15.4	106.2	154	9	62.1	194	5.67	39.1
35	28.7	197.9	75	22.8	157.2	115	15.3	105.5	155	8.88	61.2	195	5.61	38.7
36	28.6	197.2	76	22.6	155.8	116	15.1	104.1	156	8.77	60.5	196	5.55	38.3
37	28.5	196.5	77	22.4	154.4	117	14.9	102.7	157	8.66	59.7	197	5.5	37.9
38	28.4	195.8	78	22.2	153.1	118	14.7	101.4	158	8.55	59.0	198	5.44	37.5
39	28.3	195.1	79	22	151.7	119	14.5	100.0	159	8.44	58.2	199	5.39	37.2
40	28.1	193.7	80	21.9	151.0	120	14.3	98.6	160	8.33	57.4	200	5.33	36.7

5.3.3 Beam Design

Beams are structural members transversely loaded to their lengths and in the plane of their webs. They resist flexure, shear, and torsion stresses.

Irrespective of their methods of design and the design code which you will use, three aspects must be considered in the design of beams. The strength requirements which necessitate that the cross section of a member must be adequate in resisting the applied bending moments and shearing forces. The stability requirements which necessitate that the steel plate elements composing the cross section must resist the local buckling phenomena should be checked. The web should be safe against buckling created by shearing forces and bending moments. The cross section should have adequate torsion strength to resist lateral torsion buckling (LTB).

Performance considerations that require the deformation of the member are not excessive under service loads; therefore, limitations of deflections sometimes govern beam design.

Beams are structured elements subjected to bending moment and shearing force only. The normal force is neglected (zero).

- purlines,
- crane track girder (CTG),
- beam in flooring system as shown in Fig. 5.3,
- girder of rigid beam and steel frame (Fig. 5.4).

The beam design is based on the failure steps as shown in Fig. 5.5. Assume a load P is applied to the center of the beam, increase the load gradually, and draw the corresponding bending moment diagram. The beam failure passes through four steps:

1. The stress on the bottom and top fiber of the steel section is less than the yield strength.
2. The top and bottom fiber of the steel section stress is equal to yield strength and the moment is M_y.
3. The part of the section has stress equal to the yield strength.
4. The entire cross section is affected by stress equal to yield strength, so we reach the plasticity limit and the moment is M_p.

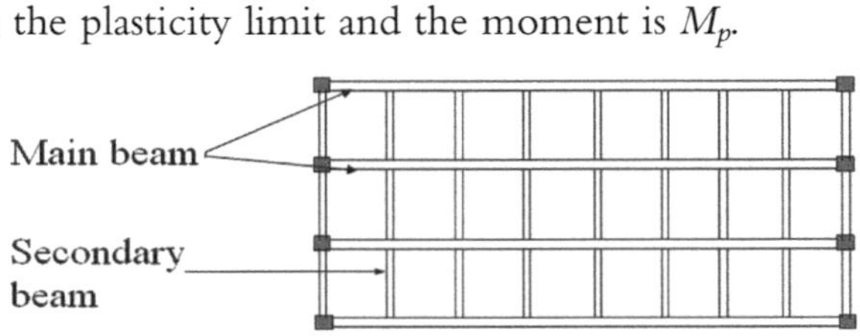

Figure 5.3 Floor structure system.

Figure 5.4 Beam system.

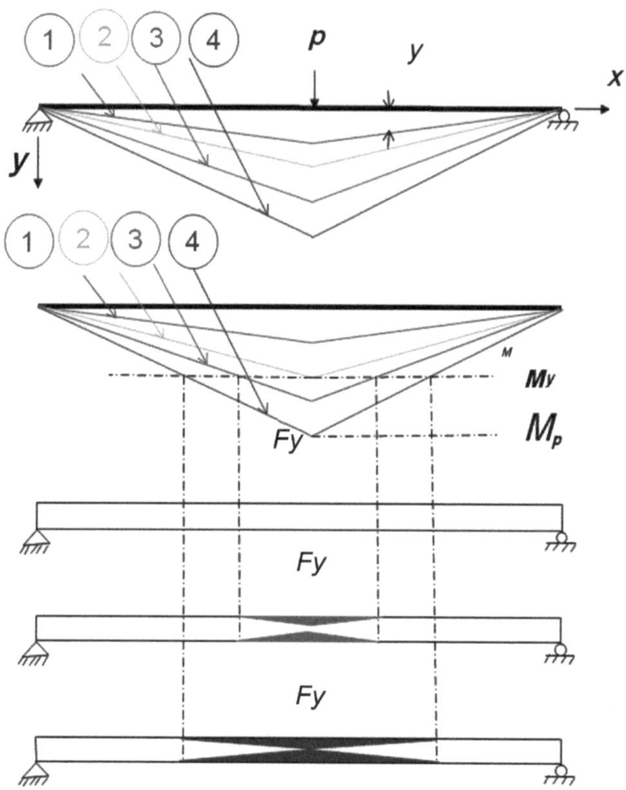

Figure 5.5 Simple beam failure stresses increasing steps.

From the AISC the following is the strength allowable in the design for the steel structure beams:

in the case of compact section,

$F_b = 0.66\ F_y$,

in the case of noncompact section,

$F_b = 0.60\ F_y$,

allowable shear stress $= 0.4\ F_y$,

It is important to define if the section is a compact or noncompact section. A compact section means that the section can achieve the plastic moment with local buckling on the web or the flange. But a noncompact section is a section that can reach the yields for the lower and bottom fiber only without causing local buckling to the web or the flange.

In general, channel and equal and nonequal angles are always noncompact sections.

5.3.3.1 Lateral Torsion Buckling

The strength of the beam is controlled by the second limit state for beams, or the lateral torsional buckling (LTB). The LTB occurs when the compression portion of a cross section is restrained by the tension portion, and the deflection due to flexural buckling is accompanied by torsion or twisting. The way to prevent LTB is to have adequate lateral bracing at adequate intervals along the beam. The limit state is the interval of the bracing.

The computed flexural stress f_b shall not exceed the allowable flexural stress F_b, given as follows (in all equations, the minimum specified yield stress F_y cannot exceed its limits):

For $L_b \leq L_c$,

$$F_b = 0.66\ F_y \tag{5.1}$$

where:

$L_c =$ smaller of

$$\frac{200 b_f}{\sqrt{F_y}} \quad \text{or} \quad \frac{137,895}{(d/A_f)F_y} \quad \text{for } I \text{ and channel shapes} \tag{5.2}$$

In which:

$b_f =$ flange width in mm,

$F_y =$ yield strength in N/mm^2,

$d =$ overall depth of section in mm,

$A_f =$ area of compression flange in mm^2.

For the above sections to be considered compact, in addition to having the width–thickness ratios of their component elements falling within the limiting value as shown in Table 5.6, the flanges of the sections must be continuously connected to the webs.

For $L_b > L_c$, the allowable flexural stress in tension is given by

$$F_b = 0.60\,F_y \tag{5.3}$$

The allowable flexural stress in compression is given by the larger value calculated from Eq. (5.4).

$$F_b = \frac{82,737A_f}{L_d} \tag{5.4}$$

where:

L = distance between cross sections braced against twist or lateral displacement of the compression flange in mm.

A_f = compression flange area in mm^2,

$C_b = 12.5M_{max}/(2.5M_{max} + 3M_A + 4M_B + 3M_C)$

where M_{max}, M_A, M_B, M_C D are the maximum moment, quarter-point moment, midpoint moment, and three-quarter point moment along the unbraced length of the member, respectively. For simplicity in design, C_b can conservatively be taken as unity.

It should be cautioned that Eq. (5.4) are applicable only to I and channel shapes with an axis of symmetry in, and loaded in, the plane of the web.

Check if the cross section is compact or not compact regarding lateral tensional buckling by applying

$$L_c \leq \frac{20b}{F_y^{0.5}}, \quad L_u \leq \frac{1380A_f}{d\,F_y}C_b$$

For purlins design the codes are not allowing the use of covering materials with design on the other hand they are not taken the effect of torsion into consideration.

5.3.3.2 Allowable Deflection

It always depends on the project specifications, but in general allowable deflection from the AISC will be as follows;

for CTG = span/800,

for floor and roof with plastering = span/360 (Table 5.7).

Table 5.6 Max width/thickness ratio for compact and noncompact section

	Compact			Noncompact section		
	St.52	St.44	St.37	St.52	St.44	St.37
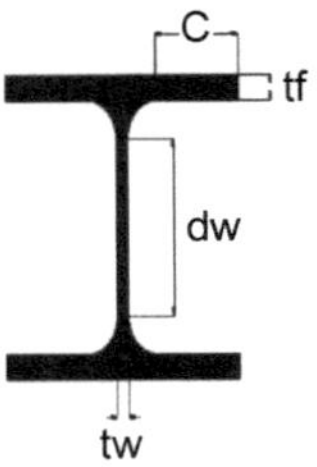	Bending					
	$d_w/t_w \leq 127/(F_y)^{0.5}$			$d_w/t_w \leq 190/(F_y)^{0.5}$		
	66.94	75.91	82	100.15	113.56	122.65
	Compression					
	$d_w/t_w \leq 58/(F_y)^{0.5}$			$d_w/t_w \leq 64/(F_y)^{0.5}$		
	30.56	34.65	37.44	33.73	38.24	41.3
	Rolled					
	$c/t_f \leq 16.9/(F_y)^{0.5}$			$c/t_f \leq 23/(F_y)^{0.5}$		
	8.9	10.1	10.91	12.12	13.74	14.8
	Welded					
	$c/t_f \leq 15.3/(F_y)^{0.5}$			$c/t_f \leq 21/(F_y)^{0.5}$		
	8.06	9.14	9.87	11.07	12.55	13.5
	Equal angles					
				$b/t \leq 23/(F_y)^{0.5}$		
				12.12	13.74	14.8
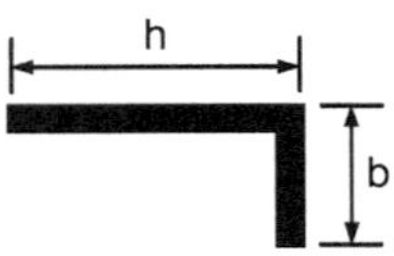	Unequal angles					
				$(b+h)/2t \leq 17/(F_y)^{0.5}$		
				8.96	10.15	11.0
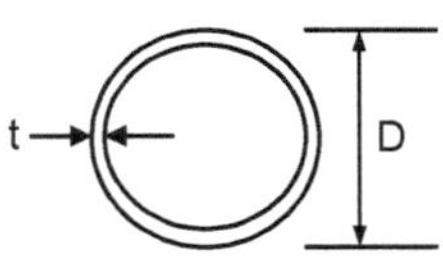	Tubular member subjected to bending or compression stress					
	$D/t \leq 165/F_y$			$D/t \leq 211/F_y$		
	45.83	58.92	68.79	58.61	75.35	88.0
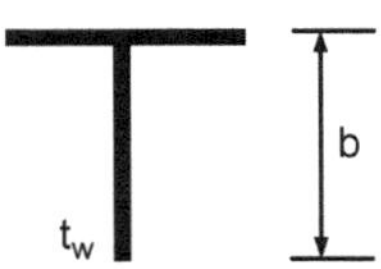	T-section subjected to compression					
				$b/t \leq 30/(F_y)^{0.5}$		
				15.81	17.93	19.36

Table 5.7 Suggested limits for calculated deflection based on BS5950

1. Vertical deflection of beam due to imposed load	
Cantilever Beam is not carrying plaster or brittle finished	Length/180 Length/200
2. Horizontal deflection of columns due to imposed load and wind load	
Top of columns in single-storey building, except portal frames. In each storey building with more than one storey.	Height/300 Height of that storey/300
3. Crane girders	
Vertical deflection due to static vertical wheel loads from overhead traveling crane Horizontal deflection	Span/600 Span/500

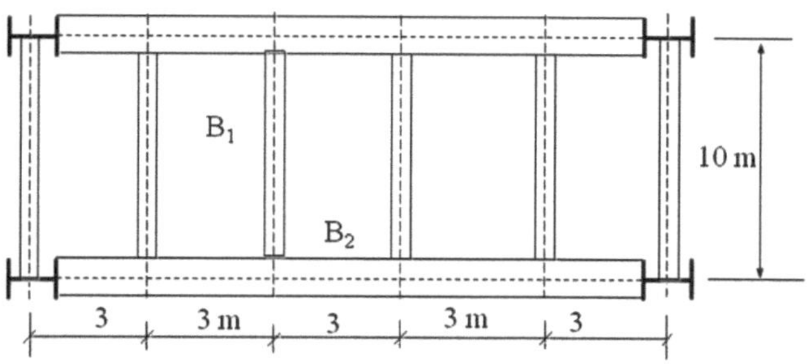

Figure 5.6 Floor geometry sketch.

Based on AISC, beams and girders supporting floors and roofs shall be proportioned with due regard to the deflection produced by the design loads. Beams and girders supporting plastered ceilings shall be so proportioned that the maximum live-load deflection does not exceed 1/360 of the span.

Example 1: For Beam Design:

The floor system of a shopping center is as shown. The floor consists of 50 mm R.C. slab over corrugated steel sheets. It is required to design the floor beams B1 and B2 (Fig. 5.6).

Solution:

$$\begin{aligned}\text{Loads/m}^2\text{ of the floor} &= 200\ \text{N/m}^2\ \text{(corrugated sheet)}\\ &= 1250\ \text{N/m}^2\ \text{(R.C. slab)}\\ &= 1000\ \text{N/m}^2\ \text{(sand + tiles)}\\ &= 4000\ \text{N/m}^2\ \text{(L.L)}\\ &= 6450\ \text{N/m}^2\end{aligned}$$

Load on the beam = 6450 × 3.0 = 19,350 N/m'
Assume beam weight = 1000 N/m'
Total load on beam = 19,350 + 1000 = 20,350 N/m'

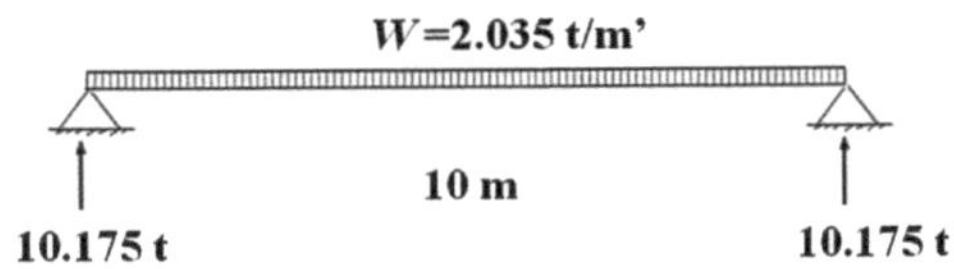

$$M_x = \frac{2.035(10)^2}{8} = 254.4 \text{ m·kN}$$

$$Q_{max} = R = 101.75 \text{ kN}$$

The compression flange is encased in R.C. slab, so the unbraced length is zero.
The cross section is compact regarding the LTB.
$F_b = 0.66\ F_y = 0.66(248) = 163\ \text{N/mm}^2$
$S = 254.4 \times 1{,}000{,}000/163 = 1{,}560{,}736\ \text{mm}^3$
Choose W18 × 55
$A = 105\ \text{cm}^2$, $h = 460$ mm, $t_w = 9.9$ mm, $t_f = 16$ mm,
$S_x = 1{,}620{,}000\ \text{mm}^3$, $I_X = 372{,}700{,}000\ \text{mm}^4$
$C = 0.5(b - t_w - 2t_f)$ check $c/t_f < 10.91$ OK
$d_w = h - 4t_f$, $d_w/t_f < 82$ OK
The section regarding the local buckling is a compact.
Check on stresses:
$f_{bc} = 254.4 \times 1{,}000{,}000/1{,}620{,}000 = 157\ \text{N/mm}^2 < 163\ \text{N/mm}^2$ OK
$Q = 101.75 \times 1000/(460 \times 16) = 13.8\ \text{N/mm}^2 < 0.4\ F_y$ OK
Check of deflection

$$\Delta = \frac{\text{span}}{360} = 27.78 \text{ mm}$$

$$\Delta = \frac{5wl^4}{384EI}$$

$$\Delta = \frac{5(0.000004 \times 3000)(10{,}000)^4}{384(205.95)(37{,}270 \times 10^4)} = 20.4 \text{ mm}$$

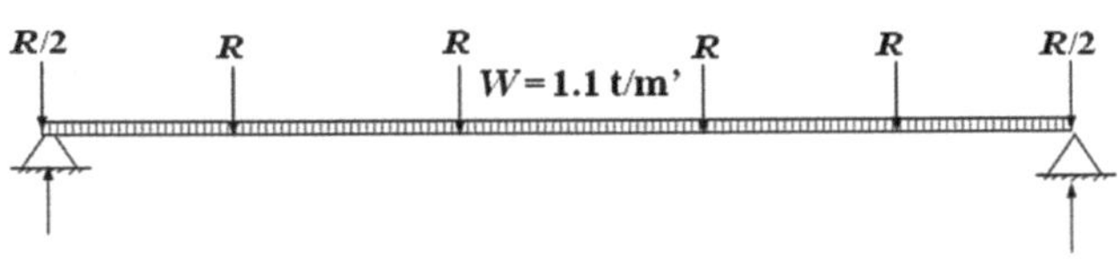

Assume own weight + wall = 11.0 kN/m'
$M_x = 1225.13$ m · kN, $Q_{max} = 286.0$ kN
$S = 1225.13 \times 1000 \times 1000/163 = 7{,}516{,}134.97$ mm^3
Try W30 × 165
$h = 785$ mm, $b = 268$ mm, $S_x = 7949$ cm^3, $I_x = 312{,}000$ cm^4,
$t_w = 18.5$ mm
$C = 0.5(b - t_w - 2t_f)$ check $c/t_f < 10.91$ OK
$d_w = h - 4t_f$, $d_w/t_f < 82$ OK
$L_u = 300$ cm s
$L_u = 200 \times b/(F_y)^{0.5} = 38{,}700$ mm > 3000 mm OK
The cross section is compact regarding the LTB.
$F_b = 0.66\ F_y = 163.7$ N/mm^2
$f_{bc} = 1225.13 \times 1{,}000{,}000/7{,}949{,}000 = 154.12$ N/mm$^2 < 0.66\ F_y$ OK
$Q = 286{,}000/(785 \times 18.5) = 19.7$ N/mm$^2 < 0.4\ F_y$ OK
Check of deflection
$\Delta = \text{span}/360 = 41.67$ mm
Convert the concentrated load to the equivalent uniform load
$W_{eq} = 4 \times 6/15 = 1.6$ t/m' = 15.7 N/mm'

$$\Delta = \frac{5(0.0157)(15{,}000)^4}{384(205.95)(312{,}000 \times 10^4)} = 16 \text{ mm} \quad \text{OK}$$

Example 2: For CTG:

Design a CTG knowing the following:

- span of girder = 19.50 m,
- span of CTG = 4.90 m,
- axis of trolly can be at 0.90 m from CL of CTG,
- distance between wheels of crane girder = 2.00 m,
- weight of trolly = 1.0 t
- own weight of crane girder = 5.0 t,
- crane capacity = 12.0 t (Fig. 5.7).

Solution:

Loads

Moving load = wt of trolly + wt to be displaced = 1.0 + 12.0 = 13.0 t

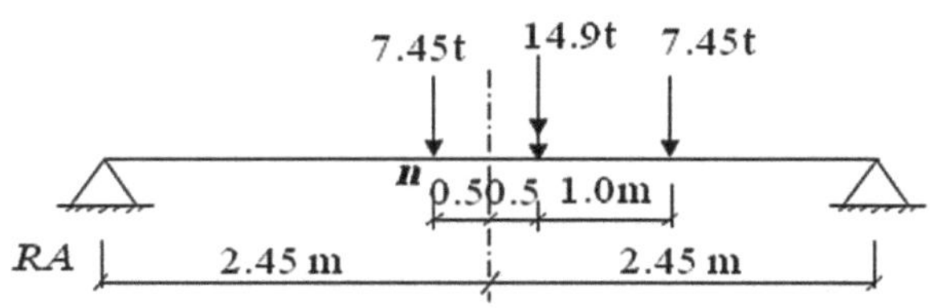

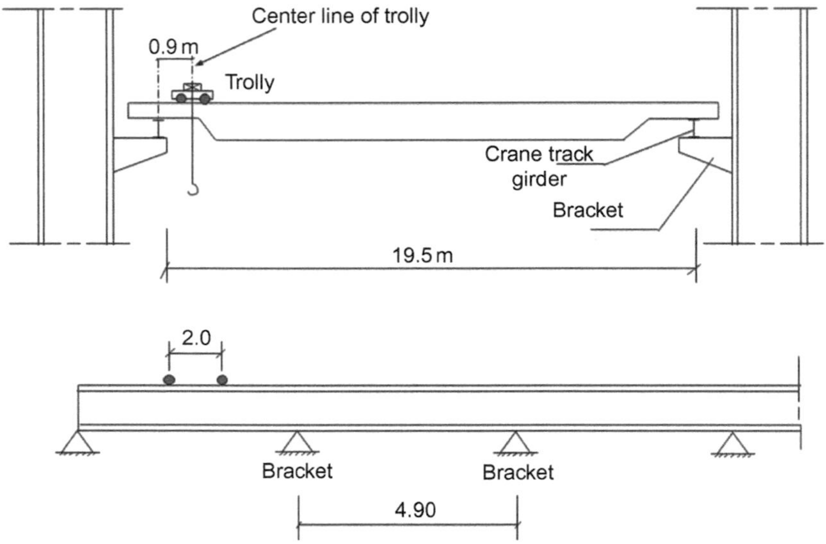

Figure 5.7 CTG configuration.

Total load taken by LTG = (13(19.5 − 0.9)/19.5) + 2.5 = 14.9 t
Load on each of the wheels on left CTG = 14.9/2 = 7.45 t
RA = [7.45(2.45 − 1.5) + 7.45(2.45 + 0.5)]/4.9 = 5.92 t
$M_x(n)_{LL} = 5.92\ (2.45 - 0.5) = 11.562$ mt
M_x (with impact) = 11.562 × 1.25 = 14.452 mt
Assume own weight of the CTG 200 kg/m'
$M_x = [0.2(4.9)^2/8] + 14.452 = 15.052$ mt
$M_y = 0.10(11.562) = 1.1562$ mt

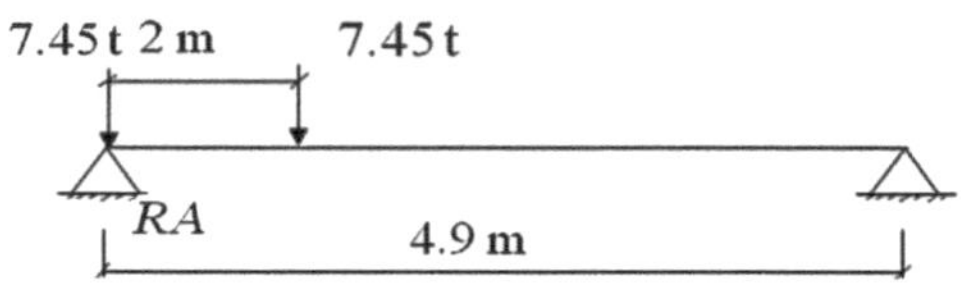

Q_y (LL) = (7.45(2.9/4.9) + 7.45) = 11.859 t
Q_y (LL) with impact = 11.859 × 1.25 = 14.824 t
$Q_x = 0.1\ (11.859) = 1.1859$ t (lateral shock)
Total $Q_y = 0.2\ (4.9)/2 + 14.824 = 15.314$ t
Assume ($S_x \cong 3\ S_y$)
$(M_x/S_x) + (M_y/0.5S_y) > 1.4 \times 1.2$, Choose HEB 280
$h = 280$ mm, $b = 280$ m, $t_f = 18$ mm, $t_w = 10.5$ mm,
$I_x = 192{,}700{,}000\ \text{mm}^4$, $I_y = 65{,}950{,}000\ \text{mm}^4$, $S_x = 1{,}376{,}000\ \text{mm}^3$,
$S_y = 471{,}000\ \text{mm}^3$

Check Local Buckling

C/t_f and d_w/t_w OK

The section regarding the local buckling is a compact section.

Lateral Torsion Buckling

$L_u = 490$ cm

$L_u = 20 \times b/(F_y)^{0.5} = 387$ cm < 490 unsafe

The section is noncompact $F_b = 0.6\ F_y$

$F_{bt} = 15.05 \times 100/1376 = 1.09$ t/cm^2

$F_{bc} = 1.09 + 1.154 \times 100/(0.5 \times 471) = 1.58 < 1.4 \times 1.2$ OK

$q_y = 15.314/(28 \times 1.0) = 0.55 < 0.4F_y$ t/cm^2

$q_x = (3/2)(Q_x/A_f) = (3/2)\ (1.185/28 \times 1.8) = 0.035 < 0.4F_y$ t/cm^2

Check of LTB

$F_{LTB} = 800\ A_f\ C_b/(L_u d) = 3.27 > f_{bc}$ OK

Check Deflection (LL Only)

$M_{max} = 11.562$ mt, $100 \times 11.562 = W_{eq} \times (490)^2/8$

$W_{eq} = 0.0385$ t/cm' $= 3.85$ t/m'

$\Delta \leq 490/800 \leq 0.612$ cm

$\Delta_{act} = 0.66$ unsafe, so it needs a bigger section.

5.3.4 Design of Beam Column Member (ASD)

For beam columns and structural elements subjected to compression (normal force) and a bending moment, safety against the following must be satisfied:

- the buckling compression stress,
- the axial compressor stress $F_c \leq 0.6\ F_y$ t/cm^2,
- the max. tensile stress $F_t \leq 0.66\ F_y$ t/cm^2,
- the safety of the LTB must be checked,
- shear stress must not exceed the limit $t_{max} \leq 0.4\ F_y$,
- the max. deflection must be satisfied.

As the beam column must be subjected to bending and compression forces, so the compact or noncompact section will be defined.

The value of (K) will be obtained from the alignment chart in Fig. 5.8.

$$G = \frac{\sum(I/L)\text{column}}{\sum(I/L)\text{girders}}$$

The case of prevented and permitted side sway is presented in Fig. 5.9.

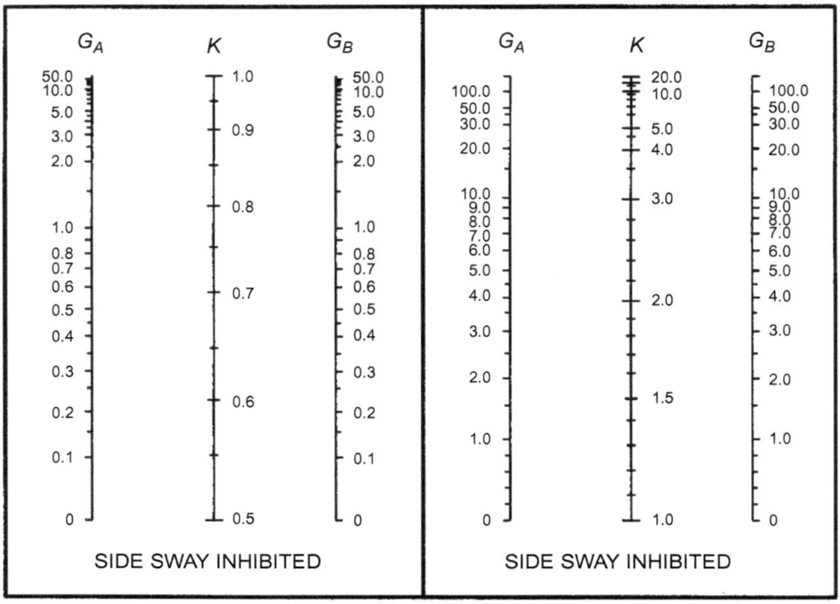

Figure 5.8 Alignment charts for buckling length factor (*k*).

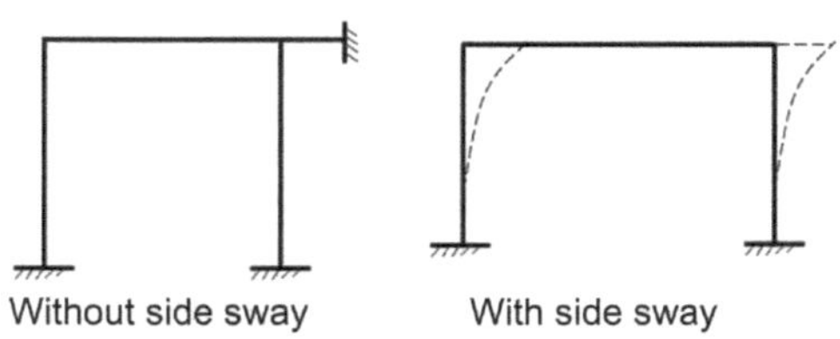

Figure 5.9 Cases of side sway and no side sway.

G is equal to 10 for hinge support and equal to 1 for fixed support.

$$\frac{f_a}{F_C} + \frac{f_{bx}}{F_{bx}}A_1 + \frac{f_{by}}{F_{by}}A_2 \leq 1.0$$

where:

f_a = Actual comp. Stress due to axial comp.
F_c = The allowable comp. Stress due to buckling.

f_{bx}, f_{by} = The actual comp. Bending stresses on moment about x and y axes, respectively.
F_{bx}, F_{by} = The allowable comp. Bending stress for the x and y axes.

$$A_1 = \frac{C_{mx}}{\left(\frac{1 - f_a}{F_{EX}}\right)}, \quad A_2 = \frac{C_{my}}{\left(\frac{1 - f_a}{F_{EY}}\right)}$$

F_{EX}, F_{EY} = the allowable buckling stress in x and y direction, respectively. $F_{EX} = 7500/(l_x)^2$
CM = moment modification factor
= 0.85 for frame permitted to sway
= 1.0 in the case of members supported at the ends.

1. Assume $I_c/I_G = 1.0$ (no crane) or
   $I_c/I_G = 1.5 - 2.0$ (with crane)
2. Calculate M_{max}, N_{max}
3. Assume $f = 1.0\ \text{t/cm}^2$ for higher N or $f = 1.3\ \text{t/cm}^2$ for higher M get approximate Cross section.
4. Check compactness (d_w/t_w, c/t_f)
5. Get l_x, l_y calculate l_x, l_y
6. F_c from the maximum (l)
7. $F_{EX} = 7500/(l_x)^2$, $f_{ca} = N/A$, $f_{bx} = M_x/Z_x$
8. $CM = 0.6 + 0.4\ (M_2/M_1)\ 0.4$
   $0.4 \leq CM \leq 0.6$ (without side sway)
   $CM = 0.85$ (with side sway)
   M_1 and M_2 as shown in Fig. 5.10.
9. $A_1 = CM/(1 - (f_{ca}/F_{EX}))$
10. Calculate F_{bx}
11. $f_{ca}/F_C + f_{bx}\cdot A_1/F_{bc} \leq 1.0$ (check buckling)
12. $f_{ca}/0.6\ F_y + f_{bx}/(0.66\ F_y$ or $0.6\ F_Y) \leq 1.0$ (check yielding).

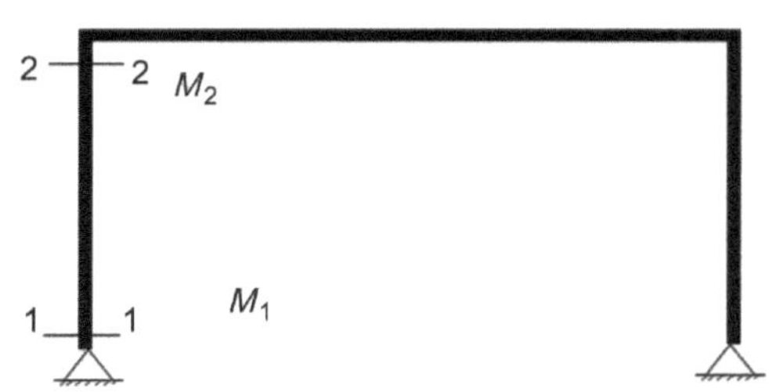

Figure 5.10 Portal frame.

5.3.5 Design of Beam Column Member (LRFD)

The design of a beam column member (LRFD) will follow the same procedure as given previously, but with the following equations:

$$\frac{P_u}{\phi P_n} \geq 0.2$$

$$\left(\frac{P_u}{\phi P_n}\right) + \frac{8}{9}\left[(M_{ux}\phi/M_{nx}) + (M_{ux}\phi/M_{nx})\right] \leq 1.0$$

$$\frac{P_u}{\phi P_n} < 0.2$$

$$\left(\frac{P_u}{2\phi P_n}\right) + \left[(M_{ux}\phi/M_{nx}) + (M_{ux}\phi/M_{nx})\right] \leq 1.0$$

Example 3: Frame design in the case of side sway:

Design the beam column where the buckling lengths are:
$L_y = 3000$ mm, $L_x = 13{,}800$ mm

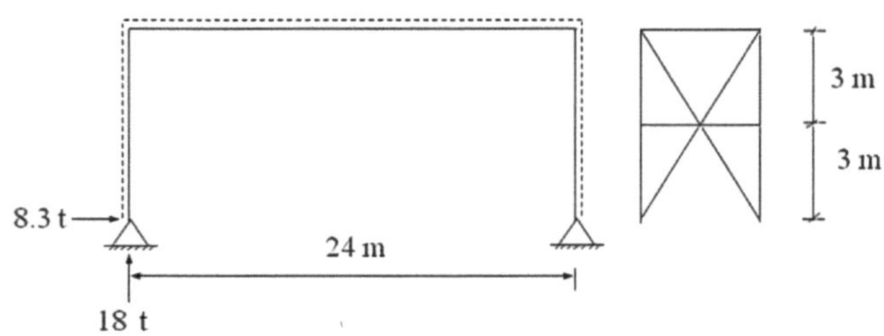

$M_x = 49.8$ mt, $N_{max} = -18.0$ t
$M_x = 55.2$ mt, $N_{max} = -22.3$ t

Solution:

Choose HEB 500
$F_y = 240$ N/mm^2
$t_f = 28$ mm, $t_w = 14.5$ mm, $r_x = 212$ mm, $r_y = 72.7$ mm,
$S_x = 4287 \times 10^3$ mm^3, $S_y = 842 \times 10^3$ mm^3
$A = 23{,}900$ mm^2
Local buckling
Web capacity $= F_y(50 - 2 \times t_f)t_w = 160.95 > 18$ t
The plastic neutral axis inside the web:
$N = 2at_wF_y$, $a = 2.5$ cms
$ad_w = h/2 + a - 2t_f = 22.5 + 2.5 - 2 \times 2.8 = 19.4$ cm

$d_w = h - 4t_f = 33.8$, $a = 0.57 > 0.5$
$d_w/t_f \leq (699/(2.4)^{0.5})/(13 \times (a-1)) = 69.9$
$d_w/t_f = 33.8/1.5 = 22.53 < 69.9$ OK
Check flange c/t_f OK
The unbraced length:
$L = 3000$ mm, $L_u = 20b/(F_y)^{0.5} = 387.3 > 300$ OK
$L_x/r_x = 13{,}800/212 = 65.1$ $L_y/r_y = 41.27$
F_{EX} (plane of bending) $= 7500/(72.63)^2 = 142{,}17$ N/mm^2
$F_c =$ from the table $= 113$ N/mm^2
$f_a = 18/239 = 7.5$ N/mm^2
$f_a = 22.3/239 = 0.09$ t/cm^2
$f_{bx} = 55.2 \times 100/4815 = 1.15$ t/cm$^2 = 11.5$ N/mm^2
Equivalency and amplification factors:
$CM = 0.85$ (in the case of side sway permitted)
$A = 1/(1 - (0.09/1.4217)) = 1.0573$ $A_1 = CM$, $A = 0.898 \geq 1.0$
$A_1 = 1.0$
$(f_{ca}/F_c) + (f_{bx} CM \cdot A/F_{bc}) \leq 1.0$
$(9/113) + (115 \times 1.0/0.66F_y) = 0.8 < 1.0$ OK
$f_{ac}/F_c + f_{bx}/F_{bc} \leq 1.0$
$(9/0.6\ F_y) + (115/0.66\ F_y) = 0.79 < 1.0$ OK

This section is safe but not economical, so it requires trying less section dimensions.

Example 4: Design for frame without side sway:

Design the beam column where the buckling lengths are:
$L_y = 5000$ mm, $L_x = 3500$ mm

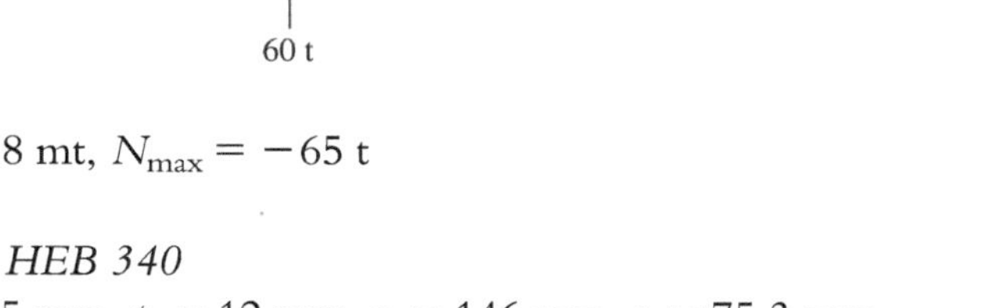

$M_x = 18$ mt, $N_{max} = -65$ t

Solution:

Choose HEB 340

$t_f = 21.5$ mm, $t_w = 12$ mm, $r_x = 146$ mm, $r_y = 75.3$ mm,
$S_x = 2156 \times 10^3$ mm^3, $S_y = 646 \times 10^3$ mm^3

$A = 17{,}100\ \text{mm}^2$
Local buckling
Web capacity $= F_y\ (34 - 2 \times t_f)\ t_w = 85.54 > 60$ t
The plastic neutral axis inside the web:
$N = 2at_wF_y,\ a = 9.61$ cm
$ad_w = h/2 + a - 2t_f = 17 + 9.61 - 2 \times 2.1 = 22.31$ cm
$d_w = h - 4t_f = 25.4,\ a = 0.90 > 0.5$
$d_w/t_f \leq (699/(2.4)^{0.5}/(13 \times a - 1) = 41.48$
$d_w/t_f = 25.4/2.15 = 11.81 < 41.48$ OK
Check flange c/t_f OK
$L = 5000$ mm, $L_u = 20b/(F_y)^{0.5} = 387.3 < 500$ OK
The section is noncompact, $F_b = 0.6\ F_y$
$L_x/r_x = 34.25,\ L_y/r_y = 66.41$
F_{EX} (plane of bending) $= 7500/(34.25)^2 = 6.39\ \text{t/cm}^2 = 639\ \text{N/mm}^2$
$F_c = 1.18\ \text{t/cm}^2 = 118\ \text{N/mm}^2$ (from the table)
$f_a = (65/171) \times 100 = 38\ \text{N/mm}^2$
$f_{bx} = 20.5 \times 10{,}000/2156 = 95.1\ \text{N/mm}^2$
Equivalency and amplification factors:
$f_a/F_c = 0.35/1.18 = 0.29 > 0.15$
$A = 1/(1 - (0.38/6.39)) = 1.05,\ CM = 0.6 - 0.4\ (18/20.5) =$
$0.25 < 0.4,\ CM = 0.4$
$A1 = CM \times A < 1.0,\ A_1 = 1.0$
$C_b = 1.75 + 1.05(18/20) + 0.3(18/20)^2 = 2.938 > 2.30$ take it 2.3
$F_{ltb} = (800 \times 30 \times 2.5) \times 2.3/(500 \times 32) = 6.98 > 0.6\ F_y$
Check of stresses:
For buckling:
$(f_a/F_c) + (f_{bx} \cdot CM \cdot A/F_{bc}) \leq 1.0$
$(38/118) + (95.1 \times 1.0/0.6F_y) = 0.98$ OK
For yielding:
$f_{ac}/F_c + f_{bx}/F_{bc} \leq 1.0$
$(38/0.6F_y) + (95.1/0.6F_y) = 0.92 < 1.0$ OK

5.4 STEEL PIPE RACK DESIGN

The purpose of this section is to provide guidelines and recommended procedures for the analysis and design of steel pipe rack design.

This design guide defines the minimum requirements for the analysis and design of pipe racks in process industry facilities. It covers general design philosophy and requirements to be used in the analysis and design of pipe racks. Criteria presented herein pertain to loads, load combinations, allowable stresses, and superstructure and foundation design.

5.4.1 Pipe Rack Design Guide

Pipe racks are frame structures that support pipes and auxiliary equipment in the process areas of industrial plants. Just like loads from the wind and or earthquakes, piping loads can vary greatly from project to project. Clearly, it is difficult to define specific criteria for the design of such structures. This guideline sets general requirements which the engineer should incorporate into pipe rack designs if possible.

This guideline is recommended for application to the following three types of steel pipe racks:

- strutted main pipe racks,
- unstrutted secondary or miscellaneous pipe racks,
- "T" supports.

The pipe rack superstructures and foundations shall be designed for the loads and load combinations as specified in Chapter 2, Loads on the Industrial Structures.

The piping work will be in contact with the pipe rack and will act as one unit. Fig. 5.11 presents a piperack under construction. The deflection should be restricted based on the design to avoid higher stress on the piping system.

Figure 5.11 Pipe rack under construction.

The deflection requirements for pipe rack beams and transverse bents shall be as follows:

The maximum allowable beam *deflection*(Δ_{max}) due to total load shall be as follows:

- $\Delta_{max} = L/240$
  where L is the span length.
- The maximum allowable *drift limits* for the pipe rack shall not exceed $H/100$
  where H = pipe rack height.

5.4.2 Pipe Rack Superstructure Design

The principal structural components of a pipe rack are the transverse beams, the columns, the longitudinal struts, and the vertical bracing. Design criteria applicable to each of these components are presented below.

In general, the pipe support framing system is designed as a rigid frame with fixed or pinned bases in the transverse direction and as braced frames in the longitudinal direction.

Additional longitudinal and/or intermediate transverse beams may be required to support electrical conduit, instrumentation lines, or other small lines. Electrical conduit and cable trays usually must be supported every 3 m unless the vendor recommends larger spans in his drawings.

Structural components of the pipe rack must be capable of resisting the axial loads, shears, and moments.

An elastic analysis shall be used to determine moments and forces in pipe rack members.

- Structural Steel Expansion

  For pipe rack design, provisions shall be made for thermal expansion of steel, with the structural steel checked for temperature change. Slotted connections (sliding connections) shall be provided in each segment of the pipe rack between vertical bracing to allow for structural steel thermal expansion. The maximum segment for the pipe rack shall be limited to 42 m in length unless calculations indicate otherwise.

  A—Transverse Beams

  - In computing the allowable bending stress F_b, the unbraced length shall be taken as the span of the beam, and the AISC factor C_b shall be used to account for end fixity. *A* C_b value of 1.0 is a very conservative and safe assumption. In no case shall the assumption of lateral support from piping be used in computing F_b.

- Generally, the horizontal members' depth should not be less than 1/24 of the span.
- If top flange lateral loads are significant, the transverse beam shall be investigated for bending about the $y-y$ axis and for torsion. This can be estimated by using $My/(0.5Sy)$.
- In axial load design, the total span of the beam should be used, modified by the appropriate effective length factor for each direction. This factor should be equal to 1.0 for the weak direction of the beam.
- Special consideration shall be given to the design of transverse beams which support large vapor lines to be hydrotested or which support large anchor or guide forces. Normally, if the local bending stresses are too high, it is required to use horizontal bracing locally.

B—Columns

For strutted pipe racks, columns shall normally be designed with pinned or fixed bases depending on the lateral drift requirements.

For normal pipe racks, column bases shall be considered pinned in the transverse direction and fixed in the longitudinal direction. The major axis of columns should normally be perpendicular to the longitudinal direction of the pipe rack (i.e., the plane formed by the column web is parallel to the longitudinal direction).

"T" support column bases, which are used in carrying one or two pipes, shall be considered fixed in both the transverse and longitudinal directions. The major axis of columns may be turned in either direction.

The column base plates for steel structure, miscellaneous piperacks, or "T" supports foundations shall be fixed to the concrete foundation by anchor bolts.

C—Longitudinal Struts

In areas where gravity loading of struts is anticipated, struts shall be designed for axial loads produced by longitudinal pipe loads plus gravity load moments and shears. Such struts should be designed for the actual load but not less than 50% of the gravity loading of the loaded transverse pipe support beam. This loading requirement will account for the usual piping and electrical conduits.

Concentrated loads for large pipes shall also be included in the design. Where gravity loading is not anticipated, struts shall be designed for axial loads only. The primary source of axial loads is longitudinal pipe loads.

D—Vertical Bracing

Vertical bracing may be used to transmit transverse and longitudinal forces to the foundations. *K*-bracing or *X*-bracing is usually used for this purpose. In modern research, it is preferred to use *X*-bracing rather than *K*-bracing from a redundancy point of view.

Braced bays in strutted pipe rack systems should be spaced at a maximum of 42 m. Longitudinal bracing should be provided in about every fourth bay.

Compression bracing for steel pipe rack systems shall normally be designed with wide flange shapes. For tension bracing, single or double angles are usually used.

Example 5: Pipe Rack Design:

Design typical pipe rack bent in oil plant. The pipe rack configuration shall be as shown in Example 1 in Chapter 2, Loads on the Industrial Structures and the plan dimensions as shown in Fig. 5.12.

Solution:

Assumptions:

Assume in the first trial that the main beams are W10 × 33 or HEA 260 for the cable tray support and W14 × 48 and W14 × 53 for the pipe supports.

The beam levels are 6.0, 7.5 and 9.0 m. The beams are rigidly connected to the columns (i.e., moment connection). The columns are W14 × 53 and fixed at the base.

The longitudinal struts (W10 × 33) located at levels 5.25, 6.75 and 8.5 m act as struts to transfer the thermal load to the vertical

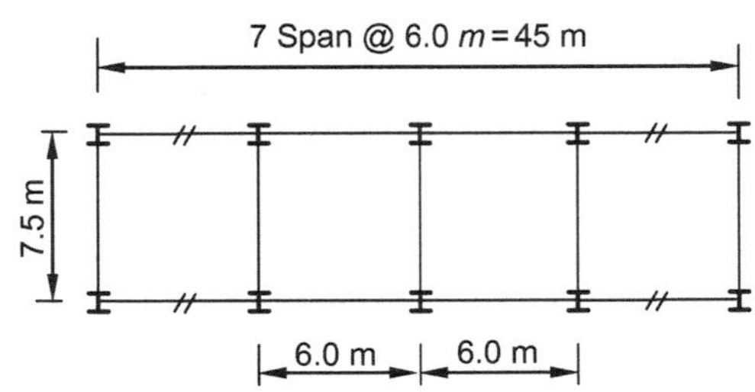

Figure 5.12 Pipe rack–plan layout.

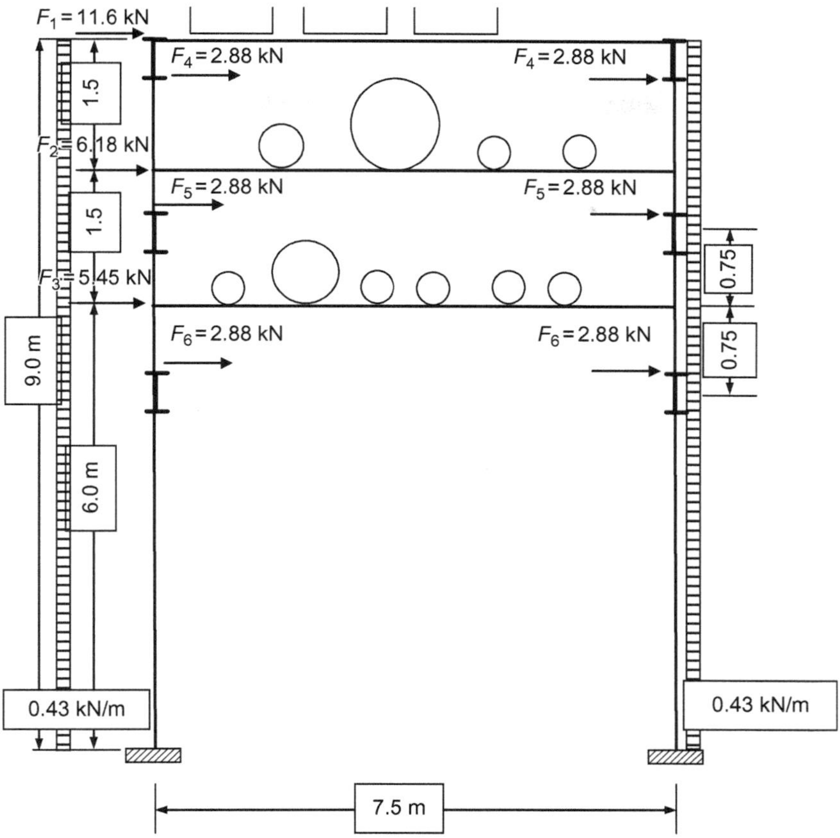

Figure 5.13 Sketch for loads on the pipe rack for Example 1.

bracing of the rack. These levels will be considered as braced in the longitudinal direction.

The loads affect the structure is as shown in Fig. 5.13. The second step is to build the model in any software package used in structure analysis and to perform the analysis and obtain the stresses and the bending moment as shown in Fig. 5.14. The unity check for the member is shown in Fig. 5.15. From this figure, it is found that the unity check is lower than unity so the chosen member is safe.

K_z (effective length factor about the column local z-axis) calculations for columns. The selected steel section is presented in Figure 5.16.

Refer to detailed calculations in Table 5.8.

Check the critical K_z for the governing bottom portion of column:

W14 × 53 $I_x = 225.18 \times 10^6$ mm^4 Column

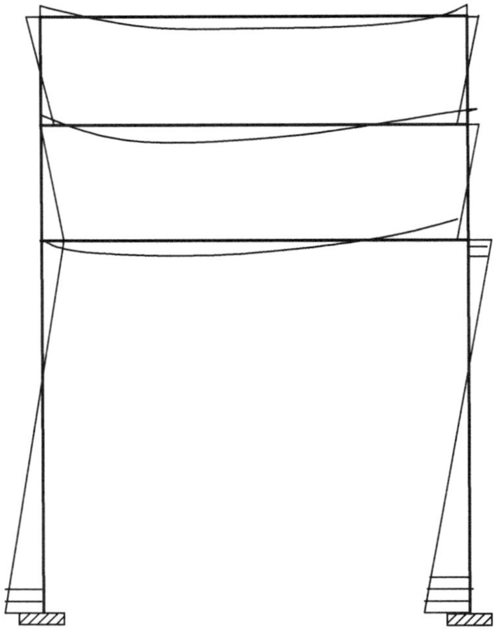

Figure 5.14 Sketch for B.M. diagram.

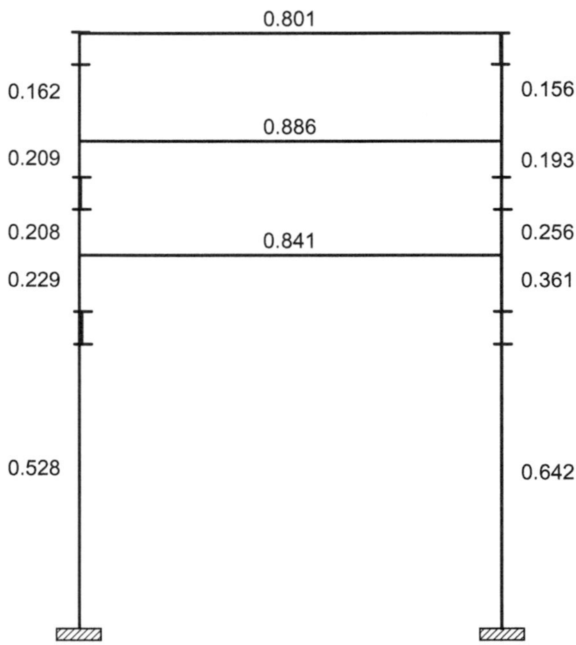

Figure 5.15 Unity check value for the structure members.

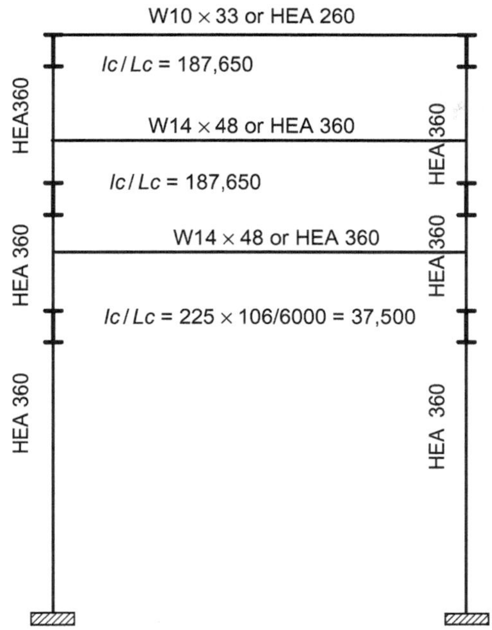

Figure 5.16 Sketch for loads on the pipe rack for Example 1.

Table 5.8 *G* and *K* calculation

Elevation (m)	Column member	*G*	K_z
9	5,10	19.78	3.3
7.5		13.09	
	3,4,8,9		2.7
6		7.85	
	1,2,6,7		1.8
0		1.0	

W14 × 53 $I_x = 225.18 \times 10^6$ mm^4 Beam at level
W10 × 33 $I_x = 71.18 \times 10^6$ mm^4 Beam at level 9 m
W14 × 48 $I_x = 201.5 \times 10^6$ mm^4 Beam at level 7.5 m
$GB = 1.0$ Fixed Base
$GT = \Sigma\ IC/LC$
$\Sigma\ IB/LB$
$\Sigma\ IC/LC = (225.18 \times 10^6/6000) + (225.18 \times 10^6/1500) = 187{,}650$
$\Sigma\ IB/LB = 225.18 \times 10^6/7500 = 30{,}024.133$
$GT = 187{,}650/30{,}024.133 = 6.25$
With $GB = 1.0$ and $GT = 6.26$ $K_z = 1.8$

At level 9

$IB/LB = 71.18 \times 106/7500 = 9490$

$IC/LC = 187{,}650$

$G = 187{,}650/9490 = 19.78$

At level 7.5

$IB/LB = 201.5 \times 106/7500 = 28{,}666.7$

$G = 187{,}650 + 187{,}865/28{,}666.7 = 13.09$

At level 6

$G = 37{,}500 + 187{,}650/28{,}666.7 = 7.85$

Pipe rack bent is designed per the attached STAAD PRO input and output file. Check STAAD PRO output for the following:

Unity check:

Ensure that unity checks for all structural members are less than 1.0.

Beam deflection:

Ensure that maximum beams vertical deflection is less than $L/240$, where L = span length.

Lateral drift:

Ensure that maximum lateral drift for the pipe rack is less than $H/100$ for load combinations with wind load and $H/100$ for earthquakes.

5.5 STAIRWAY AND LADDERS DESIGN GUIDE

The following requirements are summarized from the 1987 edition of the OSHA General Industry Requirements:

Fixed stairs are usually used to access one structure level from another where operations necessitate regular travel between levels, and for access to operating platforms at any equipment which requires regular monitoring during operations.

Fixed stairs shall also be provided where access to elevations is daily or at each shift for such purposes as gauging, inspection, regular maintenance, or if one of the following conditions occurs: the work exposes employees to acids, caustics, gases, or other harmful substances, or the carrying of tools or equipment by hand is normally required. OSHA 1910.24(b) and the industrial practice provide a guideline for fixed stair design as follows:

- Every flight of stairs having four or more risers shall be equipped with standard handrails.
- The stairways shall be designed and constructed to carry a load of five times the normal live load anticipated but never of less strength than to safely carry a moving concentrated load of 4.4 kN.

- The width of fixed stairways is recommended to be not less than 760 mm.
- Stairways shall be installed at angles to the horizontal of between 30° and 50°. Any uniform combination of rise/tread dimensions may be used that will result in a stairway at an angle to the horizontal within the permissible range. Table 5.9 gives rise/tread dimensions that will produce a stairway within the permissible range, stating the angle to the horizontal produced by each combination.
- Stairway platforms shall be no less than the width of a stairway and a minimum of 760 mm in length measured in the direction of travel.
- For handrails, standard railings shall be provided on the open sides of all exposed stairways and stair platforms for at least one side of closed stairways, preferably on the right side descending.
- Vertical clearance above any stair tread to any overhead obstruction shall be at least 2.1 m measured from the leading edge of the tread.

Every open-sided floor or platform 1.2 m or more above the adjacent floor or ground level shall be guarded by a standard railing on all open sides except where there is an entrance to a ramp, stairway, or fixed ladder. All handrails and railings shall be provided with a clearance of not less than 75 mm between the handrail or railing and any other object.

Table 5.9 Stairway rise and run dimensions

Angle to horizontal	Rise (mm)	Tread run (mm)
30°35′	165.1	279.4
32°08′	171.45	273.05
33°41′	177.8	266.7[a]
35°16′	184.15	260.35
36°52′	190.5	254
38°29′	196.85	247.65
40°08′	203.2	241.3
41°44′	209.55	234.95
43°22′	215.9	228.6
45°00′	222.25	222.25
46°38′	228.6	215.9
48°16′	234.95	209.55
49°54′	241.3	203.2

[a]Preferred.

Table 5.10 Maximum allowable girder deflections

Top–running CMAA class A, B, and C cranes	*L*/600
Top–running CMAA class D cranes	*L*/800
Top–running CMAA class E and F cranes	*L*/1000
Under–running CMAA class A, B, and C cranes	*L*/450
Monorails	*L*/450

5.6 CRANE SUPPORTS

The vertical deflection is very critical in the movement of the crane, and it mainly depends on the crane manufacturer. Table 5.10 shows the guidelines for the maximum allowable deflection of the runway girders under the maximum static load without impact load (where L = the span length).

Vertical deflection of the jib crane support beams shall not exceed $L/225$ (where L = the maximum distance from the support column to load location along the length of the jib beam) if loaded with the maximum lifted plus hoist load(s), without impact.

Lateral deflection of support runway girders for cranes with lateral moving trolleys shall not exceed $L/400$ (where L = the span length) when loaded with a total crane lateral force not less than 20% of the sum of the weights of the lifted load (without impact) and the crane trolley. The lateral force shall be distributed to each runway girder with consideration for the lateral stiffness of the runway girders and the structure supporting the runway girders.

5.7 CONNECTIONS DESIGN

The connection can be a welded or bolted connection according to the design and fabrication procedure which is constrained if the connection will be fabricated on site or in the workshop. A sample of bolted and welded connection is presented in Fig. 5.17.

The following rule should be followed when selecting the weld connection. Welding in the field is avoided if possible due to welding condition requirements. Field welding is not to be performed while it is raining, snowing, or below 0°F; in certain ambient temperatures, preheating of the material to be welded is required. AWS Code D1.1

Figure 5.17 Special bolted connection.

(2004b) specifies minimum preheat temperatures, which are designed to prevent cracking.

The connections shall conform to the following requirements:

1. Shop connections may be either bolted or welded. Field connections shall be bolted where possible. Connections may be field welded when conditions are such that a bolted connection is not suitable.
2. Bolted connections for primary members shall utilize high-strength bolts conforming to ASTM A-325-N, and bearing-type connections with threads included in the shear plane. However, slip-critical-type connections shall be used in connections subject to vibration or repeated stress reversal.
3. Standard connections shall be designed by the fabricator in accordance with the project construction specifications and loads shown on the drawings. Moment connections and special connections shall be designed by the engineer and shall be shown on the engineering drawings.
4. Moment connections shall preferably be of the bolted end plate type.

In general, the differences between the bolting and welding which will be used in selection are summarized as follows:

- Bolting is generally a faster operation than welding.
- Bolting does not have the temperature and weather condition requirements that are associated with welding.
- Unexpected weather changes may delay welding operations.

5.7.1 Bolts

Four basic types of bolts are commonly used. They are designated by ASTM as A307, A325, A490, and A449. A307 bolts are called unfinished or ordinary bolts. They are made from low-carbon steel. Two grades, A and B, are available. They are available in diameters from 6 to 100 mm. They are used primarily for low-stress connections and for secondary members. A325 and A490 bolts are called high-strength bolts. A325 bolts are made from a heat-treated medium carbon steel.

They are available on the market in three types of bolts:

Type 1: Bolts made of medium carbon steel.

Type 2: Bolts made of low carbon steel.

Type 3: Bolts having atmospheric corrosion resistance and weathering characteristics comparable to A242 and A588 steel.

A490 bolts are made from quenched and tempered alloy steel, and thus have a higher strength than A325 bolts. Just like A325 bolts, three types (Types 1−3) are available.

Both A325 and A490 bolts are available in diameters from 12 to 38 mm, in 1/8 in. increments. They are used for general construction purposes. A449 bolts are available in diameters up to 75 mm. A449 bolts are used when diameters over 38 mm are needed. They are also used for anchor bolts and threaded rods.

High-strength bolts can be tightened to two conditions of tightness: snug-tight and fully tight.

Snug-tight conditions can be attained by a few impacts of an impact wrench or the full effort of a worker using an ordinary spud wrench.

Snug-tight conditions must be clearly identified on the design drawing and are permitted only if the bolts are not subjected to tension loads; loosening or fatigue due to vibration or load fluctuations are not design considerations. Bolts used in slip critical conditions (i.e., conditions for which the integrity of the connected parts is dependent on the frictional force developed between the interfaces of the joint) and in conditions where the bolts are subjected to direct tension are required to be fully tightened to develop a pretension force equal to about 70% of the minimum tensile stress F_u of the material from which the bolts are made.

The stresses on the high strength bolt are shown in Fig. 5.18. This can be accomplished by using the turn-of-the-nut method, the calibrated wrench method, or by the use of alternate design fasteners or direct tension indicators. The bolted number and component are presented in Figs. 5.19 and 5.20, respectively.

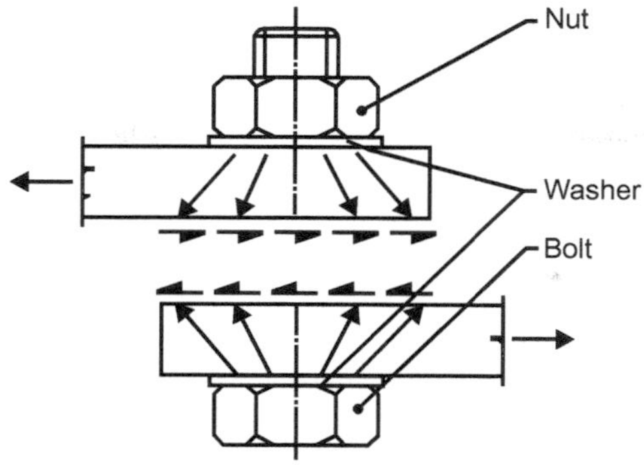

Figure 5.18 High strength bolt stresses.

Figure 5.19 Photo presents the bolt number.

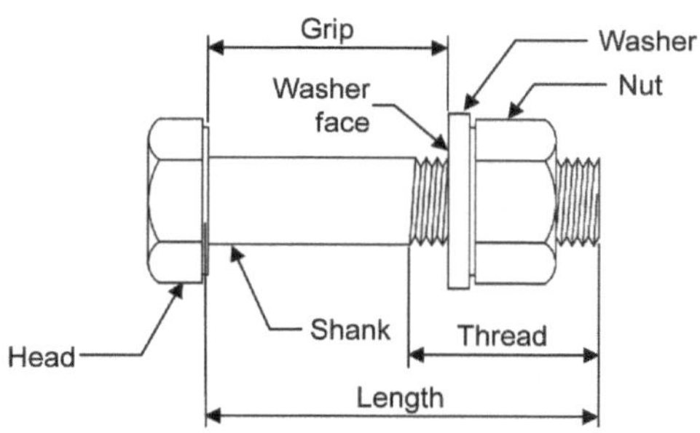

Figure 5.20 The bolt component.

In cases of bolted connections, the following tools can be used as shown in Figs. 5.21 and 5.22.

The common tools used by ironworkers include spud wrenches, pins, and correction bars of various sizes as shown in Fig. 5.21.

Figure 5.21 Tools for erecting connection.

Figure 5.22 Hydraulic wrench.

Table 5.11 Properties and strength of high strength bolts (grade 10.9)

d	*A*, mm²	A_s, mm²	*T*, tons	M_a, Kg.m	Permissible friction load of one bolt per one friction surface (p_s) ton			
					St. 37 and 42–44 ($\mu = 0.4$)		St. 50–55 ($\mu = 0.5$)	
					I	II	I	II
M12	113	84	5.29	12	1.69	2.01	2.11	2.52
M16	201	157	9.89	31	3.16	3.37	3.95	4.71
M20	314	245	15.43	62	4.93	5.90	6.17	7.36
M22	380	303	19.08	84	6.10	7.27	7.63	9.10
M24	452	353	22.23	107	7.11	8.45	8.89	10.60
M27	573	459	28.91	157	9.25	11.03	11.56	13.78
M30	706	561	35.34	213	11.30	13.48	14.13	16.86
M36	1018	817	51.47	372	16.47	19.64	20.58	24.55

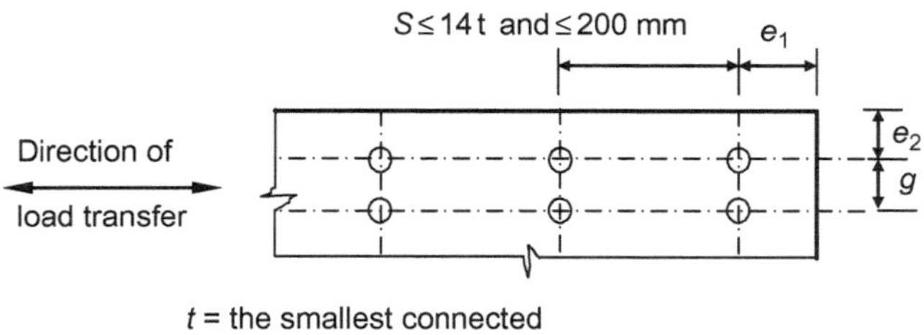

Figure 5.23 Spacing in tension or compression member.

Impact wrenches will be needed for certain installations as shown in Fig. 5.22.

Electricity or compressed air is required depending on the impact wrench being used.

From ECS2008, Table 5.11 presents the size of each bolt with its area, A_s, and the area of stresses with the torque required, M_a, to achieve the required load. The shearing force resistance P_s is presented in Case I for primary stresses due to dead load, live load or superimposed load and the dynamic effects and centrifugal force. In Case II as presented in Table 5.11 is for primary and additional stresses in Case I in addition to earthquake load, wind load, braking force or lateral shock.

Table 5.11 can be applied for grade 8.8 by reducing the values in the table by 30%.

The steel detailing guide is shown in Figs. 5.23 and 5.24.

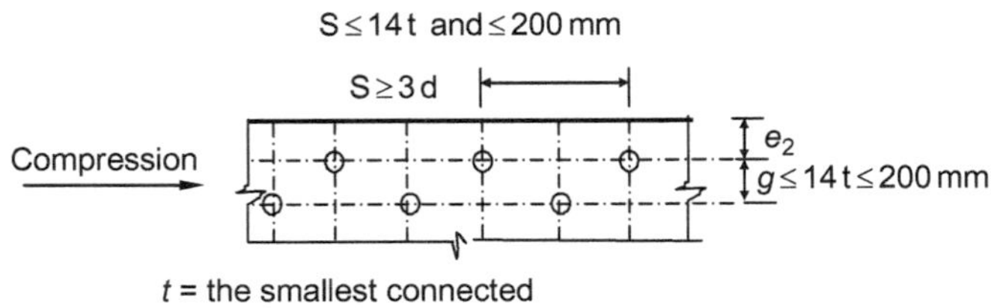

Figure 5.24 Staggered spacing in compression.

5.7.2 Welding

Welded connections are connections whose components are joined together primarily by welds. Welds can be classified according to:

- Types of welds: groove, fillet, plug, and slot welds.
- Positions of the welds: horizontal, vertical, overhead, and flat welds.
- Types of joints: butt, lap, corner, edge, and tee.

Although fillet welds are generally weaker than groove welds, they are used more often because they allow for larger tolerances during erection than groove welds. Plug and slot welds are expensive to make, and they do not provide much reliability in transmitting tensile forces perpendicular to the surfaces. Furthermore, quality control of such welds is difficult because inspection of the welds is rather arduous. As a result, plug and slot welds are normally used just for stitching different parts of the members together.

5.7.2.1 Welding Symbols

A shorthand notation giving important information on the location, size, length, etc. for the various types of welds was developed by the American Welding Society to facilitate the detailing of welds. This system of notation is reproduced in Fig. 5.25.

5.7.2.2 Strength of Welds

In ASD, the strength of welds is expressed in terms of allowable stress. In LRFD, the design strength of welds is taken as the smaller of the design strength of the base material and the design strength of the weld electrode. These allowable stresses and design strengths are summarized in Table 5.12. When a design uses ASD, the computed stress in the weld shall not exceed its allowable value. When a design uses LRFD, the design strength of welds should exceed the required strength obtained by dividing the load to be transmitted by the effective area of the welds.

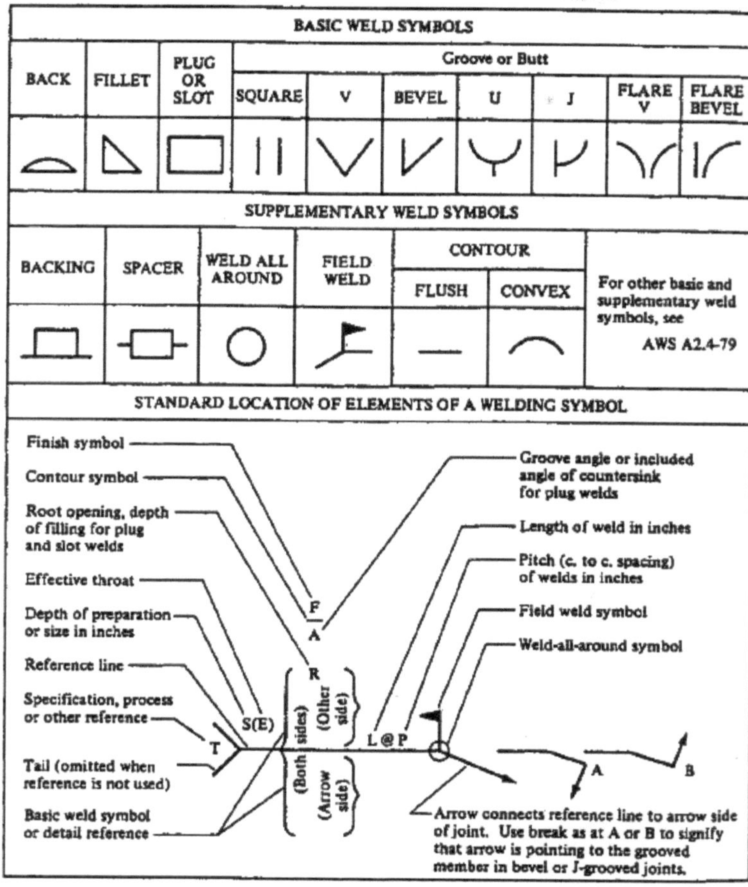

Figure 5.25 Welding symbol.

The general limitation to the welding will be as follows:

- $S \leq$ min of t_1 or t_2
- min of t_1 or $t_2 < 20$ mm, $S \geq 5$ mm
- 20–30 mm, $S \geq 6$ mm
- 30–50 mm, $S \geq 8$ mm
- 50–100 mm, $S \geq 10$ mm
- For building, $S \geq 4$ mm
- L_{min} = largest of 4 S or 5 cm
- $L_{max} = 70$ S
- No limit for "L" for beam-column connections or flange-to-web welds in welded built up girders.

Table 5.12 Strength of welds

Types of weld and stress	Material	ASD allowable stress	LRFD ϕ FBM or ϕ FW	Required weld strength level
Full penetration groove weld				
Tension normal to effective area	Base	Same as base metal	$0{:}90F_y$	"Matching" weld must be used
Compression normal to effective area	Base	Same as base metal	$0{:}90F_y$	Weld metal with a strength level equal to or less than "matching" must be used
Tension of compression parallel to axis of weld	Base	Same as base metal	$0{:}90F_y$	
Shear on effective area	Base weld electrode	0.3 × nominal tensile strength of weld metal	0.9 ($0.6F_y$) 0.8 (0.6FEXX)	
Partial penetration groove welds				
Compression normal to effective area	Base	Same as base metal	$0{:}90F_y$	Weld metal with a strength level equal to or less than "matching" must be used
Tension or compression parallel to axis of weldShear parallel to axis of weld	Base weld electrode	0:30 × nominal tensile strength of weld metal	0:75[:60FEXX]	
Tension normal to effective area	Base weld electrode	0:30 × nominal tensile strength of weld metal ≤ 0.18 × yield stress of base metal	$0.90F_y$ 0.8[0.6FEXX]	
Fillet welds				
Stress on effective area	Base weld electrode	0:30 × nominal tensile strength of weld metal	0:75[:60FEXX] $0.9F_y$	Weld metal with a strength level equal to or less than "matching" must be used
Tension or compression parallel to axis of weld	Base	0:30 × nominal tensile strength of weld metal	$0.9F_y$	

Notes:

- See AWS D1.1 for "matching" weld material.
- Weld metal one strength level stronger than "matching" weld metal will be permitted.
- Fillet welds partial-penetration groove welds joining component elements of built-up members such as flange-to-web connections may be designed without regard to the tensile or compressive stress in these elements parallel to the axis of the welds.

Example 6: Welded connection:

Fig. 5.26 presents unequal angle $80 \times 200 \times 120$ connected to the column by a welding and applies a force, $P = 6$ t to this angle.

Solution:

Straining action:

$Q = 60$ kN, $M = 60 \times 45 = 2700$ kN/m

Section 1-1 (take $S = 10$ mm)

$q = 60/(2 \times 120 \times 10) = 0.025$ kN/mm^2

$q(M_t) = 2700 \times 60/(2 \times (10 \times (120)^3/12)) = 0.0563$ kN/mm^2

$q(t) = [(0.025)^2 + (0.0563)^2]^{0.5} = 0.061 < 0.2F_u < 0.072$ kN/mm^2 OK

Section 2-2

$I_x = 2 \times (120)^3 \times 10/12 = 2{,}880{,}000$ mm^4

$F_t = 2700 \times (120/2)/2{,}880{,}000 = 0.05625$ kN/mm^2

$F_{eq} = [(0.05625)^2 + 3(0.025)^2]^{0.5} = 0.07 < 0.072 \times 1.1$ OK

5.7.2.3 Welding in Existing Structures

Welding to the existing structure when performing an extension or during retrofit projects requires careful consideration of numerous factors:

- Determine weldability—Identify the steel grade to establish a welding procedure.
- Select and design the weld—Fillet welds are preferred and avoid overwelding.
- Surface preparation—Remove contaminants such as paint, oil, and grease.
- Loads during retrofit—An engineer should determine the extent to which a member will be permitted to carry loads while heating, welding, or cutting.

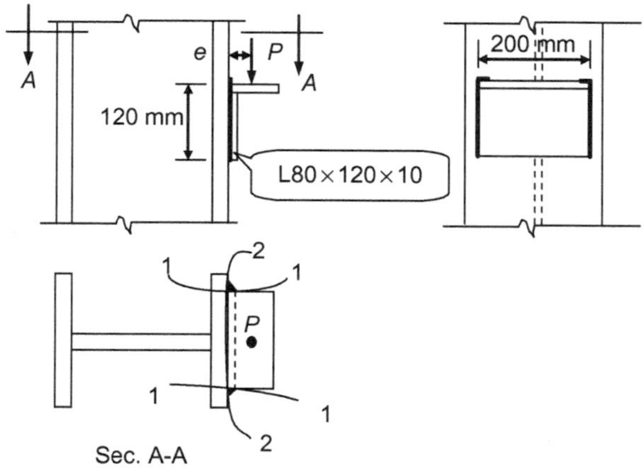

Figure 5.26 Welded connection.

- Fire hazards—Follow all governing fire codes, regulations, and safety rules to avoid fires.

5.7.3 Connection Design

The following is the guideline for the connection design in most steel structures used in the process plant, which is the pipe support and the pipe rack.

Example 7: Bracing connection design:

The normal force is $P_x = 80$ kN and the bracing consists of two equal angles 100 × 100 × 10 mm.

Solution:

This calculation was done by excel sheet. You can contact me or download from the website www.elreedyma.comli.com or www.elreedyman.tk.

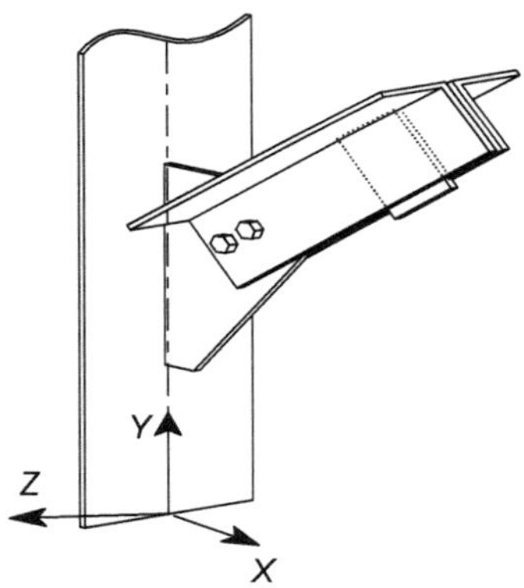

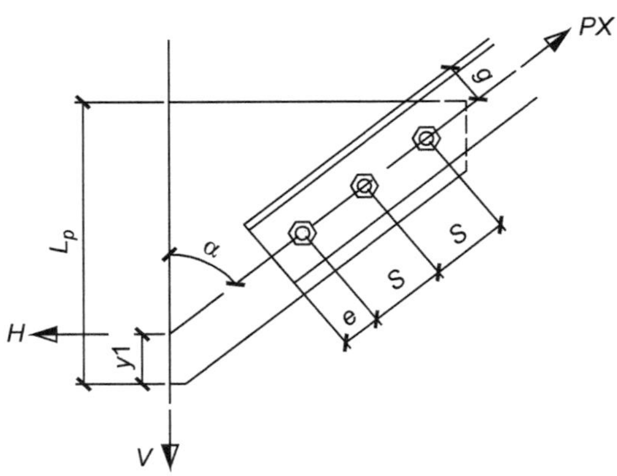

Section properties:

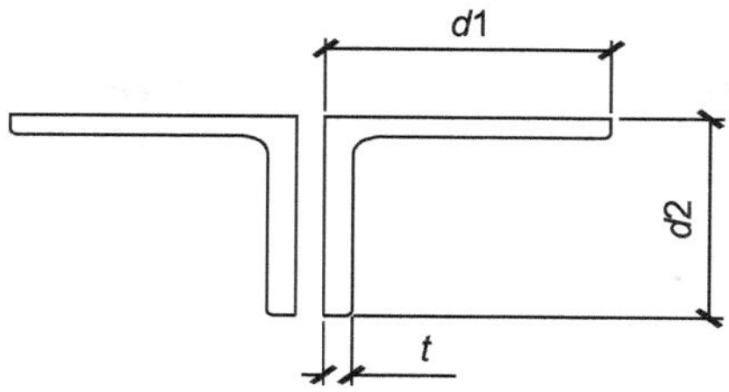

$d_1 = 100$ mm
$d_2 = 100$ mm
$t = 10$ mm
$A_g = 1920$ mm^2

The steel structure member and the connection plates grade is ASTM A 36.

$F_{us} = 400$ N/mm^2 ultimate strength
$F_{ys} = 245$ N/mm^2 yield strength

- *Connection bolts*

  Choose bolt grade ASTM A325
  $d = 20$ mm bolt diameter
  $d_h = 22$ mm hole diameter
  A_b gross $= 314$ mm^2 bolt gross area
  A_b net $= 245$ mm^2 bolt net area
  $f_v = 145$ N/mm^2 allowable shear stress
  $f_p = 480$ N/mm^2 allowable bearing stress $= 1.2F_u$
  $f_t = 303$ N/mm^2 allowable tension stress

- *Geometry*

  $e = 35$ mm edge distance
  $s = 70$ mm spacing between bolts
  $g = 55$ mm gauge of bolts
  $N_b = 2$ no. of bolts
  $t_p = 12$ mm gusset plate thickness
  $L_p = 200$ mm gusset plate length
  $Y_1 = 0$
  $L_1 = 120$ mm buckling length
  $K = 0.5$ the effective length factor
  $\alpha = 30°$
  $S_w = 8$ mm weld size for gusset plate

- *Check stress for bolts*
  - F_s (all) allowable double shear force per bolt
  - F_s (all) $= A_b$ gross $\times f_v \times 2$
  - F_s (all) $= 91$ kN
  - F_s (act) actual shear force per bolt $= P_x/N_b$
  - F_s (act) $= 40$ kN
  - F_s (all) $> F_s$ (act)
  - F_p (all) allowable bearing force per bolt
  - F_p (all) $= f_p \times \min(2t, t_p) \times d$
  - F_p (all) $= 115$ kN
  - F_p (all) $> F_s$ (act)
- *Check for brace*

1. Stress on effective area

   A_{net}, net area for brace section $= 2(A_g - t \times d_h)$

   $A_{net} = 3400$ mm^2

   A_e = effective area $= 0.75 \times A_{net} = 2550$ mm^2

   F_t (act), actual tension stress $= P_x/A_e = 31$ N/mm^2

   F_t (all), allowable tension stress $= 0.5 \times F_{u1} = 200$ N/mm^2

   F_t (all) $> F_t$ (act) OK

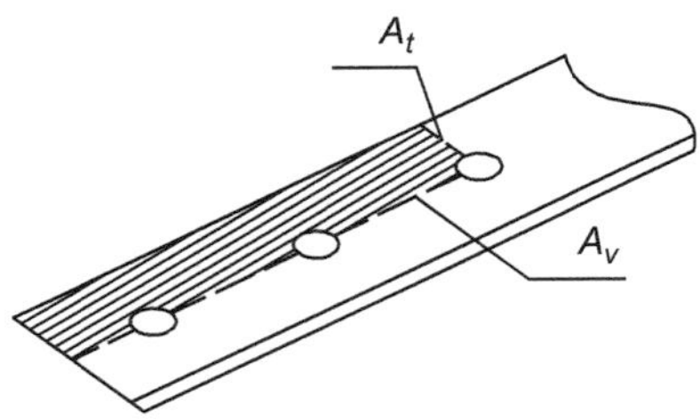

2. Block shear capacity

   Shear area $A_v = 2 \times t \times [(e + (N_b - 1) \times s - (N_b - 0.5) \times d_h]$

   $A_v = 1440$ mm^2

   Tension area $A_t = 2 \times t \times [d_1 - g - 0.5 \times d_h]$

   $A_t = 680$ mm^2

   R_{bs}, resistance for block shear

   $R_{bs} = 0.3 \times F_{u1} \times A_v + 0.5 \times F_{u1} \times A_t$

   $R_{bs} > Px$ OK

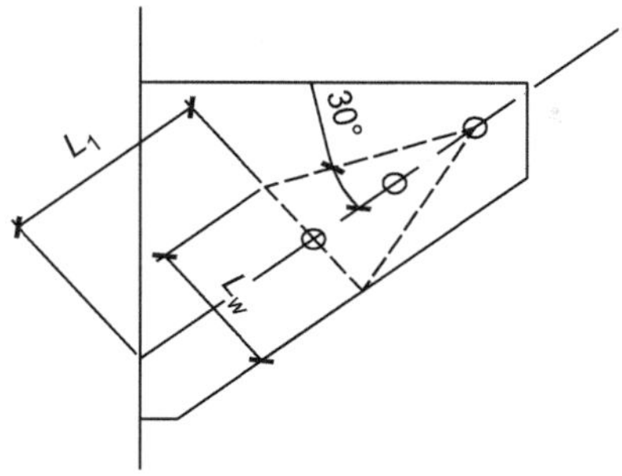

- *Check for gusset plate*

3. Stress on plate section at plate buckling

$$r = \text{radius of gyration} = \sqrt{\frac{I}{A}} = \sqrt{\frac{tp3xL/12}{tpxL}} = \frac{tp}{\sqrt{12}}$$

Slenderness ratio, $\lambda = kL_1/r$

$$C_c = \sqrt{\frac{2\pi^2 E}{F_y}}$$

$kL/r < C_c$

$F_a(\text{all}) = [1 - (kL/r)^2/2C_c^2]F_y/[5/3 + 3(kL/r)/8\ C_c$
$- (kL/r)^3/8C_c^3]$

$F_a(\text{act}) = P_x/(L_w \times t_p)$

$F_a(\text{all}) > F_a(\text{act})$

4. Stress on plate section at contact with column

ecc = Eccentricity is the distance from the point where axial force acting to the plate length L_p center

$H = P_x \sin \alpha$

$V = P_x \cos a$

$M = \text{ecc} \times H$

5. Stress on plate section at contact with column

ecc = distance from point where axial force acting to the plate length L_p center

$H = P_x \sin a$

$V = P_x \cos a$

$M = \text{ecc} \times H$

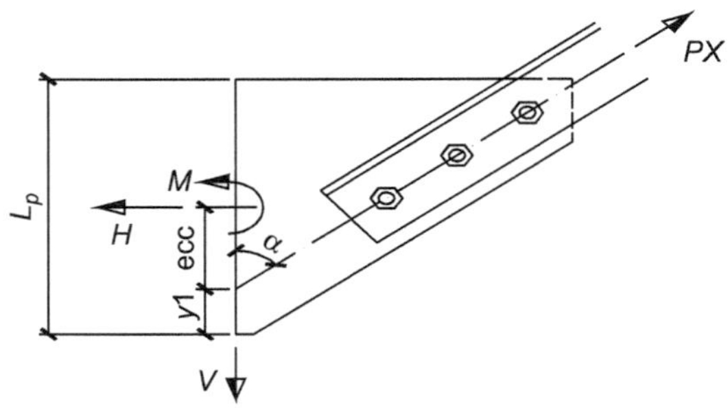

f_t(act), actual tensile stress $= H/(L_p \times t_p) + M/(t_p \times L_p^2/6)$
f_t(act) = 67 N/mm^2
f_t(all), allowable tensile stress $= 0.6 \times F_y$
f_t(all) = 147 N/mm^2
f_t(all) > f_t(act) OK
f_v(act), actual shear stress $= V/(L_p \times t_p)$
f_v(act) = 29 N/mm^2
f_v(all), allowable shear stress $= 0.4 \times F_y$
f_v(all) = 98 N/mm^2
f_v(all) > f_v(act) OK

6. Block shear capacity
   Shear area $A_v = t_p \times [(e + (N_b - 1) \times s - (N_b - 0.5) \times d_h]$
   $A_v = 864$ mm^2
   Tension area $A_t = t_p \times [d_1 - g - 0.5 \times d_h]$
   $A_t = 408$ mm^2
   R_{bs}, resistance for block shear
   $R_{bs} = 0.3 \times F_{u1} \times A_v + 0.5 \times F_{u1} \times A_t$
   $R_{bs} = 185$ kN
   $R_{bs} > P_x$ OK

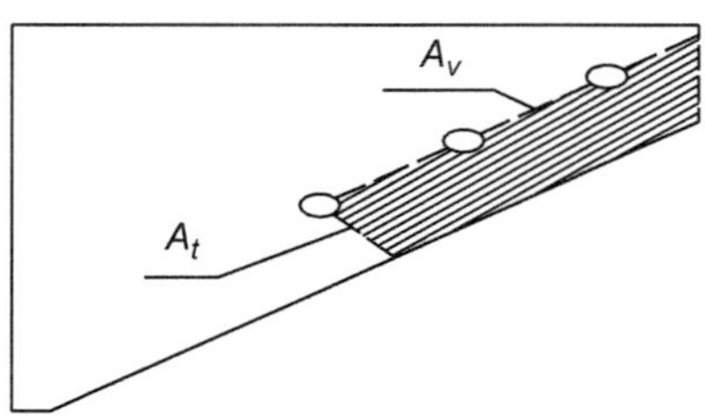

- *Check for welds*
  We will use Electrode E70
  $F_w = 0.707 \times 0.3 \times 70$ ksi

$F_w = 14.8$ ksi $= 102$ N/mm^2
Weld stress components
$f_1 = H/(2 \times L_p \times s_w) + M/2x\ (s_w \times L_p^2/6)$
$f_1 = 50$ N/mm^2
$f_2 = V/(2 \times L_p \times s_w)$
$f_2 = 21.7$ N/mm^2
$f_r = 54.5$ N/mm^2

$$f_r = \sqrt{f_1^2 + f_2^2}$$

$F_w > f_r$ OK

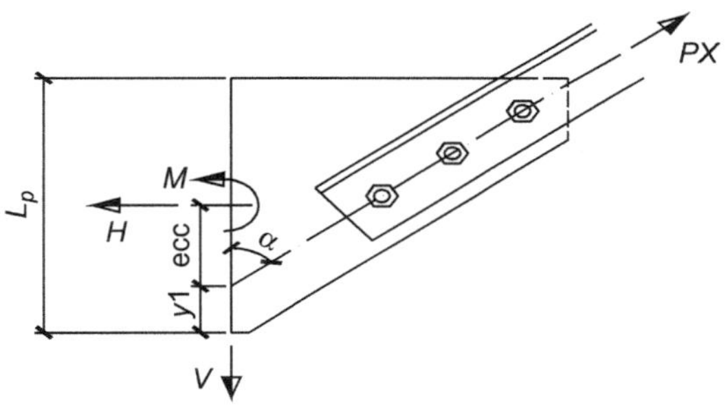

5.7.4 Base plate design

The stresses on the base plate are shown in Fig. 5.27 where d, b is the length and width of the base plate, respectively.

$$f_{1,2} = \frac{-N}{A} \pm \left[6\left(\frac{Ne}{bd^2}\right)\right]$$

$$M = F_1 \cdot X_1 + F_2 \cdot X_2$$

$$f_a = \frac{6M}{b \cdot t^2}$$

Get base plate thickness (t)

The base plate dimensions are obtained from the following equations:
$D = 0.95d + 100$ in mm
$B = 0.8b + 100$ in mm

Example 8: Base plate for fixed connection:

Design a fixed-base connection as shown in Fig. 5.28 to support a vertical force of $N = 500$ kN, a bending moment of $M = 30$ m kN, and a

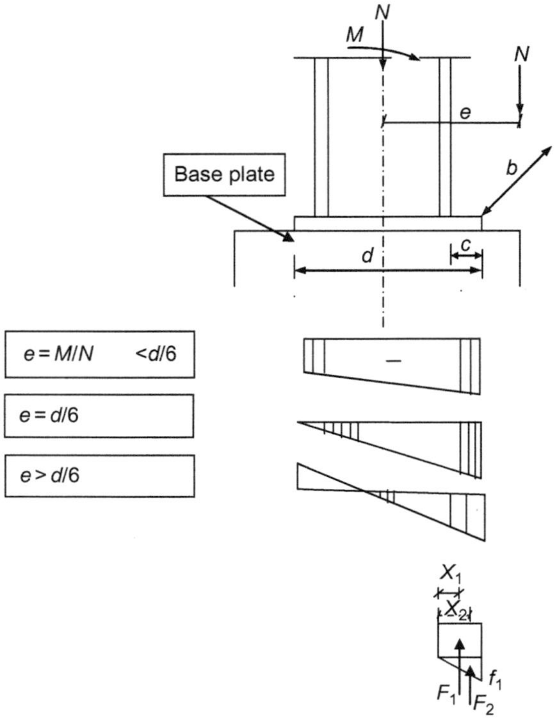

Figure 5.27 Base plate stresses for fixed support.

horizontal force of $H = 50$ kN. The column section is HEB 300 with $F_y = 275$ MPa.

Solution:

Assume the base plate is 400 × 400 mm

$e = M/N = 30/500 = 0.06\text{ m} = 60\text{ mm} < 400/6$

Therefore, the base plate is under compression as the resultant force inside the core.

Calculate stresses

$f_1 = -N/A \pm M \cdot y/I$

$f_1 = (500{,}000N/(400 \times 400)) + 30 \times 1{,}000{,}000 \times 200/(400 \times (400)^3)/12 = 5.94\text{ N/mm}^2$

$f_1 < 30$ MPa which is the concrete characteristic strength

$f_2 = (500{,}000\text{N}/(400 \times 400)) + 30 \times 1{,}000{,}000 \times 150/(400 \times (400)^3)/12 = 5.23\text{ N/mm}^2$

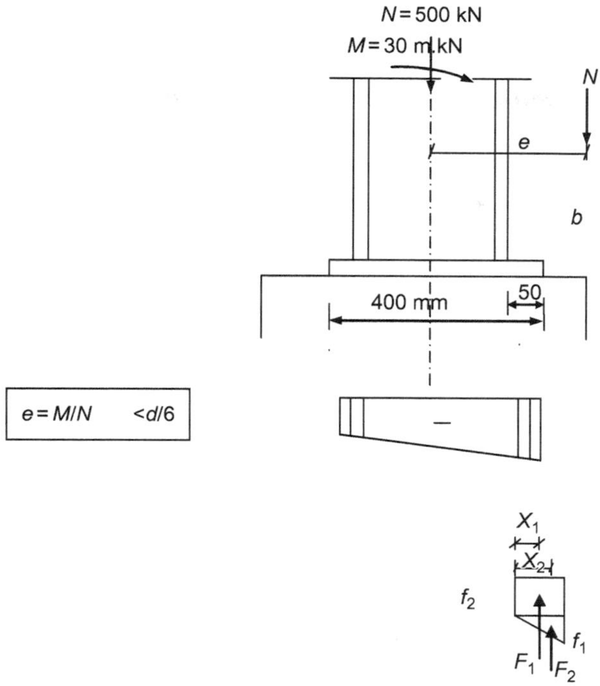

Figure 5.28 Base plate stresses for hinge support.

Calculate moment as in Fig. 5.28 where $X_1 = 25$ mm and
$X_2 = 2/3 \times 50 = 33.3$ mm
$M_1 = 5.23 \times 50 \times 25 + ½(5.94 - 5.23) \times 50 \times 33.3 = 7128.58$
$t = (6M/0.72F_y)^{0.5} = 14.7$ mm choose thickness 15 mm.

Example 9: Base plate for hinge connection:

This example for designing a base plate for hinge connection as presented in Fig. 5.29 with a normal force $N = 450$ kN and a shear force $Q = 40$ kN. The steel section is HEB 300, and the concrete characteristic strength is 35 MPa.

Solution:

$D = 0.95 \times 300 + 100 = 385$ mm
$B = 0.8 \times b + 10 = 340$ mm
$f_c = 450 \times 1000/385 \times 340 = 3.44$ MPa $< f_c'$ OK
$M = 3.44\ (50)^2/2 = 4300$ mm · N/mm
$t = (6M/(0.72F_y))^{0.5}$
$t = 12.2$ mm choose 14 mm

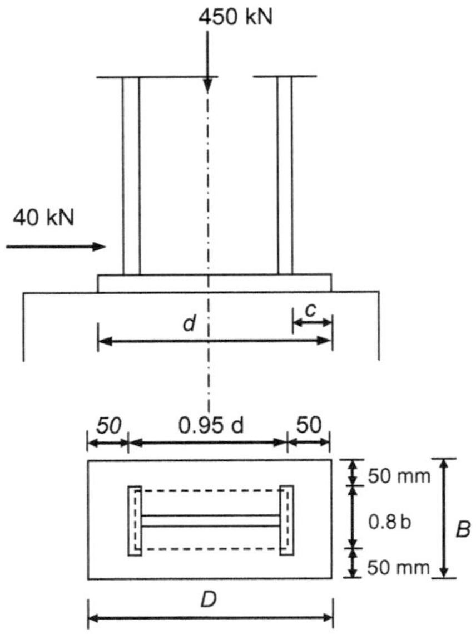

Figure 5.29 Base plate design.

In the case of welding

$S = (40/2 \times 0.2\ F_u \times 300) = 6.9$ mm

Bolts

$A_r = 40/(2 \times 10.8) = 1.85$

5.8 ANCHOR BOLT DESIGN

Definitions for the different types of anchor bolts:

Cast-in anchor: A headed bolt, anchor rod, or hooked bolt that is installed before the concrete is placed.

Postinstalled anchor: An anchor installed in hardened concrete. Expansion, undercut, and adhesive anchors are examples of postinstalled anchors.

Pretensioned anchor bolt: An anchor bolt that is designed to be tensioned to a predefined force to prevent premature failure of the anchor bolt due to fatigue.

Shear lug: A pipe or plate section welded to the bottom of a base plate that is used to resist base shear forces.

The cast-in anchor bolt is the most traditional type in industrial projects; however, the other types are very special for certain facilities and base according to the fabricator or manufacturer requirements.

5.8.1 Anchor Bolts, Nuts, and Washers

5.8.1.1 Anchor Bolts

For most ordinary structures ASTM A307 headed bolts, ASTM A36/A36M rods, or ASTM F1554 grade 36 rods should be specified for ordinary strength anchor bolts. Applications requiring high-strength materials should use ASTM A193/A193M grade B7 or ASTM F1554 grade 105 anchor bolts and rods.

Anchor bolt materials and properties are shown in Table 5.13, while the anchor bolt nut is presented in Table 5.14.

5.8.1.2 Washers

Because base plates typically have holes larger than oversized holes to allow for tolerances on the location of the anchor rod, washers are usually furnished from ASTM A36 steel plates. They may be round, square, or rectangular, and generally have holes 2 mm larger than the anchor rod diameter. The thickness must be suitable for the forces to be transferred.

Table 5.13 Anchor bolt materials and properties

Anchor material type		F_y ksi (Mpa)	f_{ut} ksi (Mpa)	Galvanize?	Ductile?
A307		Not clearly defined	60 (410)	Yes	Yes
A 36 or f1554 grade 36		36 (250)	58 (400)	Yes	Yes
F1554 grade 105		105 (720)	125 (860)	Yes	Yes
A 193 trade and 760 assed on bolt diameter (d_b)	$d_o \leq 64$ mm	105 (720)	125 (860)	No	Yes
	$64 > d_o \leq 102$ mm	95 (660)	115 (790)	No	Yes
	$d_o > 102$ mm	75 (515)	100 (690)	No	Yes

Table 5.14 Anchor bolt nut specifications

Anchor material specification	Nut material specification
ASTM A307, GRADE A	ASTM A563, GR.A
ASTM A36/A36M	ASTM A563, GR.A
ASTM F1554, GR.36	ASTM A563, GR.A
ASTM A193/A193M, GR.B7	ASTM A563, DH HEAVY-HEX OR ASTM A194/A194M
ASTM F1554, GR.105	ASTM A563, DH HEAVY-HEX OR ASTM A194/A194M

5.8.1.3 Sleeves

The sleeves are very important tools to solve small movements of anchor bolts during installation of machine skids after pouring concrete. Therefore, sleeves are recommended to use when small movement of the anchor bolt is desired after the bolt is set in concrete. There are two types of sleeves commonly used with anchors. A partial sleeve is primarily used for alignment requirements, while the full sleeve is used for alignment as well as pretensioning.

Sleeves do not affect the tensile capacity of a headed anchor because the tension in the anchor is transferred to the concrete through the head and does not rely on the bond between the anchor and surrounding concrete. Sleeved anchors can only resist shear forces when the sleeve is filled with grout.

A partial length sleeve is used where precise alignment of anchors is required during installation of equipment or where bolt groups or patterns involve six or more anchor bolts with interdependent dimensional requirements. In this situation, the sleeve should be filled with grout after installation is complete. Partial length sleeves should not be used for base plates on structural steel columns which have oversized holes.

Full-length anchor bolt sleeves are used when anchor bolts will be pretensioned in order to maintain the bolt under continuous tensile stresses during load reversal.

Pre-tensioning of anchor bolts requires the bolt to be un-bonded over a well-defined length which is stretched. When sleeves are used for pretensioned bolts, the top of these sleeves should be sealed and the sleeve should be filled with elastomeric material to prevent grout or water from filling the sleeve.

In case of vertical vessel with high height the effect of wind produces high forces in the anchor bolt. So the friction between the anchor bolt and concrete cannot be only carry the force on the anchor bolt so it is recommended to put a plate connected to the lower point for the anchor bolt as shown in Fig. 5.30.

5.8.2 Anchor Bolt Plate Design

In this case, the three anchor bolts are connected by the steel plates, and we need to design the thickness for this plate to withstand the load from the anchor bolts.

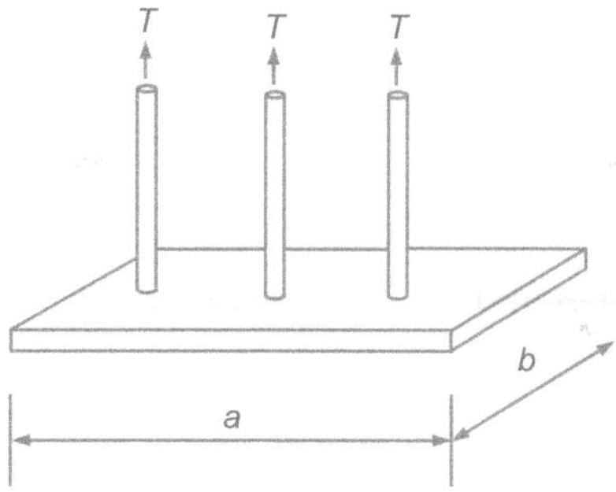

$$b = 3d$$

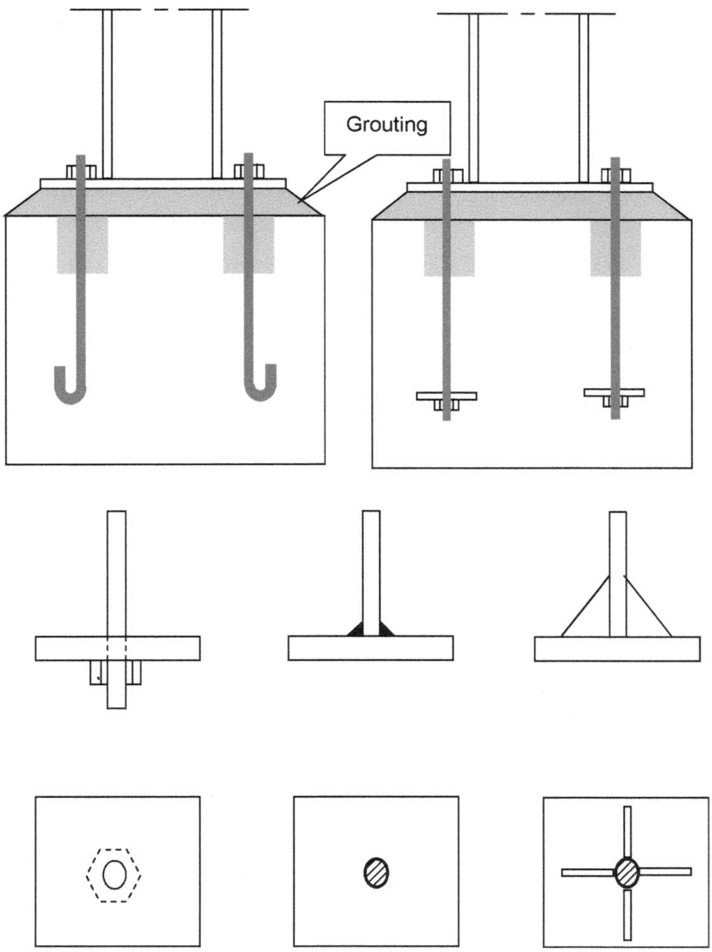

Figure 5.30 Anchor bolt detail.

where d is the anchor bolt diameter.

$$l = l_{\text{min}} = l_{\text{RC}} - 100$$

$$T_{\text{bond}} = F_b \times 3.14 \times d \times l_{\text{min}}$$

where: l_{RC} is the concrete foundation depth,
F_b is the concrete bond strength,
$T_e = T - T_b$,
$W = 3T_e/(b \cdot a) \leq f_c$,
$(M/0.72F_y) = (b_t^2/6)$ get (t).

5.8.3 Coatings and Corrosion

Corrosion is the main factor affecting the anchor bolt materials. It is important to define the elements that will be in contact with the anchor bolts and the environmental conditions surrounding it such as the water table, humidity, salt water, and caustic gases.

Galvanizing is a common option for ASTM A307 bolts and ASTM A36/A36M and ASTM F1554 threaded rods. Hot dip galvanizing is not recommended for ASTM A193/A193M grade B7 anchor bolts because of the possibility of hydrogen embrittlement that can be caused by the hot-dip galvanizing process.

Practical designs for the anchor should consider at least 3 mm as a corrosion allowance in the anchor bolt diameter if it is close to highly corrosive environments such as in buildings near the coast or if it is washed periodically as in the case of sphere tank legs due to examining the fire fighting system regularly.

Pedestal design and anchor arrangement should consider water collection and anchor environment in order to reduce the amount of contact with corrosive substances or plant wash down water. Therefore, the top of pedestals is usually to be at least 150 mm above the finished paving surface.

Painting the anchor bolt will protect the anchor but may require more maintenance. ASTM A193/A193M anchors are required to be painted. If the engineer determines prolonged contact with a corrosive substance is unavoidable, a metallurgist consultant should determine another anchor material or protective options; this precaution will be discussed in Chapter 9, Assessment for Structures in Industrial Plant.

5.8.4 Bolts Types, Details, and Layout

It is traditional to use cast-in-place anchor bolts which can come in several configurations. The following is a list of some of the most common types:

- Headed bolts or threaded rods with heavy hex nuts.
- J and L shaped anchor rods.

Current design practices do not rely on the concrete bond to develop the tension capacity of smooth rods bent into J and L shapes. While these bolts are still recognized in ACI 318-02, they have no advantage over headed bolts and can be harder to fit into small foundation pedestals.

Determining the number, type, projection, and diameter of anchor bolts should be done as follows:

- Structural steel base plates require a minimum of four anchor bolts for stability during construction per the latest OSHA safety requirements.
- The layout of the anchor bolts and required foundation pedestal size should be established based on the design forces and minimum dimensions.
- The minimum anchor bolt size is 20 mm for most items. However, there is an exception for very small equipment when suggested by the manufacturer or for small miscellaneous steel items such as ladders, stair stringers, small base pipe supports, etc.
- For most structures and equipment, ordinary strength anchor bolts can be used (ASTM A307, A36/A36M or F1554 gr. 36). If the anchor bolt loads require anchor bolts in excess of 50 mm, high-strength anchor bolt material should be considered.

5.8.5 Anchor Bolt Projection

Anchor bolt projection above the top of concrete is computed as follows:

For single nuts

$$P = t_g + t_p + 1.5\ d$$

For double nuts

$$P = t_g + t_p + 2.5\ d$$

where:

P = anchor bolt projection above concrete,

t_g = thickness of grout,

t_P = thickness of base plate,

d = diameter of anchor bolt.

Table 5.15 Types of anchor bolts

Equipment type	Bolt	Anchor bolt shape		
		J	K	X
Stair at 45° inclination	M16	×		
Ladders	M12	×		
Steel structure	Variable	×		
Horizontal vessel and heat exchanger	M20/M24		×	
Vertical vessel on steel leg support	Variable			×
Pie racks	M24/M30	×		
Stacks, pumps and compressors, and vertical vessel with skirt	Variables		×	
For high precision installation	Variables			×

The thread length required at the top of the anchor bolt must be sufficient to accommodate two nuts and about ½ of the anchor bolt diameter projecting above the top nut. Normally, a thread length is about three bolt diameters to provide some tolerance for errors in the elevation of the anchor bolt placement during construction.

The minimum edge distance shall be in accordance with the ACI 318 Code and should be in accordance with ASCE recommendations, which is four times the diameter in the case of normal steel and six times the diameter in the case of anchors with high strength. Minimum embedment and anchor spacing should be in accordance with the recommendations of the ASCE Anchor Bolt Report.

Table 5.15 presents the different types of anchor bolt shapes with different equipment which is traditionally in industrial facilities.

The anchor bolts approximate dimensions as shown in Figs. 5.31, 5.32, and 5.33 are based on concrete characteristic strength 30 N/mm^2 and these dimensions as a guideline is presented in Table 5.16 these dimensions should be verified before use according to every project separately. Table 5.17 presents the height of the threaded part of different diameters.

5.8.5.1 Edge Distance

From a practical point of view, the clear distance for anchor bolts or anchor bolt sleeves to the edge of the concrete shall be a minimum of 100 mm. This clear distance is intended to prevent corrosion

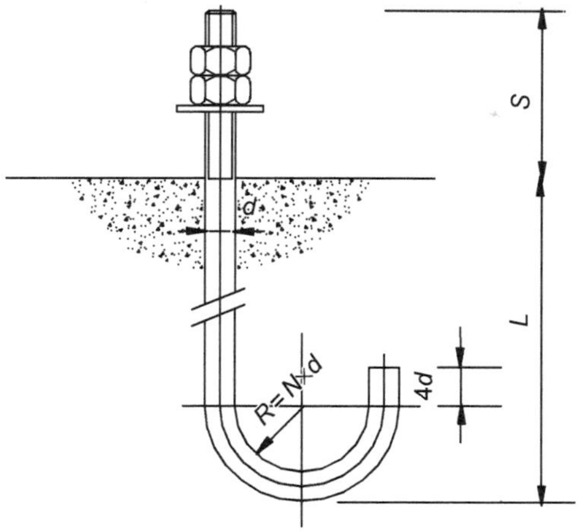

Figure 5.31 Anchor bolt type (J).

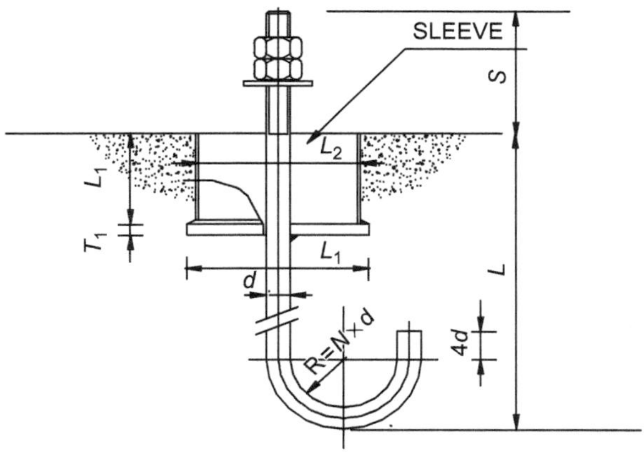

Figure 5.32 Anchor bolt type (K).

and to make sure that the anchor bolts are not in contact with the rebar cage.

ACI 318 requires that cast-in headed fasteners, which will be torqued, have minimum edge distances of 6d.

For constructability reasons, the ASCE Anchor Bolt Report recommends a minimum edge distance of 4d for ASTM A307 or A36 or their equivalent and 6d for high strength bolts.

The minimum bolt spacing is 8d.

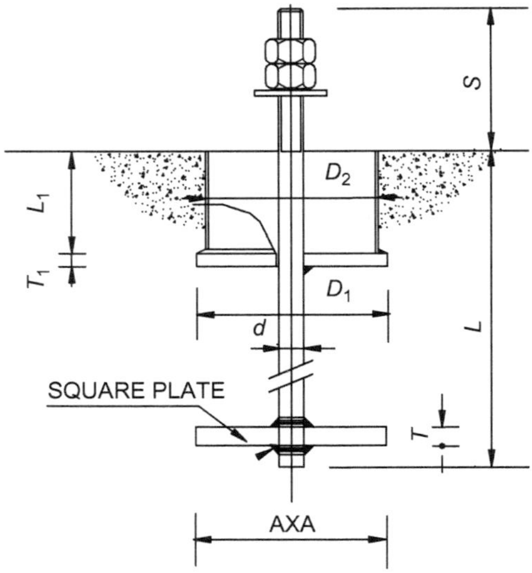

Figure 5.33 Anchor bolt type (X).

Table 5.16 Approximate dimensions for anchor bolt type J, K, and X in mm

d	N	L	H_1	T_1	D_2	D_1	T^a	B^a	L^a
10	2.5	320	100	8	80	110			
12	2.5	360	100	8	80	110			
14	2.5	390	100	8	80	110			
16	2.5	400	100	8	80	110			
18	2.5	400	100	8	80	110			
20	2.5	420	100	8	80	110	15	80	380
22	2.5	460	150	8	80	110	20	80	380
24	2.5	460	150	8	80	110	20	80	380
26	2.5	500	150	10	100	130	20	80	460
28	2.5	500	150	10	100	130	30	80	460
30	3.5	520	150	10	100	130	30	80	430
32	3.5	560	150	10	100	130	30	100	530
34	3.5	560	200	10	100	130	30	100	530
36			200	10	100	130	30	100	540
38			200	14	100	180	30	100	600
42			200	14	150	180	40	100	630
46			200	14	150	180	40	100	690
50			250	14	150	180	40	150	750
52			250	14	150	180	40	150	780
56			250	14	150	240	40	150	840
60			300	14	200	240	45	150	900
64			300	14	200	240	50	150	960

[a]Anchor bolt type X has a higher bolt size than 20 mm.

Table 5.17 Height of threaded part, mm

Anchor bolt diameter	Height of threaded part	
	One nut and washer	Two nuts and washer
10	15	25
12	20	30
14	25	40
16	25	40
18	30	50
20	30	50
22	35	55
24	35	60
26	40	70
28	40	75
30	45	75
34	50	85
36	55	90
38	60	100
42	65	105
46	70	115
50	75	12
52	80	130
56	85	140
60	90	150
64	95	160

5.8.5.2 Embedment Depth

In cases of industry, there will be tall vessels and structures subjected to heavy wind or seismic loads. This will result in large diameter anchor bolts and also higher embedment length as it will transfer the loads from the anchor bolts to the concrete.

There is no minimum embedment depth specified in ACI 318 as long as there is enough effective embedment depth to resist uplift forces. If ductility is required, greater embedment may be necessary. The ASCE Anchor Bolt Report recommends a minimum embedment depth to be 12d where d is the anchor bolt diameter.

5.8.6 Calculation of Vessel Anchor Bolts

Based on ASCE, the vertical vessel anchor bolt maximum tension is calculated by assuming elastic distribution of forces and moment, which is based on the moment of inertia of the bolt group.

$$T_b = \frac{4M}{N \times BC} - \frac{W}{N}$$

where:

T_b = tension force on the anchor bolt,

M = maximum moment on vessel,

N = number of anchor bolts,

BC = bolt circle diameter,

W = minimum weight of vessel.

This equation is conservative, as this equation assumes that the moment is resisted only by the bolt group. This equation may provide a large embedment length and can control the design in cases of tall vessels and stacks. Another sophisticated solution by iteration method defines the natural axis location and determines the compressive area and the tensile forces in the bolt.

5.8.7 Anchor Bolt Strength Design

A ductile anchorage design can be defined as one where the yielding of the anchor (or the reinforcement or the attachment that the anchor attaches to) controls the failure of the anchorage system. This will result in large deflections, redistribution of loads, and absorption of energy prior to any sudden loss of capacity of the system resulting from a brittle failure of the concrete (ASCE Anchor Bolt Report).

Anchors embedded in concrete and pulled to failure, fail either by pullout of the concrete cone or by tensile failure of the anchor itself. The former is a brittle failure and the latter is a ductile failure. A brittle failure is sudden and without warning, possibly causing catastrophic results. In contrast, a ductile failure will cause the steel to yield, elongate gradually, and absorb a significant amount of energy, often preventing structures from collapsing. Consequently, when the design of a structure is based upon ductility or energy absorption, one of the following mechanisms for ductility shall be used.

The anchors shall be designed to be governed by tensile or shear strength of the steel, and the steel shall be a ductile material.

5.8.7.1 Ultimate Strength Design

When the bolt is subject to both tension and shear, the following equation will apply:

$$\left(\frac{T_u}{\phi_1 P_n}\right) + \left(\frac{V_u}{\phi_2 V_n}\right) \leq 1 \qquad (5.5)$$

where:

T_u is the factor tensile load per bolt,

$\phi_1 = 0.9$ strength reduction factor for tension load,

$V_u =$ factored shear force per bolt,

$\phi_2 = 0.85$ strength reduction factor for shear load.

$$\phi_1 P_n = \phi_t F_t A_s$$

where:

$F_t =$ the smaller of 1.0 times bolt yield stress or 0.9 times bolt tensile strength,

$A_t =$ bolt tensile stress area.

$$\phi_2 V_n = \phi_2\ {}_{\mu} f_y A_s$$

where:

$\mu =$ friction coefficient from ACI 349 appendix B

$= 0.55$ when the bottom of the base plate is raised above the concrete surface as on a grout bed

$= 0.7$ when the bottom of the base plate is on the concrete surface

$= 0.9$ when the top of base plate is at or below the concrete surface,

$F_y =$ anchor bolt yield stress,

$A_s =$ bolt tensile stress area.

5.8.7.2 Allowable Stress Design

AISC ASD provides allowable stress design requirements for ASTM A307 and is valid for A36, A325, A449 and A490 bolts.

Based on ASCE this equation is considered conservative:

$$\left(\frac{f_t}{F_t}\right) + \left(\frac{f_v}{F_v}\right) = 1$$

$F_t =$ calculated tensile stress,

$F_v =$ calculated shear stress,

$F_t =$ allowable tensile stress $= 0.33F_u$,

$F_v =$ allowable shear stress $= 0.22F_u$ in the case of threaded excluded from the shear plane,

$= 0.17F_u$ in the case of threaded included in shear plane.

5.8.7.3 Calculate Required Embedment Length

Design an embedment using a straight reinforcing bar welded to an embedment plate as shown in Fig. 5.34.

$$l_d = \left(\frac{3}{40}\right)\frac{f_y}{\sqrt{f'_c}}\frac{\alpha\beta\gamma}{[(c + k_{tr})/d]}d$$

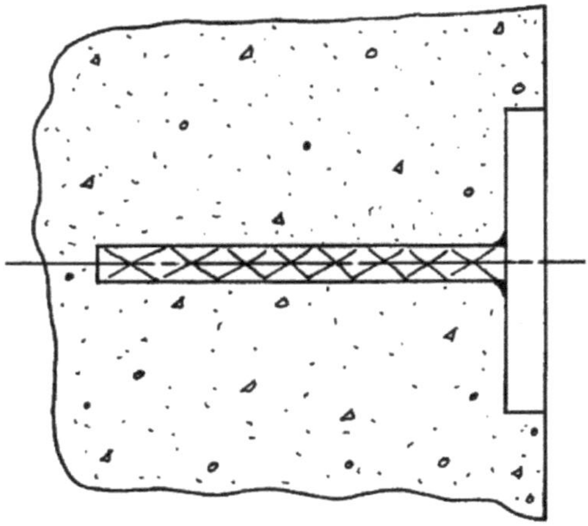

Figure 5.34 Reinforcing bar anchor.

Assume no transverse reinforcement ($k_{tr} = 0$), use $ld = 24$ in. no adjacent anchors or edges ($[c + k_{tr}]/db = 2.5$, max.), more than 300 mm of fresh concrete to be cast below the anchor ($\alpha = 1.3$), uncoated anchor ($\beta = 1.0$), 19 mm bar ($\gamma = 0.8$).

Determine required embedment length for the stud to prevent concrete cone failure.

The design pullout strength of the concrete, P_d, must exceed the minimum specified tensile strength ($A_s f_{ut}$) of the tensile stress component as shown in Fig. 5.35.

$$P_d > A_s f_u t$$

$$P_d = \phi 4 \sqrt{f'_c} A_{CP}$$

$$A_{cp} = \pi \left[\left(L_d + \frac{d_h}{2} \right)^2 \right] \phi 4 \sqrt{f'_c} \geq A_s f_{ut}$$

5.8.8 Anchor Design Considerations

If the base plate is designed with oversized bolt holes and there is shear load in excess of the amount that can be transmitted through friction, it is recommended that:

1. either shear lugs be used or
2. a mechanism to transfer load from the base plate to the bolt without slippage is incorporated (such as welding washers in place).

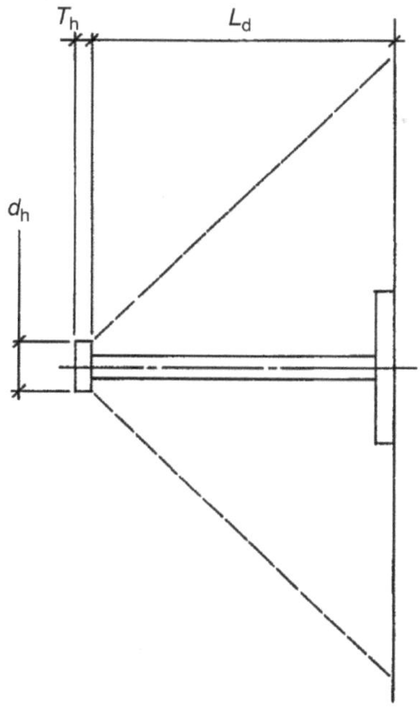

Figure 5.35 Embedment stud.

If no tensile force is effectively applied to the anchors, the anchors need not be designed for tension. Where the tensile force is adequately transferred to properly designed rebar, there is no requirement to check for concrete breakout strength of the anchor or anchors in tension.

If the anchors are welded to the attachment so as to distribute the force to all fasteners, the strength may be based on the strength of the row of anchors furthest from the edge.

5.8.9 Pretensioning

Pretensioning of anchor bolts can be used to effectively fix the base plate to the foundation. This primary application for pretensioning is for large vibrating equipment, tall process columns, and cantilever stacks. Pretensioning anchor bolts will virtually eliminate fatigue stresses in the anchor bolts and is also effective in eliminating shear forces on the bolts. Proper anchor bolt details, careful installation procedures, and field quality control are essential to ensure that the anchor bolts are successfully tensioned.

Table 5.18 Pretension load and torque recommendation[a]

Nominal bolt diameter, mm	Torque (N m)	Pretension load (kg)
M16	60	3311
M24	100	7447
M30	160	18,247
M52	245	37,136

[a]Based on 30ksi bolt internal stress.

The advantages of a pretension anchor bolt are that it can prevent stress reversals on anchors susceptible to fatigue weakening, may increase dampening for pulsating or vibrating equipment, will decrease the drift for process towers under wind or seismic load, and will increase the frictional shear resistance for process towers and other equipment.

The disadvantage of this type of anchor bolt is that it can be a costly process to accurately install and there is no recognized code authority that gives guidance on the design and installation of pretensioned anchors. There is little research in this area, but it is dependent on the requirement of the manufacturer.

The pretensioning should always be done when recommended by the equipment manufacturer. The manufacturer's instructions should be followed carefully.

When not otherwise specified, anchors for turbines and reciprocating compressors should be torqued to the values shown in Table 5.18. When pretensioning is required, high strength anchors should be used. Pretensioned anchor bolts should also be considered for tall process columns or cantilever stacks, which can develop significant bolt tension under wind or seismic loads.

REFERENCES

Industry Codes and Standards

American Association of State Highway and Transportation Officials (AASHTO).

AASHTO Standard Specifications for Highway Bridges.

American Concrete Institute (ACI).

ACI 318/318R—*Building Code Requirements for Structural Concrete and Commentary.*

ACI 350R—*Environmental Engineering Concrete Structures.*

ACI 530/ASCE 5—*Building Code Requirements for Masonry Structures* American Institute of Steel Construction (AISC).

AISC Manual of Steel Construction—Allowable Stress Design (ASD).

AISC Manual of Steel Construction—Load and Resistance Factor Design (LRFD).

Specification for Structural Joints Using ASTM A325 or A490 Bolts.

ANSI/AISC 341-02—Seismic Provisions for Structural Steel Buildings □ American Iron and Steel Institute (AISI).

AISI SG 673, Part I—*Specification for the Design for Cold-Formed Stee l Structural Members.*

AISI SG 673, Part II—*Commentary on the Specification for the Design for Cold-Formed Steel Structural Members.*

AISI SG 913, Part I—*Load and Resistance Factor Design Specification for Cold-Formed Steel Structural Members.*

AISI SG 913, Part II—*Commentary on the Load and Resistance Factor Design Specification for Cold-Formed Steel Structural Members.*

American Petroleum Institute (API)—API Standard 650—*Welded Steel Tanks for Oil Storage.*

American Society of Civil Engineers (ASCE).

SEI/ASCE 7-02—*Minimum Design Loads for Buildings and Other Structures.*

SEI/ASCE 37-02—*Design Loads on Structures during Construction.*

ASCE Guidelines for Seismic Evaluation and Design of Petrochemical Facilities.

ASCE Guidelines for Wind Loads and Anchor Bolt Design for Petrochemical Facilities.

ASCE Design of Blast Resistant Buildings in Petrochemical Facilities.

American Society of Mechanical Engineers (ASME).

ASME A17.1—*Safety Code for Elevators and Escalators.*

American Society for Testing and Materials (ASTM).

ASTM A36/A36M—*Standard Specification for Carbon Structural Steel.*

ASTM A82—*Standard Specification for Steel Wire, Plain, for Concrete Reinforcement.*

ASTM A185—*Standard Specification for Steel Welded Wire Fabric, Plain, for Concrete Reinforcement.*

ASTM A193/A193M—*Standard Specification for Alloy-Steel and Stainless Steel Bolting Materials for High-Temperature Service.*

ASTM A307—*Standard Specification for Carbon Steel Bolts and Studs, 60,000 psi Tensile Strength.*

ASTM A325—*Standard Specification for Structural Bolts, Steel, Heat Treated, 120/105 ksi Minimum Tensile Strength.*

ASTM A325M—*Standard Specification for High-Strength Bolts for Structural Steel Joints [Metric].*

ASTM A354—*Standard Specification for Quenched and Tempered Alloy Steel Bolts, Studs, and Other Externally Threaded Fasteners.*

ASTM 490/490M—*Standard Specification for Heat-Treated Steel Structural Bolts, 150 ksi Minimum Tensile Strength.*

ASTM A615/A615M—*Standard Specification for Deformed and Plain Billet-Steel Bars for Concrete Reinforcement.*

ASTM A706/A706M—*Standard Specification for Low-Alloy Steel Deformed Bars for Concrete Reinforcement.*

ASTM A992/A992M—*Standard Specification for Steel for Structural Shapes for Use in Building Framing.*

ASTM F1554—*Standard Specification for Anchor Bolts, Steel, 36, 55, and 105 ksi Yield Strength.*

American Welding Society (AWS).

AWS D1.1—*Structural Welding Code—Steel.*

American Forest and Paper Association.

National Design Specification for Wood Construction (NDS).

NDS Supplement—Design Values for Wood Construction.

Crane Manufacturers Association of America (CMAA).

CMAA No. 70—*Specifications for Top Running Bridge and Gantry Type Multiple Girder Overhead Electric Traveling Cranes.*

CMAA No. 74—*Specifications for Top Running and Under Running Single Girder Overhead Electric Traveling Cranes Utilizing Under Running Trolley Hoist.*
Precast/Prestressed Concrete Institute (PCI).
PCI MNL 120—*Design Handbook—Precast and Prestressed Concrete.*
Steel Joist Institute (SJI).
SJI Standard Specifications and Load Tables.
Federal Standards and Instructions of the Occupational Safety and Health Administration (OSHA), including any additional requirements by state or local agencies that have jurisdiction in the state where the project is to be constructed, shall apply.
U.S. Department of Labor, Occupational Safety and Health Administration (OSHA).
OSHA 29 CFR 1910—*Occupational Safety and Health Standards.*
OSHA 29 CFR 1926—*Safety and Health Regulations for Construction.*
Lui, E.M., 1999. Structural steel design. In: Wai-Fah, Chen (Ed.), Structural Engineering Handbook. CRC Press LLC, Boca Raton, FL.
From PIP REIE 686, recommended Practices for machinery Installation and installation design.

CHAPTER 6

Blast Resistance Building Design

6.1 INTRODUCTION

Blasts are not always caused by combustion; they can also result from any rapid release of energy that creates a blast wave, such as a bursting pressure vessel from which compressed air expands, or a rapid phase transition of a liquid to a gas. The loads resulting from a blast are created by the rapid expansion of the energetic material, creating a pressure disturbance or blast wave radiating away from the explosion source. Blast pressure is more properly called *overpressure* because it is relative to ambient conditions. Shock waves are high-pressure blast waves that travel through the air at a velocity faster than the speed of sound. These waves are characterized by an instantaneous increase in pressure followed by a rapid decay.

On the other hand, pressure waves are lower amplitude and travel below the speed of sound. These waves are characterized by a more gradual increase in pressure than a shock wave, with a decay of pressure much slower than a shock wave. In most cases, shock waves have a greater potential for damage and injury than pressure waves do.

The calculation of the pressure due to a blast load will be done by a loss prevention engineer, as it depends on the type of hydrocarbon and its components. The engineer's study will result in a circle of pressure ranges, as shown in the following photo depicting the pressure due to the explosion of one of the sphere tanks (Fig. 6.1).

The effect of the ground-reflected wave before it affects the building is illustrated in Figure 6.2 and the reflecting pressure that affect the face of the building is presented in Figure 6.3.

The effect of the blast load in the building is presented clearly in Fig. 6.4. The step 1 the building is stable and at step 2 the shock wave affect the structure and there are drift happened on the structure, then in step 3 it return to its original position and then at step 4 the negative phase effect will affect the structure to drift in the opposite direction and finally at step 5 the building will be in its original static condition with the final damage effect on it. Noting that all these steps are in seconds. Figure 6.5 presents the pressure versus time.

Onshore Structural Design Calculations.
DOI: http://dx.doi.org/10.1016/B978-0-08-101944-3.00006-1

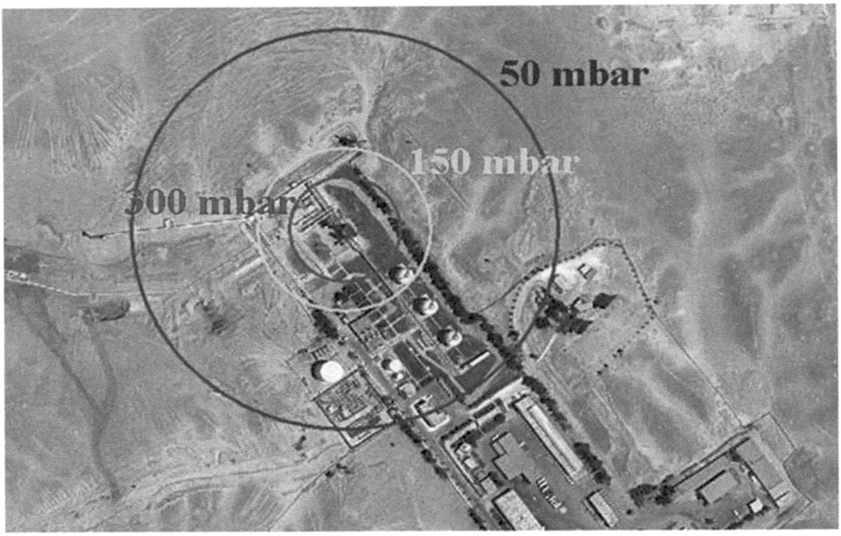

Figure 6.1 Blast pressure contour.

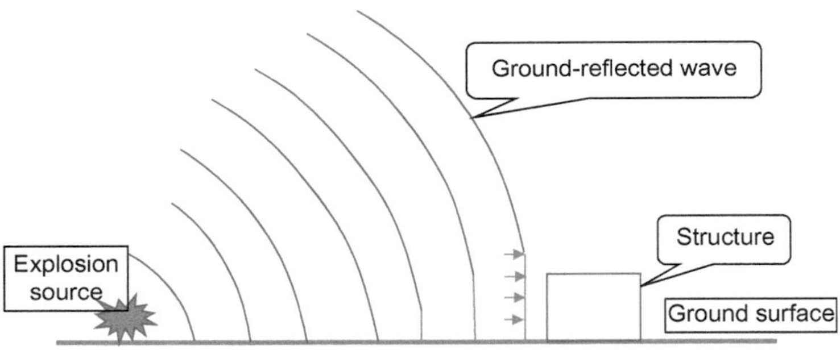

Figure 6.2 Reflected wave on the building.

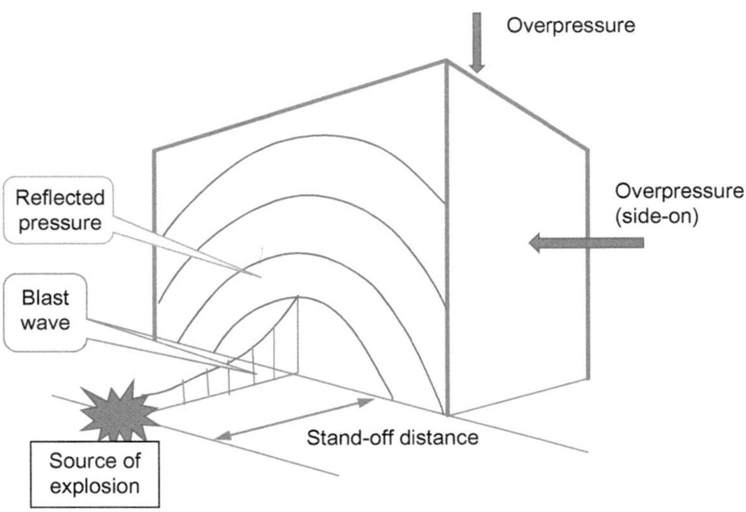

Figure 6.3 Blast pressures on building surfaces.

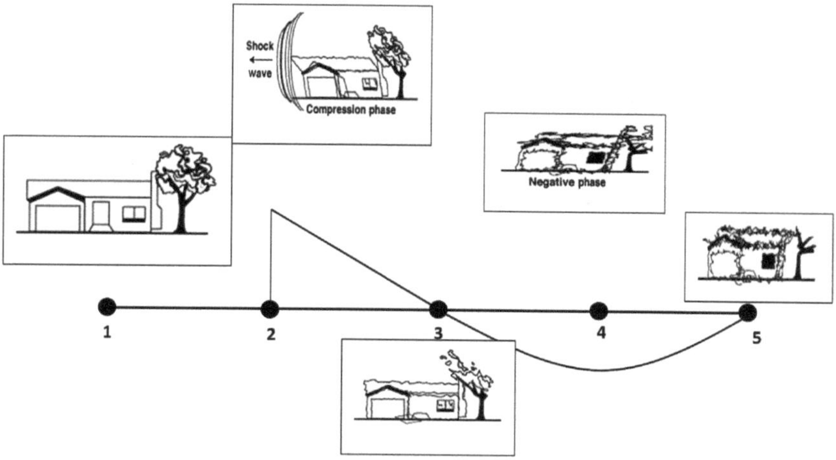

Figure 6.4 Life cycle blast load effect on the building.

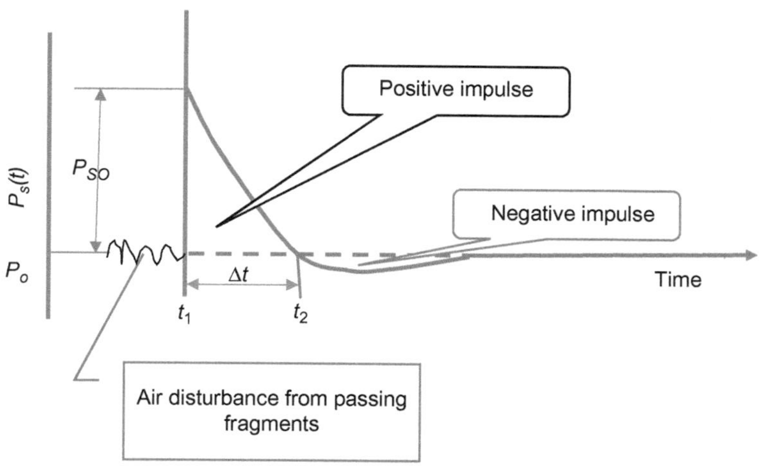

Figure 6.5 Pressure versus time. P_{SO}, peak static over pressure; P_o, ambient (atmospheric) pressure; $P_s(t)$, static overpressure at time (t); t_1, time of arrival of shock front; t_2, end of positive phase; Δt, time duration of positive phase ($\Delta t = t_2 - t_1$); I_s, overpressure impulse.

The effect of the blast wave at the front wall in the building in elevation and plain view is presented in Fig. 6.6. The vortex around the building is clear in the building as described in the *TNO green book*.

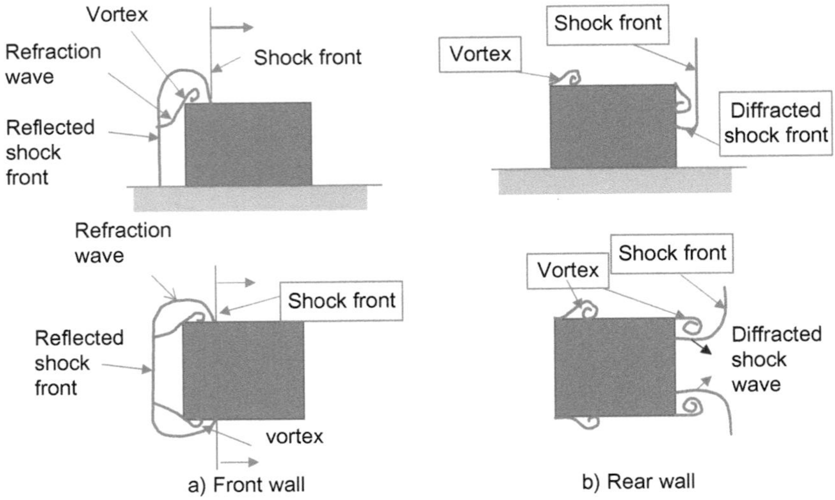

Figure 6.6 Blast wave at front wall and rear wall.

The effect to the rear wall is presented in Fig. 6.6, and it is a clear that there is a diffracted shock front which starts at the top of the building, as in the *TNO green book*.

The practical value that should be considered in designing the resistance building to blast load as follow:

- As per SG-22 and CIA 1992 [stand-off distance 30 m (100 ft)].
- High pressure, short period over pressure 69 kpa (10 psi) with time duration, t_d 20 ms.
- Low pressure, long duration, 3 psi, 100 ms.
- Some owners, provide blast design for building 15−365 m, most owners, 30−120 m.
- From historical data, review vapor cloud explosion from the distance 60−300 m with range from P_r 14−83 kPa with $t_d = 35-33$ ms with side on over pressure 6.9−38 kPa
- Conventional building can withstand 1 psi, 6.9 kPa, or 30 psi-ms.

Before going into detail for blast resistance, the following definitions should be known, as they are standard for any reference for blast load effects.

- Angle of incidence: The angle between the direction of the blast wave travel and a line perpendicular to the surface of a structure at the point of interest.
- Blast loads: The transient dynamic loads from the blast effects of an explosion, usually stated in terms of peak pressure and impulse or duration.

- Dynamic increase factor (DIF): A multiplier applied to the static strength of a material to reflect the increased effective strength due to fast strain rates caused by rapidly applied blast loads.
- Ductility ratio: A measure of the degree of plasticity in a member at maximum dynamic response, equal to either the maximum displacement or the displacement at yield. This value is a key measure of dynamic response.
- Duration: The length of time from the start of the initial positive phase of the blast pressure to the return to ambient pressure.
- Dynamic reaction: The support reaction of a structural component to the dynamic blast loading, taking into account inertia effects.
- Fragment resistance: The resistance to high-speed fragments that result from the breakup of equipment or structures that are close to the explosion source.
- Impulse: A measure used, along with the peak blast pressure, to define the ability of a blast wave to do damage. Impulse is calculated as the integrated area under the positive pressure versus duration curve, in units of psi-ms (MPa-ms).
- Negative phase: The portion of the pressure-time history typically following the p_o.
- Period: The fundamental natural period of a structural component if modeled as a single degree-of-freedom (SDOF) system.
- Positive phase: The portion of the pressure-time history in which the pressure is above ambient pressure.
- Rebound: The deformation in the direction opposing the initial blast pressure. This occurs after a component has reached a peak deformation and returns in the direction of its initial position.
- Reflected pressure: The rise in pressure above ambient pressure produced by a shock wave or pressure wave striking a surface facing the direction of blast wave propagation.
- Response range: The degree of structural damage permitted for blast resistant buildings.
- Strength increase factor (SIF): A multiplier applied to the nominal strength properties of a material to reflect its actual strength above minimum specified values.
- Support rotation: The angle formed between the axis of a member loaded between its endpoints and a straight line between one endpoint and the point of maximum deflection. This value is a key measure of dynamic response.

6.2 BUILDING CONFIGURATION

The building that withstands the blast load should have some configuration, as the single-story construction shall be used where possible. Where multistory construction is required, the number of stories shall be minimized, and special design considerations shall be given to the interstory response to the blast loading.

The floor plan and elevation preferably shall have clean rectangular profiles without reentrant corners, and recessed areas shall be minimized. A box-type structure is preferable, in which the shorter side is exposed to the larger reflected blast load, and the longer side is exposed to the lower side-on blast load value. Roof overhangs, canopies, and reentrant corners should be avoided where possible to avoid additional blast wave reflections. Architectural items such as canopies and signs should be designed with light construction materials, such as canvas, to avoid creating a debris hazard for the structure.

6.3 BUILDING MATERIALS

The materials that will be used in the resistance building to blast load shall have minimum requirements and special characteristics; for example, cast-in-place reinforced concrete construction is typically used to provide resistance to severe blast loads. The wall thicknesses for structures in or immediately adjacent to large process areas are typically 205 mm (8 in.) to 305 mm (12 in.) but can be thicker in special cases.

Reinforced concrete is especially appropriate for short duration loading that produces an impulsive response. Its large mass relative to the surface area is especially effective in resisting these types of loads. Reinforced concrete construction is typically used where a protective structure is needed around an existing structure, when the existing structure is close to a blast source and thus must resist large blast loads.

Brittle structures, such as unreinforced masonry, have little ductility and will fail under very low blast loads. These failures are sudden and should be avoided in all situations. For this reason, unreinforced masonry construction is not permitted in the design of blast resistant structures. The blast capacity and ductility for reinforced masonry is less than the R.C. structure. As there is a lower compressive strength of masonry, there is limited steel reinforced and limited mortar bond strength. A masonry wall with a reinforcement can be used as illustrated in Fig. 6.7.

Figure 6.7 Masonry block with steel bars.

Metal frame, metal clad construction is commonly used in petrochemical plants for warehouses, maintenance shops, and process support office buildings. This type of construction is appropriate for relatively low blast overpressures and should typically be located several hundred feet from major process units. Moment-resisting frames are typically used in a metal building to resist the lateral load applied on its long side. Cross bracing is used between frames to resist loads applied to the end or short walls of the building. Frame spacing must typically be closer, on the order of 6 m (20 ft), than for conventional construction, where frame spacing may exceed 10 m (30 ft).

6.3.1 Dynamic Material Strength (Steel)

In case of a blast load, the structure will be affected by a high load within a small period of time; this means that the relation between stress and strain is different rather than the static case in case of the lower load rate increasing. According to TM 5-1300, the stress strain curve in case of steel under a dynamic load will be as shown in Fig. 6.8.

In case of a concrete structure, the stress–strain curve in case of a dynamic load will be as shown in Fig. 6.9.

The DIF for steel hot rolled sections, steel reinforcement and concrete with different grades is presented in Fig. 6.10.

Dynamic yield stress, F_{dy}, shall be calculated as follows:

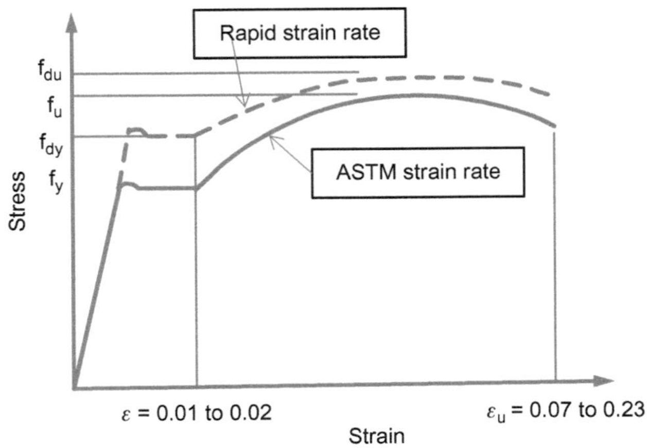

Figure 6.8 Effect of strain rate on stress–strain curve of steel.

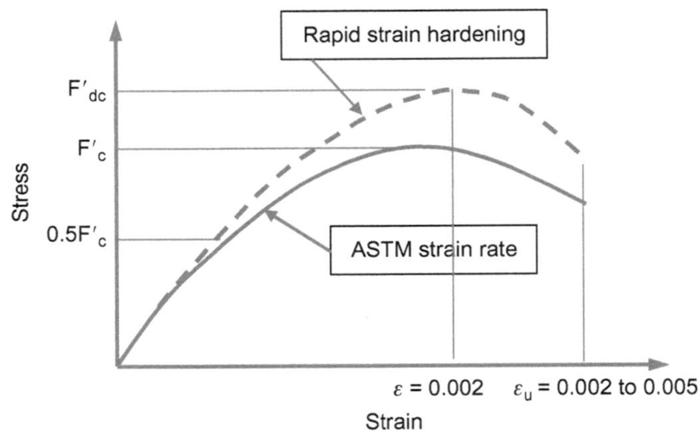

Figure 6.9 Effect of strain rate on stress–strain curve of concrete.

$$F_{dy} = F_y \times \text{SIF} \times \text{DIF}$$

where:

F_y = specified minimum static yield stress,
DIF = dynamic increase factor,
SIF = strength increase factor,
μ = ductility ratio,
θ = support rotation (°),
F_{du} = dynamic ultimate strength.

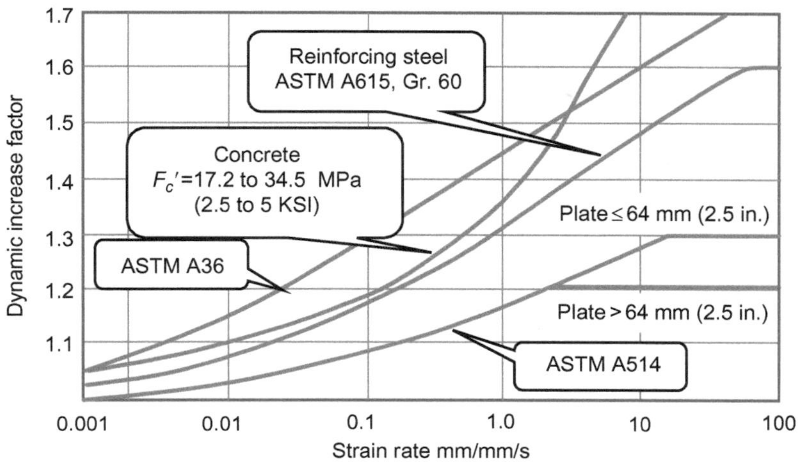

Figure 6.10 Effect of strain rate on dynamic material strength.

Table 6.1 Dynamic design stress for structural steel

Type of stress	Max. ductility ratio	Dynamic design stress (F_{ds})
All	$\mu \leq 10$	F_{dy}
All	$\mu > 10$	$F_{dy} + (F_{du} - F_{dy})/4$

Table 6.2 Dynamic design stress for reinforced concrete

Type of stress	Type of reinforcement	Maximum support rotation	Dynamic design stress (F_{ds})
Bending	Tension and compression	$0 < \theta \leq 2$	F_{dy}
		$0 < \theta \leq 5$	$F_{dy} + (F_{du} - F_{dy})/4$
Direct shear	Diagonal bars	$5 < \theta \leq 12$	$(F_{dy} + F_{du})/2$
Diagonal tension	Stirrups		F_{dy}
Compression	Column	All	F_{dy}

Tables 6.1 and 6.2 present the dynamic design stress for steel and reinforced concrete structures, respectively.

SIFs for different structure materials are presented in Table 6.3. Table 6.4 displays the values of DIF for reinforcing bars, concrete, and masonry.

For steel with different grades and aluminum sections, the values of DIF are displayed in Table 6.5.

Table 6.3 SIFs for structural materials

Structural material	SIF
Structural steel-yield strength of 345 MPa (50 ksi) or less	1.1
Steel reinforcement of Grade 60 (414 MPa)	1.1
Prestressed reinforcement	1.0
Cold formed steel cladding panels:	
• Yield strength of 228 MPa (33 ksi) or less	1.2
• Yield strength of 345 MPa (50 ksi) or more	1.1
Concrete and masonry	1.0
Other materials	1.0

Table 6.4 DIFs for reinforced concrete/masonry

Stress type	DIF			
	Reinforcing bars		Concrete	Masonry
	(F_{dy}/F_y)	(F_{du}/F_u)	(F_{dc}/F_c)	(F_{dm}/F_m)
Flexure	1.17	1.05	1.19	1.19
Compression	1.10	1.00	1.12	1.12
Diagonal tension	1.00	1.00	1.00	1.00
Direct shear	1.10	1.00	1.10	1.00
Bond	1.17	1.05	1.00	1.00

F_{dc}, dynamic concrete strength; F_c, concrete strength; F_{dm}; dynamic masonry strength; F_m, masonry strength.

Table 6.5 DIFs for steel and aluminum

Materials type	DIF		
	Yield stress		
	Bending/ shear (F_{dy}/F_y)	Tension/ compression (F_{dy}/F_y)	Ultimate stress (F_{du}/F_u)
A36	1.29	1.19	1.10
A572, A588, A992	1.19	1.12	1.05
A514	1.09	1.05	1.00
A653	1.10	1.10	1.00
Prestressed reinforcement	1.00	1.00	1.00
Stainless steel type 304	1.18	1.15	1.00
Aluminum, 6061-T6	1.02	1.00	1.00

6.4 LOAD COMBINATION

The blast resistant structure shall be designed in accordance with the load combinations *for the following blast load* condition:

$$U(t) = D + aL + B(t)$$

where:

$U(t)$ = total applied time dependent load or its effect,
D = static dead load,
$B(t)$ = time dependent blast load or its effect (horizontal and vertical),
L = conventional static live load,
a = reduction factor applied to conventional live loads to reflect the portion of live load expected to occur simultaneously with the blast load; zero shall be used for the reduction factor if doing so will result in a more severe condition.

Table 6.6 provides the guidance for the building performance according to the criticality of the building. The building performance should be agreed upon with the owner, as it will be the first step in designing the blast building.

Table 6.6 Building performance requirements

Building classification	Example and typical performance requirements	Building response range
Critical/ essential	Central control center, main electrical station. Continued use; reusable with cosmetic repairs.	Low: • Localized building/component damage. • Building can be reused; however, repairs may be required to restore integrity of structural envelope. If elastic limit is specified, then no repairs are necessary. • Total cost of repairs is moderate.
Normally occupied	Administration and engineering offices, laboratories. Damage-limited for occupant protection; not reusable without major repairs/ replacement.	Medium: • Widespread building/component damage. • Building cannot be used until repair. • Total cost of repairs is significant.
Other	Operator shelters, warehouse. Collapse-limited, not repairable, abandon/ replace.	High: • Building/component has lost structural integrity and may collapse from additional environmental loads (i.e., wind, snow, and rain). • Total cost of repair approaches replacement cost of new building.

Table 6.7 Deformations limits for structural steel

Element type	Response range					
	Low		Medium		High	
	μ	θ	μ	θ	μ	θ
Beams, girts, purlins	3	2	10	6	20	12
Frame members	1.5	1	2	1.5	3	2
Single sheet metal panels	1.75	1.25	3	2	6	4
Open web joists	1	1	2	1.5	4	2
Plates	5	3	10	6	20	12
Building side sway limits	$H^{a}/50$		$H/35$		$H/25$	

[a]H is the building height.

Table 6.8 Deformations limits for reinforced concrete structure

Element type	Controlling stress	Ductility ratio (μ)[a]	Support rotation, θ^{b}		
			Low	Medium	High
Beams	Flexural	N/A	1	2	4
	shear[c]:				
	Concrete only	1.3			
	Concrete + stirrups	1.6			
	Stirrups only	3.0			
	Compression	1.3			
Slabs	Flexure	N/A	2	4	8
	Shear	1.3			
Beam-columns	Flexure	N/A	1	2	4
	Compression	1.3			
	Tension	3			
	Shear	1.3			
Shear walls, diaphragms	Flexure	3	1	1.5	2
	Shear	1.5			

[a]Ductility ratio $= 0.05/(\rho - \rho')$, 10 where ρ and ρ' are tension and compression reinforcement ratios, respectively.
[b]Stirrups are required for support rotations greater than 2°.
[c]Shear controls if shear resistance is less than 120% of flexural resistance.

Based on Table 6.6 one can define the building classification and the building response range category then, from Tables 6.7–6.9 to determine the response range that you will use in design based on the type of the structure members for steel, reinforced concrete and reinforced masonry, respectively.

Table 6.9 Deformation limits for reinforced masonry

Element type	Ductility ratio (μ)	Support rotation, θ (°)		
		Low	Medium	High
One-way	1	0.5	0.75	1
Two-way	1	0.5	1	2

6.5 COMPONENT DESIGN

Ultimate strength methods shall be used for designing structural components for blast resistance. In-plane and secondary bending stresses shall be considered in the design. Dynamic strength properties shall be used to reflect increased material strength under rapidly applied loads. Components shall be adequately laterally braced to prevent premature buckling failure during the positive and rebound response. The steel structure connections shall be designed for 120% of the member's controlling resistance (flexure or shear, whichever is less), and the shear capacity shall be at least 120% of the flexural capacity. In case of the design of compression elements, such as load-bearing walls and columns, bending effects including p-delta and slenderness shall be considered.

In steel structures, the structural steel components shall be designed by *AISC LRFD, or any equivalent LRFD* related to your country. In the case of AISC, the resistance factor (φ) shall be 1.0 for load combinations that include blast loads. The materials with specified yield strength of 345 MPa (50 ksi) or less shall be used for flexural design. The materials with specified yield strength greater than 345 MPa can be used if ductile behavior is not required. The oversize holes shall not be used in connections that are part of the lateral force-resisting system. The column base plates shall be designed to develop the peak member reactions applied as a static load. Dynamic material properties can be used for the design of base plates. Flexural members shall be laterally braced on both faces to provide consistent moment capacity for both positive and rebound responses.

In cases of reinforced concrete structures, the reinforced concrete components shall be designed using ultimate strength, *ACI 318,* or other relative codes based on your country codes and regulation. The strength reduction factor (φ) in case of ACI shall be 1.0 for load combinations that include blast loads. The deformation limits for shear shall be used if the member shear capacity is less than 120% of the flexural capacity. The minimum concrete strength used for the design of the concrete structures shall be 28 MPa, and the maximum reinforcing bar size shall be 32 mm (No. 10).

The concrete walls and slabs shall be reinforced on each side in the main direction. For steel bars with a (f_y) of 240 N/mm^2, there shall be a minimum of 1% on both sides of the concrete cross section; for (f_y) of 460 N/mm^2, there shall be a minimum of 0.6% on both sides of the concrete cross section. Slenderness effects shall be included for load-bearing walls and for members with significant axial loads. At least 50% of the bottom main reinforcement shall extend over the face of the support providing a good anchorage between the supports. Special attention shall be paid to ensure continuity and a minimum of local stress concentration. Adequate lapping of reinforcement is required. Due to its nonductile behavior, prestressed concrete shall not be used. The wall and roof thickness shall be between 250 and 400 mm.

Fig. 6.11 presents the control room layout, which in most cases is identical in most oil and gas fields. Fig. 6.12 illustrates a cross section of the control building.

Fig. 6.13 presents the section in the control room presenting the roof isolation and floor, which is used to lay the cables under the building, and the false ceiling.

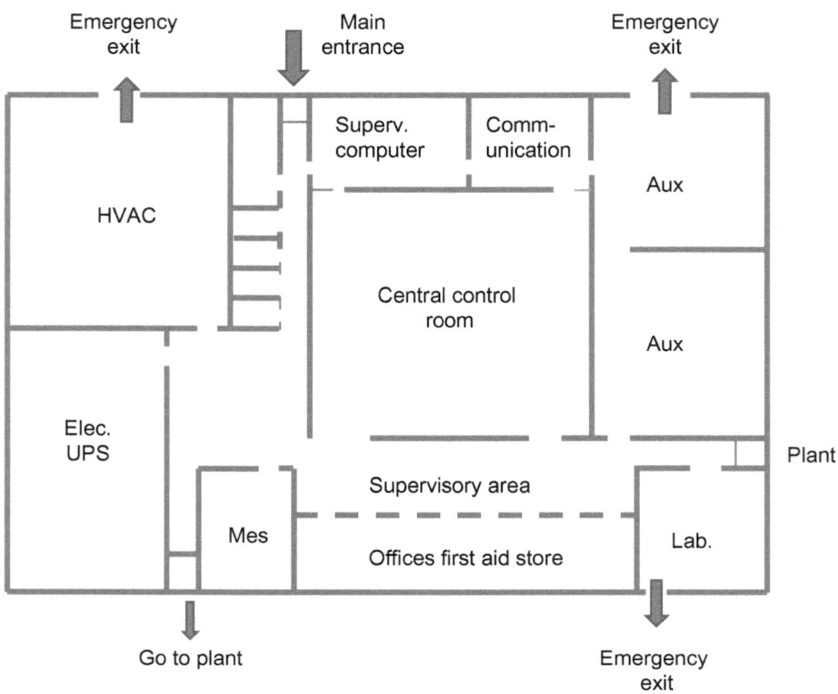

Figure 6.11 Typical control room layout.

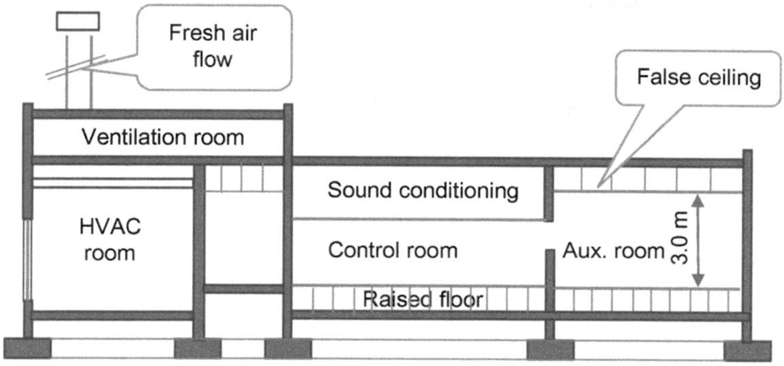

Figure 6.12 Control building cross section.

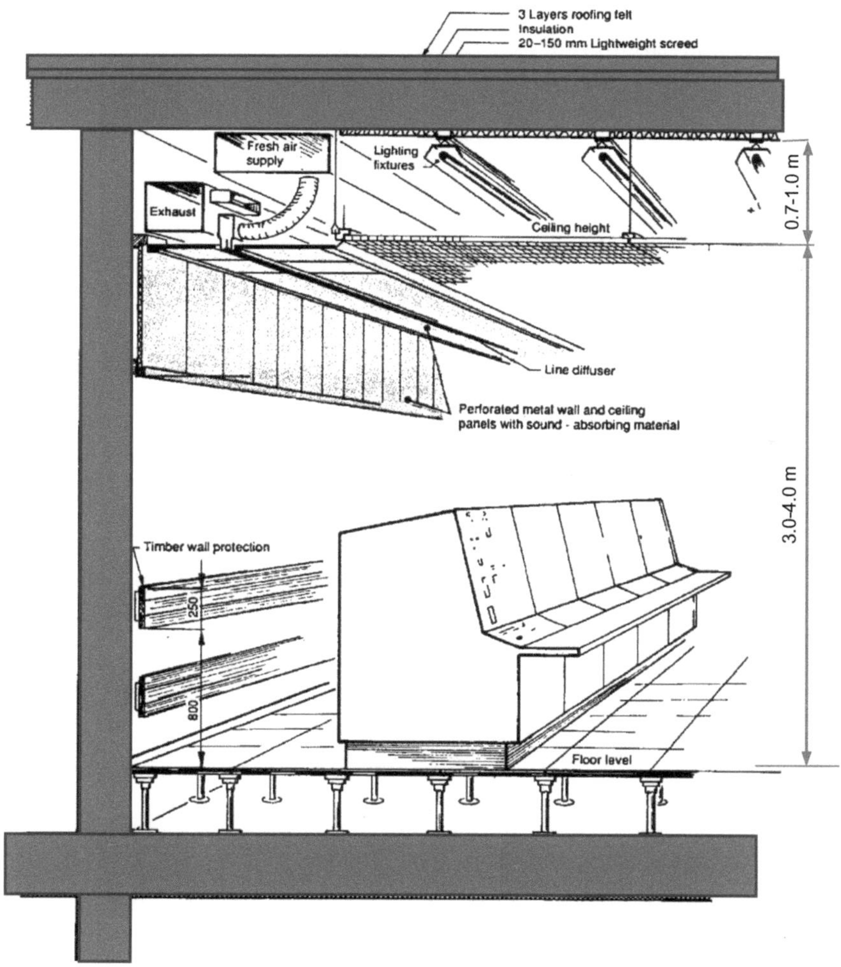

Figure 6.13 Section in a typical control room.

There are many suspended items such as ceilings, Heat Ventilation and Air Conditioning (HVAC) ductwork, and light fixtures suspended from the roof inside the building shall be secured to structural framing members. Anchorage shall be designed to resist a peak, statically applied force equal to the mass of the item times the maximum acceleration of the roof or five times the weight of the item, whichever is less.

6.5.1 Blast Door

The doors issued in blast resistant buildings have their own characteristics, and there are special manufacturers for these types of doors.

Large buildings require more than one egress door in accordance with local building codes; at least two doors shall be designated as egress doors for the purpose of limiting the damage to these doors if subjected to blast loads. The designated egress doors shall not be located on the same side of the building.

Doors shall be outward opening and shall seat against the frame in response to the positive phase blast wave. The blast door manufacturer's calculations or test data shall be provided to verify adequate blast resistance and door performance for the design load conditions. Power-operated doors shall be used for exit doors that exceed the maximum opening force. Table 6.10 presents the required performance from the door, and defines the ductility limit and support rotation based on that.

Table 6.10 Blast door recommended performance response criteria

Performance category	Description	Door condition after blast	μ_{max}	θ_{max}
I	Primary exit	To be operable	1.0	1.2
II	Prevent entrapment	To be operable but significant deformation to the door is permitted	3	2.0
III	Prevent blast from entering building	The door will be inoperable	10	8
IV	Prevent a door from becoming debris hazard	Inoperable and outward rebound forces and resulting hardware failure is acceptable	20	12

6.5.2 Windows

The windows blast resistant buildings have their own special specifications that should be considered in the purchasing and architectural design considerations.

In general, windows normally shall not be used in blast resistant buildings. If windows are required, they shall be designed to provide protection in response to the applied blast load consistent with the overall building performance requirements as specified. The design of the windows shall include the window frames and anchorage. Properly designed laminated glass, tempered glass, and polycarbonates shall be permitted for use in windows. The performance of window products shall be substantiated by test data.

The following rooms shall have no windows in the outer walls:

- control room,
- computer room,
- instrument and computer room,
- electrical equipment and battery room,
- heating, ventilating, and air-conditioning machine room,
- shift supervisor's office,
- first aid compartment/room,
- social amenities, with exception of the mess room.

6.5.3 Openings

Large openings in the building envelope, such as intake ducts, shall be designed to prevent entry of excessive blast pressures. Blast valves, blast attenuators, or other devices shall be used to limit excessive blast pressure entry into the structure. Performance of the blast valve or attenuator shall be substantiated by test data and/or calculations.

Blast valves shall be provided for openings greater than 1000 cm^2 (150 $in.^2$) in any surface in which the peak applied pressure is greater than 0.07 MPa (10 psi). Blast attenuators can be used for openings greater than 1000 cm^2 (150 $in.^2$) in any surface if the peak applied pressure is greater than 0.035 MPa (5 psi).

6.5.4 Suspended and Extended Mounted Element

Equipment and furnishings such as ceilings, HVAC ductwork, and light fixtures suspended from the roof inside the building shall be secured to structural framing members. Anchorage shall be designed to resist a statically applied force equal to the mass of the item times the maximum acceleration of the roof or five times the weight of the item, whichever is less.

To avoid the potential for hazardous debris, large nonstructural features such as canopies and signs on the building exterior shall not be permitted. Small items such as instruments, fire alarms, lights, strobes, and beacons can be mounted on the exterior walls. Roof- and wall-mounted equipment (e.g., HVAC equipment) shall be avoided. Roof- and wall-mounted equipment shall be securely anchored, and the supporting structural components shall be specifically designed for actual equipment dynamic loads if subjected to the blast. Equipment and other items mounted on the exterior walls or roof shall be designed similarly to the structural components if they are to withstand the applied blast loads. The reactions from externally mounted items shall be considered in the design of the supporting structural components.

Instrumentation or electrical equipment shall not be mounted on the interior face of walls subjected to blast loads without the owner's prior written approval. All fixed floor-supported items (e.g., lockers, electrical cabinets, racks) shall have a minimum clearance from exterior walls equal to the maximum calculated lateral blast load deflection. The maximum deflection shall be the sum of both the overall building sidesway and the deflection of any wall component(s), and shall be calculated based on the maximum blast loads. Supports and anchorage for equipment shall be designed to resist a lateral force equal to 20% of the equipment weight.

6.5.5 Walls and Ceiling

Wall tiles shall be avoided or glued firmly to the wall with special tile adhesive to prevent them from coming loose in the event of an explosion.

The aluminum grid ceiling in the control room shall be constructed in such a way that the panels are secured in position and cannot fall down during an explosion. This also applies to the lighting fixtures, etc.

All surface finishing materials for walls and ceilings in the control room shall be fire-resistant or noncombustible and shall have no loose fibers. In these spaces, the surface of all walls, ceilings, and floors, including the space under the cavity floor, where dust is liable to accumulate shall be sealed with epoxy paint or PVC coating.

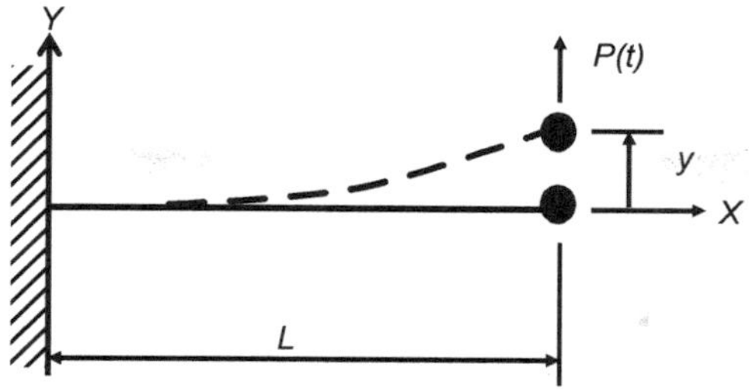

Figure 6.14 Cantilever under dynamic excitation.

6.6 FUNDAMENTAL OF DYNAMIC ANALYSIS

One of the basic methods of dynamic analysis is the examination of dynamic equilibrium to formulate an equation of motion. If at a time period, t, the mass is displaced upward from the static equilibrium position to a distance y, then it has a velocity, $\dot{y}$ and acceleration $\ddot{y}$ (Fig. 6.14).

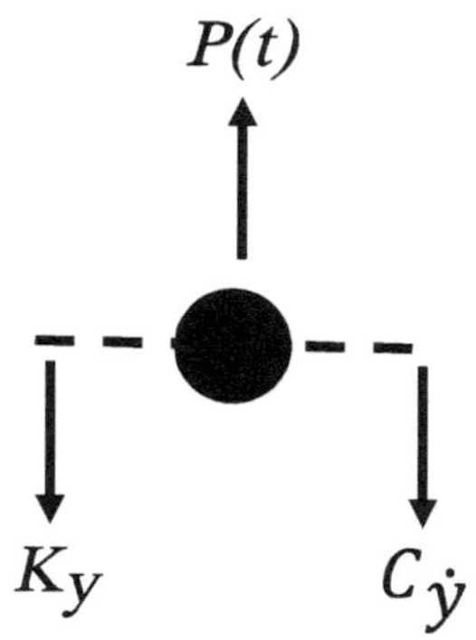

- External force $P(t)$ acting on mass (M) at time (t).
- Stiffness force K_y $(K = 3EI/L$ for uniform cantilever).
- Damping force C_y (a dissipative force assumed to act in the opposite direction of the velocity).
- The resultant of these forces will cause the mass to accelerate according to Newton's second law of motion.

$$M_{\ddot{y}} + C_{\dot{y}} + K_y = P(t) \tag{6.1}$$

$$M_{\ddot{y}} = P(t) - K_y - C_{\dot{y}} \tag{6.2}$$

$$M_a + C_v + K_y = F(t) \tag{6.3}$$

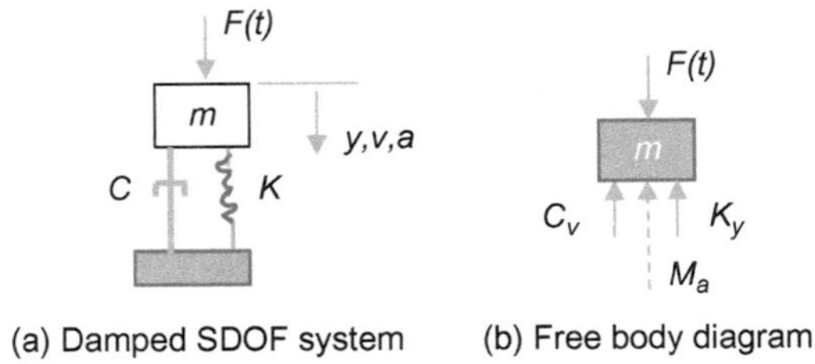

Figure 6.15 Dynamic system.

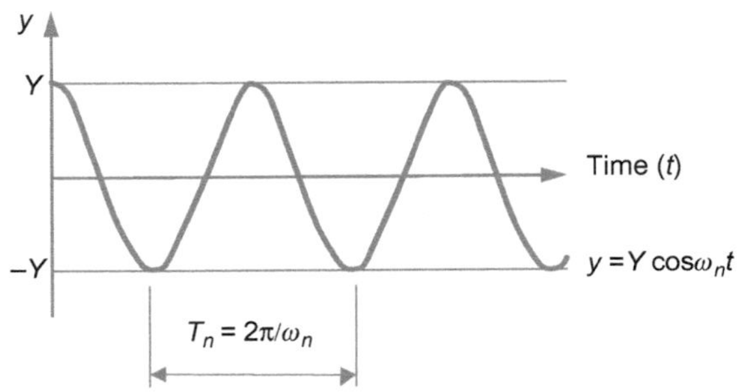

Figure 6.16 Simple harmonic motion.

where:

M = mass,

A = acceleration,

C = viscous damping constant,

v = velocity,

k = stiffness,

$F(t)$ = applied forces as a function of time.

The case of damped SDOF system and free body is presented in Fig. 6.15 and the simple harmonic motion in case of free body without damping is presented in Figure 6.16.

Equations in case of undamped free vibration:

$$M_{\ddot{y}} + K_y = 0 \tag{6.4}$$

$$\frac{K}{M} = \omega_n^2 \tag{6.5}$$

$$y = Y \cos\ \omega_n t \tag{6.6}$$

6.7 DYNAMIC ANALYSIS METHODS

1. Equivalent static method:
   Used in the past, this method is not recommended for general use except for cases where the structure is far from the blast source, so the blast load resembles a gust speed.
2. Equivalent SDOF method:
   The majority of dynamic analysis performed in blast resistant design of petrochemical facilities is using SDOF approximations.
3. Finite element analysis method:
   It is costly and unnecessary unless the structure has unusual features, or the member natural frequency to the support natural frequency is in the range 0.5–2.0.

6.7.1 Analysis Output

From this analysis, we should obtain the following:

- maximum relative displacement for each member,
- relative rotation angle at plastic hinge,
- deflection and reaction due to rebound.

6.8 EQUIVALENT SDOF

To start the dynamic analysis for hand calculation for a one story building, it needs to be converted to a SDOF structure. Equivalent SDOF for typical structures is illustrated in Fig. 6.17.

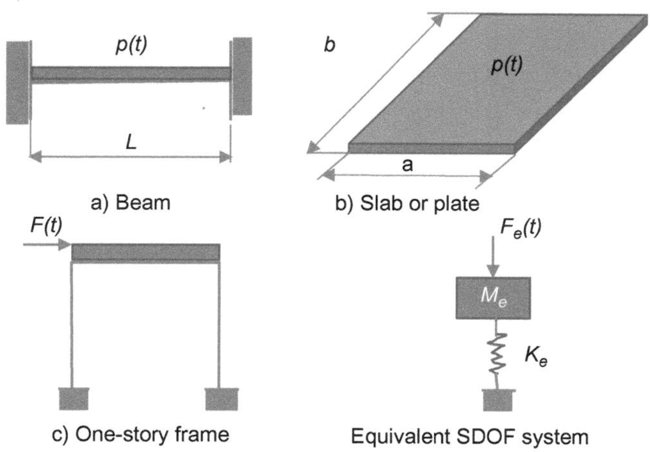

Figure 6.17 Equivalent SDOF for typical structure.

$$M_a + R = F_t \tag{6.7}$$

where R = less of K_y or R_u.

- Transformation factors

  The procedure for obtaining an equivalent SDOF approximation for a structural component is based on a deformed shape under blast load; the control point is selected at the maximum response, such as a plastic hinge location within the span.

  The transformation factors are as follows:
  equivalent stiffness, $K_e = K_L\ _K$,
  equivalent mass, $M_e = K_M M$,
  equivalent force, $F_e = K_L F$,
  equivalent resistance, $R_e = K_L R$.

where:

K_L is load or stiffness transformation factor,
K_M is mass transformation factor.

$$M_e\, a + R_e = F_e \tag{6.8}$$

$$K_{LM} Ma + K_y = F \tag{6.9}$$

where:

$$K_{LM} = K_M / K_L$$

Tables 6.11–6.19 present the transformations for one-way members in case of simply supported, fixed end, and simple-fixed end boundary conditions, respectively, as presented by Biggs (1964).

Transformation factors for one-way remembers are presented in Table 6.11.

1. Simply supported

   1.a Uniform load

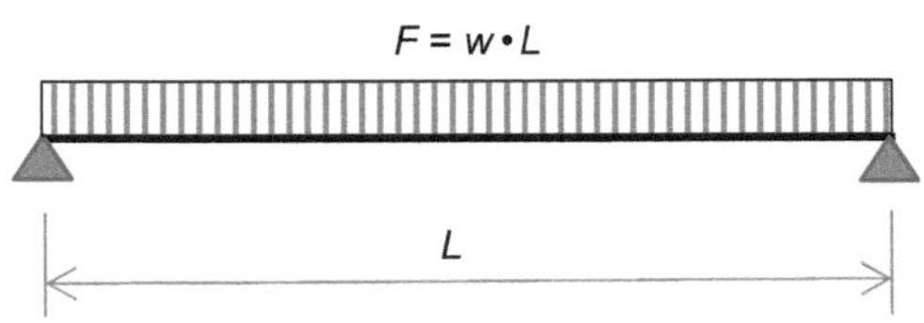

where:

w is the uniform load/m'.
M_{PC} is the ultimate moment capacity at the midspan.
The lumped mass is presented in Tables 6.12–6.19.

Table 6.11 Transformation factor for simple support beam with uniform load

Strain range	Load factor (K_L)	Lumped mass factor[a] (K_M)	Uniform mass factor (K_M)	Bending resistance (R_b)	Spring constant (K)	Dynamic reaction
Elastic	0.64	–	0.5	$8M_{PC}/L$	$384EI/L^3$	$0.39R + 0.11F$
Plastic	0.5	–	0.33	$8M_{PC}/L$		$0.38R_u + 0.12F$

[a]Equal portions of the concentrated mass are lumped at each concentrated load.

1.b Concentrated load at mid span

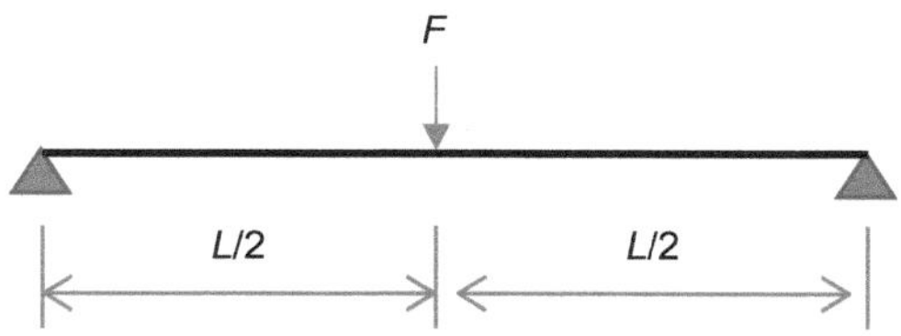

Table 6.12 Transformation factor for simple support beam with concentrated load in middle of span

Strain range	Load factor (K_L)	Lumped mass factor[a] (K_M)	Uniform mass factor (K_M)	Bending resistance (R_b)	Spring constant (K)	Dynamic reaction
Elastic	1.0	1.0	0.49	$4M_{PC}/L$	$48EI/L^3$	$0.78R - 0.28F$
Plastic	1.0	1.0	0.33	$4M_{PC}/L$		$0.75R_u - 0.25F$

[a]Equal portions of the concentrated mass are lumped at each concentrated load.

1.c Two concentrated loads

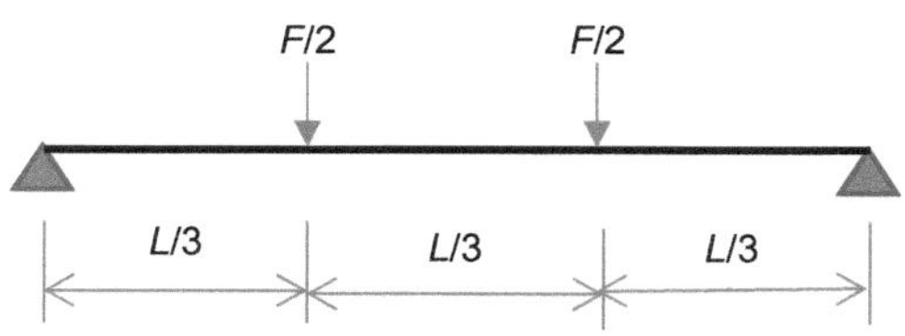

Table 6.13 Transformation factor for simple support beam with two concentrated loads

Strain range	Load factor (K_L)	Lumped mass factor[a] (K_M)	Uniform mass factor (K_M)	Bending resistance (R_b)	Spring constant (K)	Dynamic reaction
Elastic	0.87	0.76	0.52	$6M_{PC}/L$	$56.4EI/L^3$	$0.525R - 0.025F$
Plastic	1.0	1.0	.56	$6M_{PC}/L$		$0.52R_u - 0.02F$

[a]Equal portions of the concentrated mass are lumped at each concentrated load.

2. Fixed end supported

2.a Uniform load

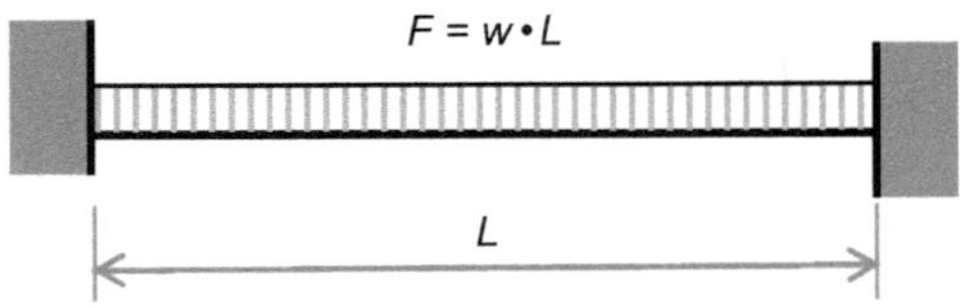

Table 6.14 Transformation factor for two fixed ends beams with uniform load

Strain range	Load factor (K_L)	Lumped mass factor[a] (K_M)	Uniform mass factor (K_M)	Bending resistance (R_b)	Spring constant (K)	Dynamic reaction
Elastic	0.53	–	0.41	$12M_{Ps}/L$	$384EI/L^3$	$0.36R + 0.14F$
Elastic-plastic	0.64		0.50	$8(M_{ps} + M_{pc})/L$	$384EI/5L^3$	$0.39R + 0.11F$
Plastic	0.50	–	0.33	$8(M_{ps} + M_{pc})/L$	0	$0.38R_u + 0.12F$

[a]Equal portions of the concentrated mass are lumped at each concentrated load.

where:

w is the uniform load/m',

M_{PC} is the ultimate moment capacity at the midspan, and M_{ps} is the ultimate moment capacity at the support.

The lumped mass

2.b Concentrated load at mid span

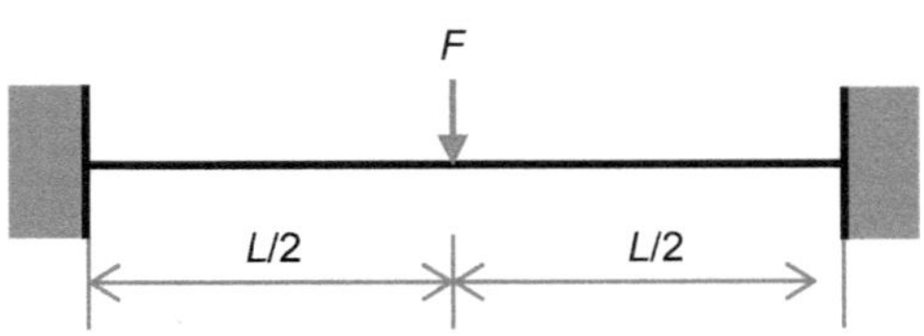

Table 6.15 Transformation factor for two fixed ends beam with one concentrated load in the middle of the span

Strain range	Load factor (K_L)	Lumped mass factor[a] (K_M)	Uniform mass factor (K_M)	Bending resistance (R_b)	Spring constant (K)	Dynamic reaction
Elastic	1.0	1.0	0.37	$4(M_{PC}+M_{ps})/L$	$192EI/L^3$	$0.71R-0.21F$
Plastic	1.0	1.0	0.33	$4(M_{PC}+M_{ps})/L$		$0.75R_u-0.25F$

[a]Equal portions of the concentrated mass are lumped at each concentrated load.

2.c Two concentrated loads

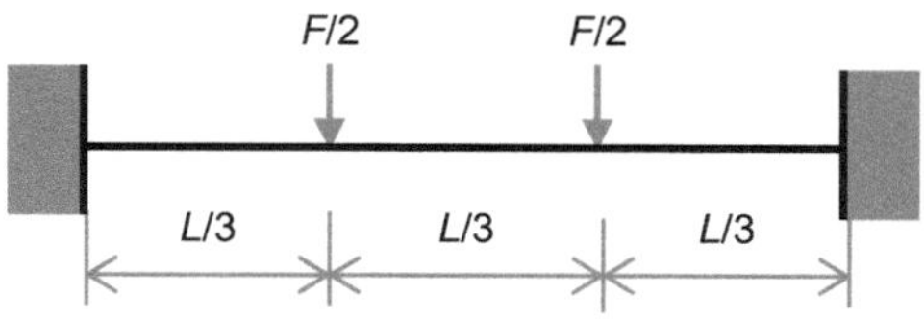

Table 6.16 Transformation factor for two fixed ends beam with two concentrated loads

Strain range	Load factor (K_L)	Lumped mass factor[a] (K_M)	Uniform mass factor (K_M)	Bending resistance (R_b)	Spring constant (K)	Dynamic reaction
Elastic	0.87	0.76	0.52	$6M/L$	$56.4EI/L^3$	$0.53R-0.03F$
Plastic	1.0	1.0	0.56	$6M/L$		$0.52R_u+0.02F$

[a]Equal portions of the concentrated mass are lumped at each concentrated load.

3. One fixed end and one hinge support

3.a Uniform load

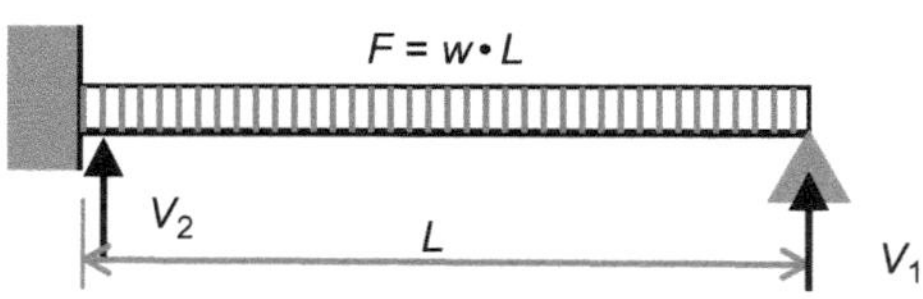

where:

w is the uniform load/m',

M_{PC} is the ultimate moment capacity at the midspan, and M_{ps} is the ultimate moment capacity at the support.

The lumped mass

3.b Concentrated load at mid span

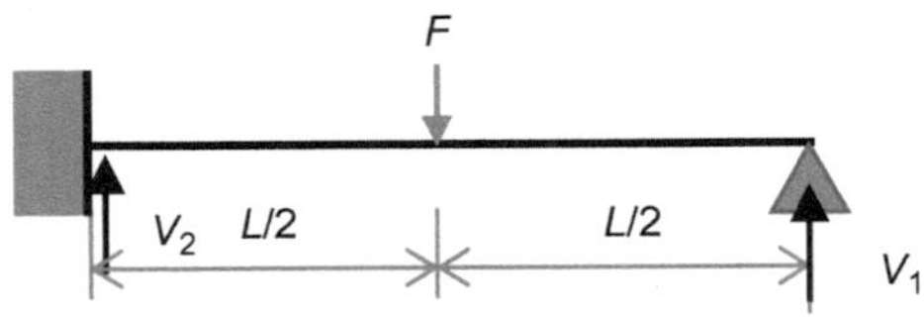

3.c Two concentrated loads

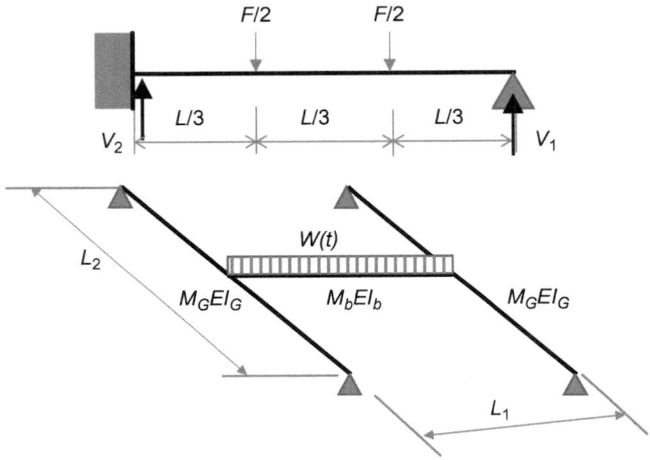

Figure 6.18 Structure system.

Mass Distribution of a Multimember System (Fig. 6.18)
To present the beam SDOF:

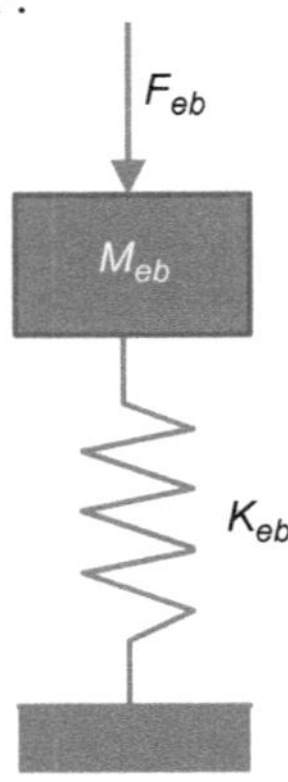

$$K_{eb} = K_{L1}\ K_b \tag{6.10}$$

$$F_{eb} = K_{L1}\ P(t)\ L_1 \tag{6.11}$$

$$M_{eb} = K_{mb}\ m_b\ L_1 \tag{6.12}$$

Table 6.17 Transformation factor for one fixed and one hinge supports beam with uniform load

Strain range	Load factor (K_L)	Lumped mass factor[a] (K_M)	Uniform mass factor (K_M)	Bending resistance, (R_b)	Spring constant (K)	Dynamic reaction
Elastic	0.58	–	0.45	$8M_{Pc}/L$	$185EI/L^3$	$V1 = 0.26R + 0.12F$ $V2 = 0.43R + 0.19F$
Elastic–Plastic	0.64		0.50	$4(M_{ps} + 2M_{pc})/L$	$384EI/5L^3$	$0.39R - 0.11F \pm M_{ps}/L$
Plastic	0.50	–	0.33	$4(M_{ps} + 2M_{pc})/L$	0	$0.38R_u - 0.12F \pm M_{ps}/L$

[a]Equal portions of the concentrated mass are lumped at each concentrated load.

Table 6.18 Transformation factor for one fixed and one hinge support beam with concentrated load in the middle of the span

Strain range	Load factor (K_L)	Lumped mass factor[a] (K_M)	Uniform mass factor (K_M)	Bending resistance (R_b)	Spring constant (K)	Dynamic reaction
Elastic	1.0	1.0	0.43	$16M_{PC}/3L$	$107EI/L^3$	$V1 = 0.25R + 0.07F$ $V2 = 0.54R + 0.14F$
Elastic–Plastic	1.0	1.0	0.49	$2(2M_{PC} + M_{ps})/L$	$48EI/L^3$	$0.78R–0.28F \pm M_{ps}/L$
Plastic	1.0	1.0	0.33	$2(2M_{PC} + M_{ps})/L$		$0.75R_u - 0.25F \pm M_{ps}/L$

[a]Equal portions of the concentrated mass are lumped at each concentrated load.

Table 6.19 Transformation factor for one fixed and one hinge support beam with two concentrated loads

Strain range	Load factor (K_L)	Lumped mass factor[a] (K_M)	Uniform mass factor (K_M)	Bending resistance (R_b)	Spring constant (K)	Dynamic reaction
Elastic	0.81	0.67	0.45	$6M/L$	$132EI/L^3$	$V1 = 0.17R + 0.17F$ $V2 = 0.33R + 0.33F$
Elastic–Plastic	0.87	0.76	0.52	$2(3M_{PC} + M_{ps})/L$	$56EI/L^3$	$0.525R - 0.0.025F \pm M_{ps}/L$
Plastic	1.0	1.0	0.56	$2(3M_{PC} + M_{ps})/L$		$0.52R_u - 0.02F \pm M_{ps}/L$

[a]Equal portions of the concentrated mass are lumped at each concentrated load.

For the girder SDOF:

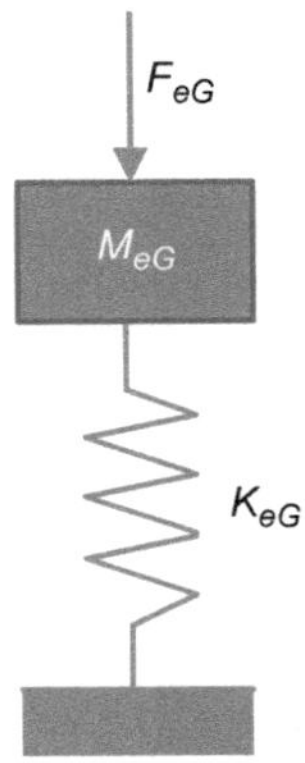

$$K_{eG} = k_{L2}\, k_G \tag{6.13}$$

$$F_{eG} = K_{LG}(\text{beam reaction force}) \tag{6.14}$$

$$M_{eG} = 0.5\, m_b L_1 + K_{mG}\, M_G \tag{6.15}$$

6.9 DYNAMIC ANALYSIS SOLUTION

The main target of the design is to calculate the maximum deformation of the member; define the ductility factor and the support rotation; then compare it with the limit that was defined previously according to the required building performance after the blast load effect. There are three methods, which are:

- Graphical methods
- Pressure impulse curve
- Numerical integration method

6.9.1 Graphical Solution

The graphical solution method for elasto-plastic SDOF is presented in Fig. 6.19 as per Biggs (1964).

6.9.2 Pressure-Impulse Curve

Another method to predict the damage as a function of affected pressure and the impulse is presented in Fig. 6.20.

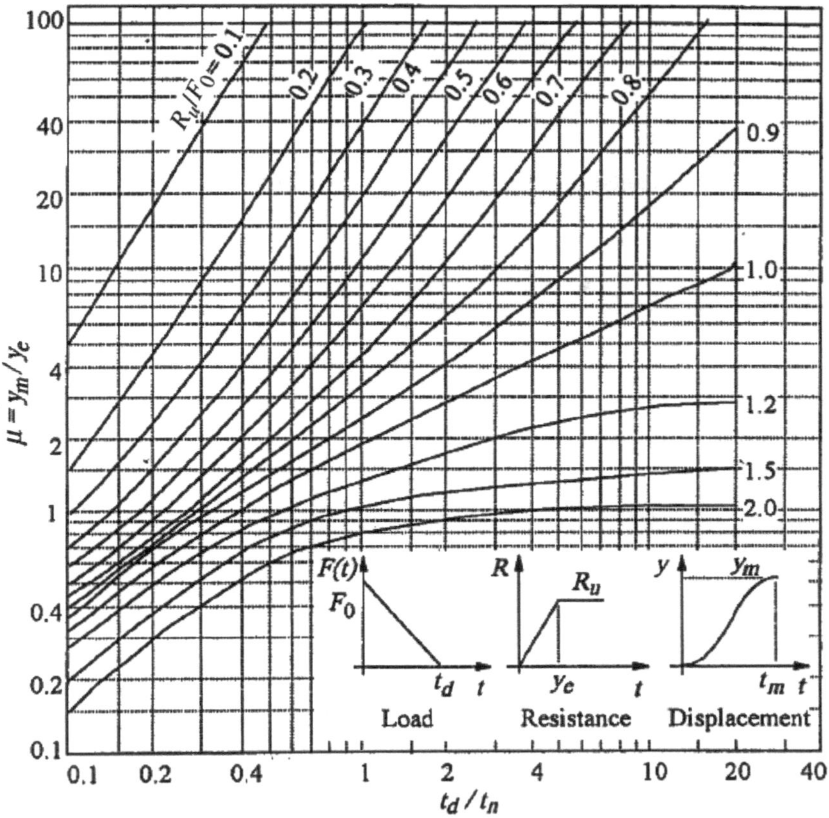

Figure 6.19 Graphic solution for elasto-plastic SDOF.

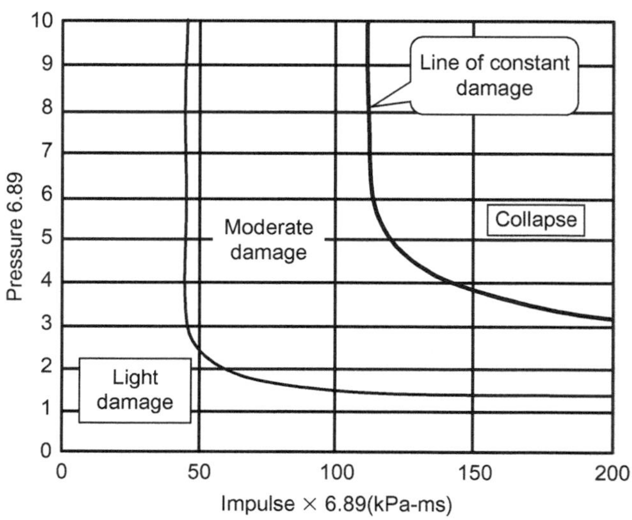

Figure 6.20 Relation between affecting pressure, impulse, and the expected damage to the building.

Closed form solution:

$$\text{If } t_d/t_n < 0.1$$

$$\mu_d = 0.5\left[\left(\frac{1.2\,\pi f}{R_o}\right)^2 + 1\right] \tag{6.16}$$

$$\text{If } t_d/t_n > 10$$

$$\mu_d = \frac{1}{[2(1 - F_1/R_o)]} \tag{6.17}$$

In between the above two extremes $t = t_d/t_n$

$$\frac{F_o}{R_m} = \frac{\sqrt{(2\mu_{d-1})}}{\pi(\tau)} + \frac{(2\mu_d - 1)(\tau)}{2\mu_d(\tau + 0.7)} \tag{6.18}$$

6.9.3 Numerical Integration

This method can be used if the simple graphical method or closed form are not appropriate. It is dependent on solving the dynamic equilibrium equation for the displacement at each step. A rule of thumb is to use a value less than or equal to 1/10th of either the natural period of the structure or load duration, whichever is smaller.

In 1959, Newmark presented a family of single-step integration methods for the solution of structural dynamic problems for both blast and seismic loading.

$$m\ddot{u}_{i+1} + c\dot{u}_{i+1} + ku_{i+1} = F_{i+1} \tag{6.19}$$

$$\dot{u_{i+1}} = \dot{u}_i + [(1 - \gamma)\Delta t]\ddot{u}_i + (\gamma\Delta t)\ddot{u}_{i+1} \tag{6.20}$$

$$u_{i+1} = u_i + (\Delta t)\dot{u}_i + \left[(0.5 - \beta)(\Delta t)^2\right]\ddot{u}_i + \left[\beta(\Delta t)^2\right]\ddot{u}_{i+1} \tag{6.21}$$

The parameters define how the acceleration is assumed over the time step, as per ASCE assume linear variation of acceleration:

$$\gamma = 0.5 \quad \text{and} \quad \beta = 0.1667$$

$$\Delta u_i \equiv u_{i+1} - u_i \tag{6.22}$$

$$\Delta \dot{u}_i \equiv \dot{u}_{i+1} - \dot{u}_i \tag{6.23}$$

$$\Delta \ddot{u}_i \equiv \ddot{u}_{i+1} - \ddot{u}_i \tag{6.24}$$

$$\Delta F_i = F_{i+1} - F_i \tag{6.25}$$

$$\Delta \dot{u}_i = (\Delta t)\ddot{u}_i + (\gamma \Delta t)\Delta \ddot{u}_i \tag{6.26}$$

$$\Delta u_i = (\Delta t)\dot{u}_i + \frac{(\Delta t)^2}{2}\ddot{u}_i + \beta(\Delta t)^2 \Delta \ddot{u}_i \tag{6.27}$$

$$\Delta \ddot{u}_i = \frac{1}{\beta(\Delta t)^2}(\Delta u_i) - \frac{1}{\beta(\Delta t)}\dot{u}_i - \frac{1}{2\beta}\Delta \ddot{u}_i \tag{6.28}$$

$$\Delta \dot{u}_i = \frac{\gamma}{\beta(\Delta t)}\Delta u_i - \frac{\gamma}{\beta}\dot{u}_i + \Delta t\left(1 - \frac{\gamma}{2\beta}\right)\ddot{u}_i \tag{6.29}$$

$$m\Delta \ddot{u}_i + c\Delta \dot{u}_i + k\Delta u_i = \Delta F_i \tag{6.30}$$

$$m\left[\frac{1}{\beta(\Delta t)^2}\Delta u_i - \frac{1}{\beta(\Delta t)}\dot{u}_i - \frac{1}{2\beta}\ddot{u}_i\right] + c\left[\frac{\gamma}{\beta(\Delta t)}\Delta u_i - \frac{\gamma}{\beta}\dot{u}_i + \Delta t\left(1 - \frac{\gamma}{2\beta}\right)\ddot{u}_i\right] + k\Delta u_i = \Delta F_i \tag{6.31}$$

$$\left[m\frac{1}{\beta(\Delta t)^2}\Delta u_i + c\frac{\gamma}{\beta(\Delta t)} + k\right]\Delta u_i = \Delta F_i + \left[\frac{1}{\beta(\Delta t)} + \frac{\gamma}{\beta}c\right]\dot{u}_i + \left[\frac{1}{2\beta}m + \Delta t\left(\frac{\gamma}{2\beta} - 1\right)c\right]\ddot{u}_i \tag{6.32}$$

$$\hat{k} = m\frac{1}{\beta(\Delta t)^2} + c\frac{\gamma}{\beta(\Delta t)} + k \tag{6.33}$$

$$\Delta\hat{F}_i = \Delta F_i + \left[\frac{1}{\beta(\Delta t)}m + \frac{\gamma}{\beta}c\right]\dot{u}_i + \left[\frac{1}{2\beta}m + \Delta t\left(\frac{\gamma}{2\beta} - 1\right)c\right]\ddot{u}_i \tag{6.34}$$

As the effective K value is known, the value of Δt is known by dividing the triangle blast load force time rectangle drawing to incremental time.

$$\hat{k}\Delta u_i = \Delta\hat{F}_i \tag{6.35}$$

$$\Delta u_i = \frac{\Delta\hat{F}_i}{\hat{k}} \tag{6.36}$$

$$\Delta\dot{u}_i = \frac{\gamma}{\beta(\Delta t)}\Delta u_i - \frac{\gamma}{\beta}\dot{u}_i + \Delta t\left(1 - \frac{\gamma}{2\beta}\right)\ddot{u}_i \tag{6.37}$$

$$\Delta\ddot{u}_i = \frac{1}{\beta(\Delta t)^2}\Delta u_i - \frac{1}{\beta(\Delta t)}\dot{u}_i - \frac{1}{2\beta}\ddot{u}_i \tag{6.38}$$

$$u_i = u_{i-1} + \Delta u_i \tag{6.39}$$

$$\dot{u}_i = \dot{u}_{i-1} + \Delta\dot{u}_i \tag{6.40}$$

$$\ddot{u}_i = \ddot{u}_{i-1} + \Delta\ddot{u}_i \tag{6.41}$$

This calculation based on having the load value for each time variation, so the value of Δt is known and from the triangle load we can define the load at each time interval. Then the mass and stiffness will be calculated for each members. At the first step in zero, we have velocity and acceleration is equal to zero.

Values of K in Eq. (6.3) is known and ΔF in Eq. (6.34) is known, so you can obtain the increment increase or decrease in displacement Δu as in Eq. (6.36); consequently, you can obtain the increment in acceleration in velocity and acceleration as in Eqs. (6.37) and (6.38), respectively. Consequently, at every step, we can calculate all the other velocities and accelerations at every load step. This calculation is done by excel sheet; you can contact the author or visit www.elreedyma.comli.com or www.elreedyman.tk to obtain it.

6.10 BLAST WAVE LOAD PARAMETERS

The blast load affects the building based on the direction of the building to the blast load source. Mainly the load applied to the building is presented by a triangle of load and time, with general two shapes as per Fig. 6.21. For the front wall shall be affected by a shock wave, so the load shape (A) is the traditional one and for the side wall as it takes a times until reach a peak which affect by pressure wave so it will match with shape (B) noting that this shape can be idealized as shown in Fig. 6.22

The blast wave attenuates as it propagates outward from the explosion epicenter. Consequently, the values of peak overpressure and impulse decrease with distance while the duration tends to increase. Values for these blast wave parameters can be determined from published data in the form of scaled values (overpressure, impulse, or duration) as a function of scaled distance. *UFC 3-340-02* provides data on high energy condensed phase explosives while *Baker, TNO* 1985, and *CCPS Explosion Guidelines*

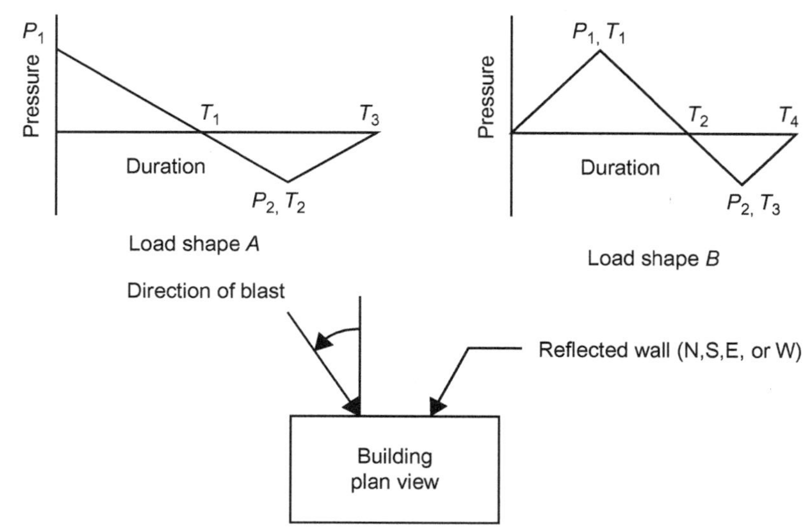

Figure 6.21 Blat load configuration.

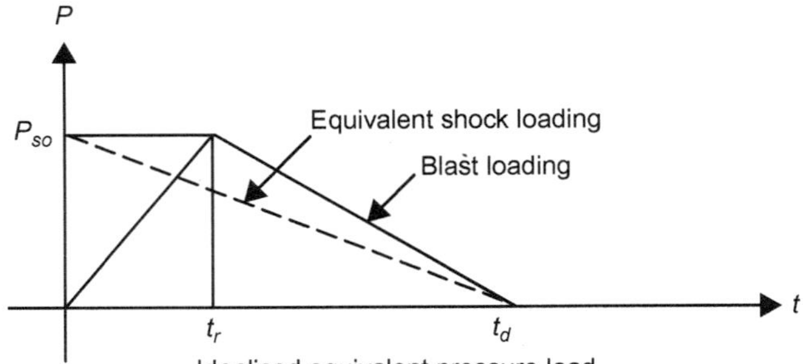

Figure 6.22 Idealized equivalent pressure load. *P*, pressure in kPa; *t*, time in s or m; t_r, rise time in s or m; P_{so}, peak incident over-pressure in kPa; t_d, positive phase duration in s or m.

provide values for vapor cloud explosions according to their respective models. As per ASCE 10, these sources do not provide data on the negative phase of the blast wave from a vapor cloud explosion.

On the other hand, the negative phase pressures are relatively small and oppose the primary lateral force, so it is usually conservative to ignore them in the design.

In addition to peak overpressure, duration, and impulse, other blast wave parameters that may enter into the determination of the blast loads for a structure include:

1. peak reflected pressure, P_r,
2. peak dynamic (blast wind) pressure, q_o,
3. shock front velocity, *U*,
4. blast wave length, L_w.

Usually, these secondary parameters can be determined from the primary blast wave parameters, as will be discussed.

The peak side-on over-pressure (P_{so}) shall be obtained via a hazard assessment or via the principal. The range of P_{so} is typically between 5 and 65 kPa.

The positive phase duration or duration shall be obtained from the hazard assessment or the principal. The value of the t_d is generally less than 150 ms. The negative phase duration may be ignored.

When the free field blast wave from an explosion strikes a surface, it is reflected as illustrated in Fig. 6.3.

The effect of this blast wave reflection is that the surface will experience a pressure much more than the incident side-on value. The magnitude of the reflected pressure is usually determined as an amplifying ratio of the incident pressure:

$$P_r = C_r P_{so} \tag{6.42}$$

where C_r is the reflection coefficient.

The reflection coefficient depends on the peak overpressure, the angle of incidence of the wave front relative to the reflecting surface, and the type of blast wave.

For peak overpressures up to 20 psi (138 kPa), the expected range for most accidental vapor cloud explosions, *Newmark* provides a simple formula for the blast wave reflection coefficient at normal (0°) incidence as follows:

$$C_r = \frac{P_r}{P_{so}} \cong (2 + 0.0073\ P_{so})\ (P_{so} \text{ in kPa}) \tag{6.43}$$

Or:

$$\cong (2 + 0.05\ P_{so})\ (P_{so} \text{ in psi})$$

The dynamic pressure (q_o) is to be calculated with the following empirical formula:

$$q_o \cong 0.0032 \times (P_{so})^2 \text{ in kPa} \tag{6.44}$$

The net dynamic pressure on a structure is $q_o \times C_d$.

The draft coefficient C_d depends on the shape and the orientation of the building. For a rectangular building, a C_d of 1.0 may be taken for the front wall and a C_d of -0.4 for the side/rear wall and roof.

The net dynamic pressure on a structure is the product of the dynamic pressure and a drag coefficient, C_d. The drag coefficient depends on the shape and orientation of the obstructing surface. For a rectangular building, the drag coefficient may be taken as $+1.0$ for the front wall and -0.4 for the side and rear walls and roof.

Shock Front Velocity, U

In the free field, the blast wave from an explosion travels at or above the acoustic speed for the propagating medium. *UFC 3-340-02* provides plots of shock front velocity versus scaled distance for high energy TNT explosives. There are no similar plots available for pressure wave propagation. However, for design purposes, it can be conservatively assumed that a pressure wave travels at the same velocity as a shock wave. In the low-pressure range, and for normal atmospheric conditions, the shock/pressure front velocity in air can be approximated using the following relationship from *Newmark:*

$$\begin{aligned} U &= 345(1 + 0.0083\ P_{so})^{0.5} \text{ m/s} \\ &= 1130(1 + 0.058\ P_{so})^{0.5} \text{ ft/s} \end{aligned} \tag{6.45}$$

6.11 BLAST DESIGN PROCEDURE

The procedure of blast study is presented in the following flow chart in Fig. 6.23. Note that the multi-degree of freedom analysis will be applied in cases of two or more story buildings; and in this case it needs software to do dynamic analysis until apply time history load, and to apply nonlinear analysis at the same time.

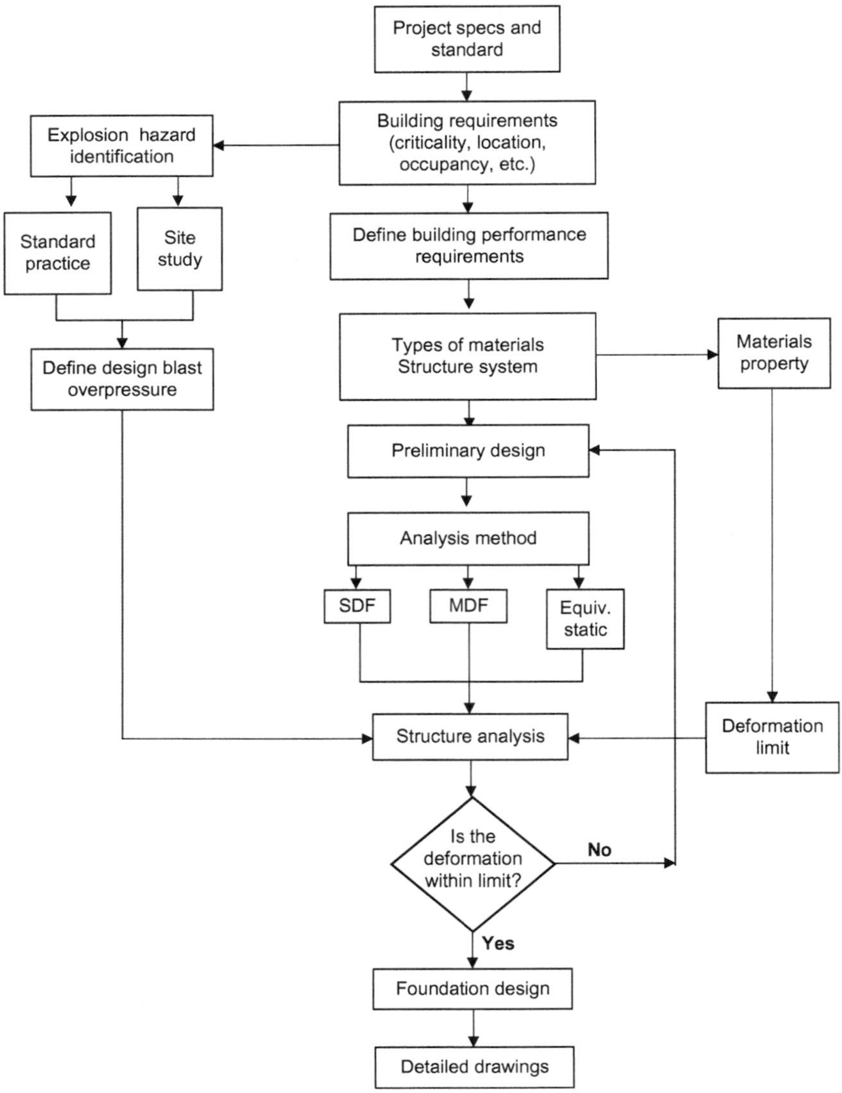

Figure 6.23 Blast load effect building design procedure.

Table 6.20 Comparison between blast and seismic loading

Parameters	Blast	Seismic
Load time variation	Shorter duration and high impulse	Longer duration and less impulse
Force direction/ acceleration	Positive overpressure in one direction and negative suction	Several direction changes under extreme acceleration
Deformation, damage	Large inelastic deformation for local member and global structure	Global large inelastic deformations
Equivalent static method	Not acceptable	Acceptable
Material straining	Strength increase	No increase in materials
Critical area	Connection	Connection

The blast load and load due to earthquakes are similar in some features, but there are main differences between them. Table 6.20 presents a comparison between the blast load and the seismic load.

6.12 STATIC ANALYSIS FOR FOUNDATIONS

Static application of the peak dynamic reactions from the wall and roof components can be used to design supporting members and to calculate overturning and sliding effects.

Foundations shall be designed using vertical and lateral soil capacities. For shallow foundations including footings and mats, vertical soil capacity shall be 80% of the ultimate net soil-bearing capacity.

For piles and other deep foundations, vertical soil capacity shall be 80% of the ultimate static capacities in compression and in tension. Passive resistance of grade beams may be used to resist lateral loads if compacted fill is placed around the building perimeter. Frictional resistance of spread footings and floor grade slabs shall be based on the coefficient of friction determined by the geotechnical study. The normal force shall be the sum of the dead loads and the applicable vertical load. However, if two or more of these resistances are used to support the lateral blast loads, the lateral capacity shall be limited to 67% of the combined ultimate static resistance.

To optimize the design, the foundation components can be analyzed dynamically for the calculated reaction-time history of the supported components.

The required dynamic material properties of the foundation soils, including resistance and stiffness, shall be determined on the basis of an appropriate geotechnical investigation. Deformation limits shall not be used for dynamic response of foundations. Based on the results of the dynamic analysis, it shall be determined whether the predicted maximum response is acceptable for the permissible damage level of the building.

6.13 EXAMPLE

A building is required to be designed against a blast load. The building dimensions are as follows:

length = 20 m,
width = 28 m,
height = 4.5 m.

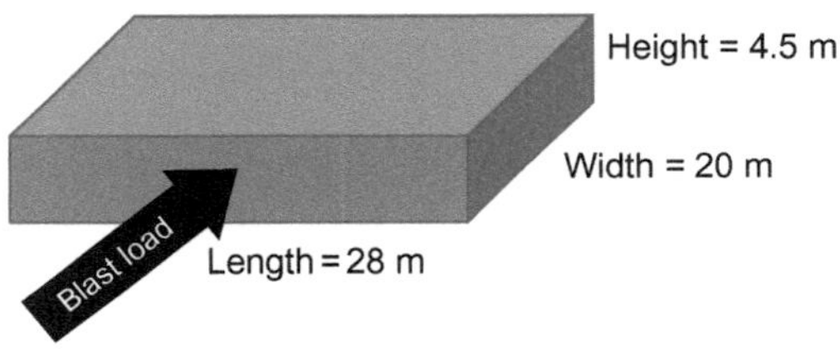

- A blast wave has been given to be applied normal to the long side of the building.
- Peak side on over pressure, $P_{so} = 41$ kPa.
- Shock front velocity:
  $U = (1130(1 + 0.058P_{so})^{0.5}) \times 0.3048 = 1312$ ft/s $= 400$ m/s Eq. (6.45)
- Length of pressure wave:
  $L_w = U(t_d) = 400 \times 0.05$ s $= 20$ m Eq. (3.6)
- Peak dynamic wind pressure:
  $q_o = 0.022(P_{so})^2 = 0.8$ psi $= 6$ KPa Eq. (3.4)

6.13.1 Front Wall Loading

The front wall is assumed to span vertically.

Reflected pressure:
$P_r = [2 + 0.05(P_{so})]P_{so} = (95$ kPa)

- Clearing distance:
  $S =$ minimum of H or $B/2 = 4.5$ m
- Reflected overpressure clearing time:
  $t_c = 3(S/U) < t_d = 3\ (4.5)/400 = 0.034 < 0.05$ OK

- Drag coefficient:
  $C_d = 1.0$
- Stagnation pressure:
  $P_s = P_{so} + C_d(q_o) = 47$ kPa
- Front wall impulse:
  $I_w = 0.5(P_r - P_s)t_c + 0.5P_s t_d = 2$ kPa-s
  Effective duration, $t_e = 2I_w/P_r = 0.042$ s

6.13.2 Side Wall Loading

- The side wall is the same as the front wall, spanning vertically from foundation to roof. As the highest load on the front wall, the side wall analysis is necessary to check in plane and out-of-plane shear wall forces. The calculation for a segment 1 m.
  Drag coefficient:
  $C_d = -0.4$
  Equivalent load coefficient:
  $L_w/L_1 = 20/1 = 20$, so essentially $C_e = 1.0$.
  Equivalent peak pressure:
  $P_s = C_e\, P_{so} + C_d\, q_o = 39$ kPa
  Rise time:
  $T_r = L_1/U = (1\text{ m})/400\text{ m/s} =$ take it 0.0
  Duration $t_d = 0.05$ s.

The value of effective over pressure C_e is as in Table 6.21, extracted from figure TM5-1300 (Fig. 6.24).

Table 6.21 Effective over pressure value factors

L_w/L	C_e
0.45	0.2
0.65	0.3
0.8	0.4
1.0	0.45
2	0.6
3	0.7
4	0.8
5	1.0
6	1.0
7	1.0
8	1.0
9	1.0
10	1.0

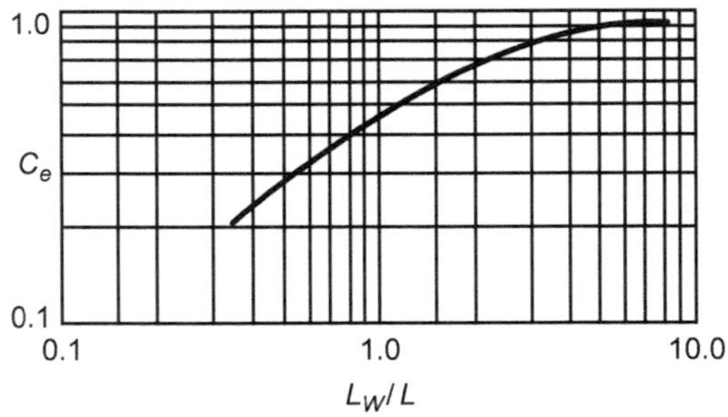

Figure 6.24 Effective over pressure values (from TM 5-1300).

6.13.3 Roof Loading

- The roof is a slab spanning between roof beams. For the roof design, a section 1 m wide by 2.5 m long will be used.
- $L_1 = (2.5\text{ m})$
- $C_d = -0.4$
- Equivalent load coefficient:
  $L_w/L_1 = 20/2.5 = 8$; therefore, $C_e = 0.92$
  Equivalent peak pressure:
  $P_s = C_e\ P_{so} + C_d\ q_o = 35\text{ kPa}$
  Rise time:
  $T_r = = 2.5/400(\text{m/s}) = 0.006\text{ s}.$
  Time of duration, $t_d = 0.05\text{ s}$
  Total positive phase duration:
  $T_o = t_r + t_d = 0.006 + 0.05 = 0.056\text{ s}$

6.13.4 Rear Wall Load

- The rear wall is proportioned the same as the front and side walls. The rear wall analysis would be necessary to determine the net loading on the overall building. The analysis will be for a wall segment 1 m wide.
- $C_d = -0.4$
- Equivalent load coefficient:
  $L_w/S = 20/4.5 = 4.4$, therefore $C_e = 0.88 = 20\text{ m}/4.5\text{ m} = 4.44$
- Equivalent peak over pressure:
  $P_b = C_e\ P_{so} + C_d\ q_o = 34\text{ kPa}$

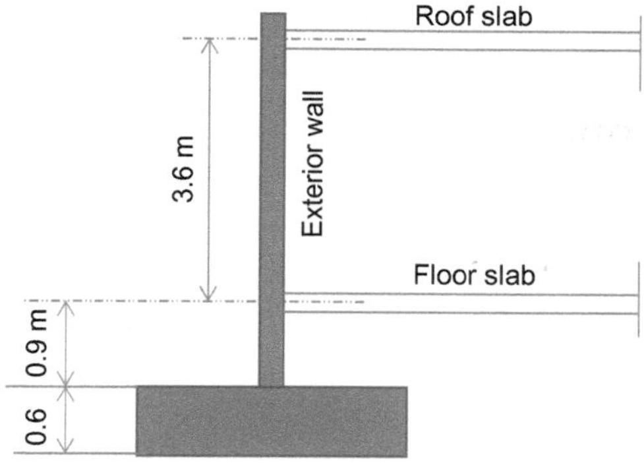

Figure 6.25 Wall elevation.

- Time of arrival:

  $t_a = L/U = 20/400 = 0.051$ s

  $t_r = S/U = 0.011$ s

  $t_d = 0.05$ s

  $t_o = t_r + t_d = 0.011 + 0.05 = 0.061$ s (Fig. 6.25).

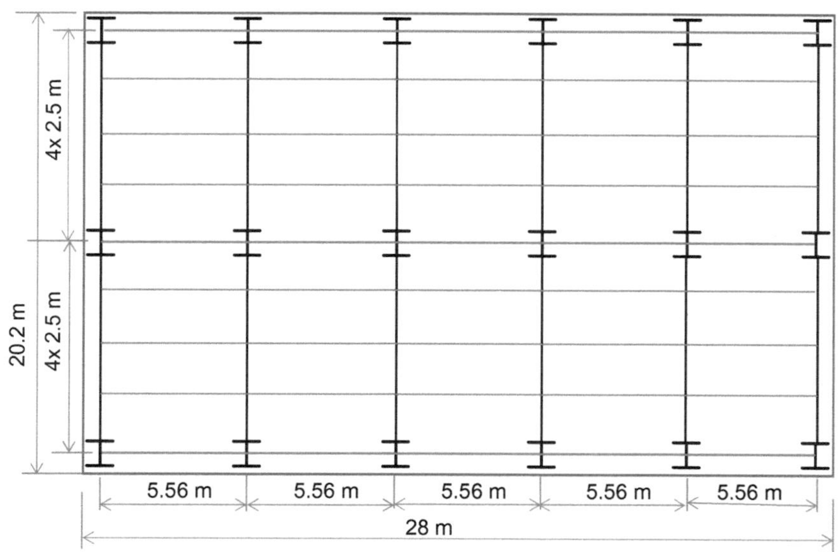

- ***Material properties***

  Structural steel: A36 $f_y = 248$ MPa

  Reinforcing steel grade 60: $F_y = 414$ MPa

  Concrete strength: $fc' = 27.6$ MPa

Table 6.22 Reinforced concrete dynamic design stress value

Type of stress	Type of reinforcement	Maximum support rotation	Dynamic stress design (F_{ds})
Bending	Tension and compression	$0<\theta\leq 2$	F_{dy}
		$2<\theta\leq 5$	$F_{dy}+(F_{du}-F_{dy})/4$
		$5<\theta\leq 12$	$(F_{dy}+F_{du})/2$
Diagonal tension	Stirrups		F_{dy}
Direct shear	Diagonal bars	$0<\theta\leq 2$	F_{dy}
		$2<\theta\leq 5$	$F_{dy}+(F_{du}-F_{dy})/4$
		$5<\theta\leq 12$	$(F_{dy}+F_{du})/2$
Compression	Column	All	F_{dy}

Steel modulus: $E_s = 200,000$ MPa
Concrete modulus: $E_c = 20,000$ MPa
$n = E_s/E_c = 8.04$
Soil density: 18.1 kN/m^3
Roof dead load: $w = 1197$ Pa
Neglect live load in case of blast
Acceleration of gravity = 9.8 m/s^2

- *Design load*
  As the previous example:
  Case A: explosion occurs on the long side of the building
  Case B: explosion occurs on the short side of the building
- Building performance requirements will be corresponding to the low response range as per Table 6.22
  -The low response range (Table 6.22)

6.13.5 Load on Exterior

- The walls are 3.6 m from floor slab to roof. The shortest wall is 20.2 m, so all walls will be considered as one-way beams pinned (hinge) at the base by a crossed reinforcing configuration (Figs. 6.26 and 6.27).

6.13.5.1 Front Wall Load

- $P_r = = 95.1$ kN/m^2
- $t_r = 0,\ t_e = 0.042$
- Peak load
- $P_o = 3.6 \times 1 \times (95.1) = 342$ kN

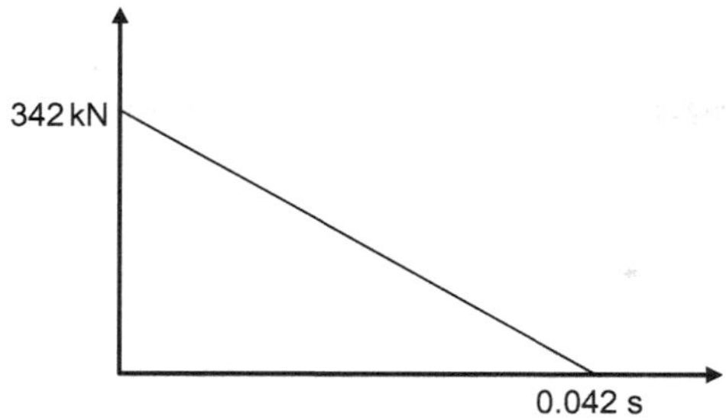

6.13.5.2 Side Wall Load

- Equivalent peak pressure
- $P_a = 40\ \text{kN/m}^2$
- $t_r = 0,\ t_d = 0.05$

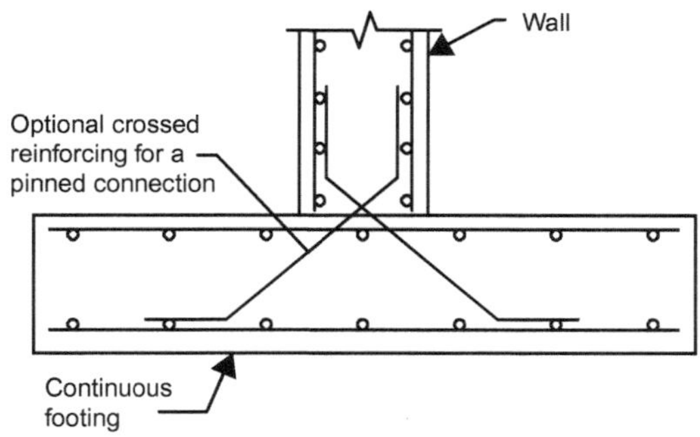

Figure 6.26 RC wall-to-foundation connection.

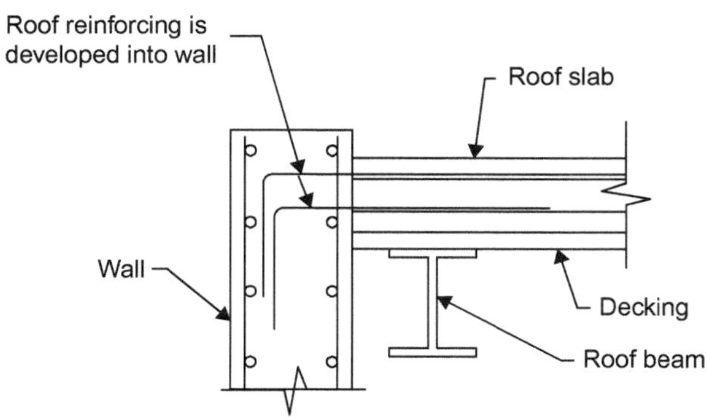

Figure 6.27 RC wall-to-roof slab connection.

- Peak load
- $P_o = 3.6 \times 1 \times 4 = 144$ kN

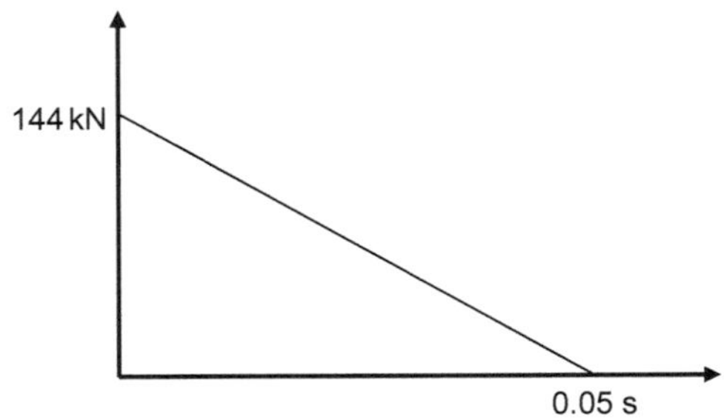

6.13.5.3 Rear Wall Load

- Equivalent peak pressure
- $P_a = 35$ kN/m^2
- $t_a = 0.051$, $t_r = 0.011$, $t_d = 0.05$
- Peak load
- $P_o = 3.6 \times 1 \times 35 = 126$ kN

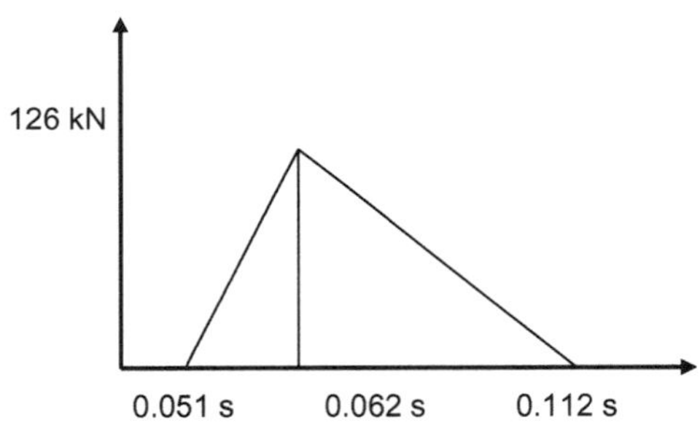

6.13.6 Trial Size

- 250 mm concrete wall thickness.
- (16 mm diameter every 140 mm distance 7 ϕ 16/m′) Vertical reinforcement will be outside the horizontal bars.
  Calculate bending resistance
- $f_{dy} = (\text{SIF})(\text{DIF})f_y = (1.1)(1.17)280 = 360$ MPa
- $f_{dc} = (\text{SIF})(\text{DIF})f_c = 1 \times 1.19 \times 30 = 35.7$ MPa
- $b = 1000$ mm

- $d = (250 - 35 - (16/2)) = 207$ mm
- $A_s = (1407\ \text{mm}^2)/\text{m}$
- $\rho = A_s/b_d = 1407/(1000 \times 200) = 0.007 > 200/f_y d$ OK
- $a = A_s f_{dy}/(0.85 f'_{dc}\ b) = 16.7$ mm
- $M_p = M_n = A_s f_{dy}(d - a/2) = 97.08$ kN m
- $R_b = 8M_p/L = 215$ kN

  Calculate shear resistance
- $f'_{dc} = (\text{SIF})(\text{DIF}) f'_c = (1.0)(1.0)\ 30 = (30\ \text{MPa})$
- $V_n = (1/6)\ (f'_c)^{0.5}\ b \cdot d = 182$ kN
- The critical section is d from the support.
- $R_s = V_n\ L/(0.5L - d) = 409$ kN
- As $R_b < R_s$ so the bending control $R_u = R_b = = 215$ kN
- Allowable response, $\theta_a = 2.0°$ (low range).

  Compute SDOF equivalent system
- $I_g = b(h)^3/12 = (100) \times (25)^3/12 = (130{,}208.3\ \text{cm}^4)$
- Cracked moment of inertia:

  $n\ A_s = 8.04 \times 1407 = 9704\ \text{mm}^2$

$$C = \frac{-nA_s + \sqrt{nA_s(nA_s + 2bd)}}{b}$$

$C = 2.23$ in. $= 56.9$ mm

$I_{cr} = bC^3/3 + nA_s\ (d - C)^2 = 293{,}055{,}128\ \text{mm}^4$

Average moment of inertia:

$I_a = (I_g + I_{cr})/2 = 611\ \text{in.}^4 = 238.7 \times 10^6\ \text{mm}^4$

$I_a = (I_g + I_{cr})/2 = 797.569231 \times 10^6\ \text{mm}^4$

Effective stiffness: $K = 384\ E_c I/5L^3 = 26{,}257.42$ N/mm

- yield deflection: $y_e = R_u/k = 90.9 \times 1000/26{,}257.42 = 3.46$ mm,
- beam mass = weight/(9.81 m/s^2)

  $= 23.5 \times 0.25 \times 1.0$ m unit width $\times 3.6/9.8 = 2.16\ \text{kN} \cdot \text{s}^2/\text{m}$

  $= 2.16\ \text{N} \cdot \text{s}^2/\text{mm}$,
- because of expected response use an average of values for K_{LM},
- elastic $K_{LM} = 0.5/0.64 = 0.78$,
- plastic $K_{LM} = 0.33/0.5 = 0.66$,
- average $K_{LM} = (0.78 + 0.66)/2 = 0.72$,
- equivalent mass: $M_e = K_{LM}$ (beam mass) $= 1.55$ N-s^2/mm,

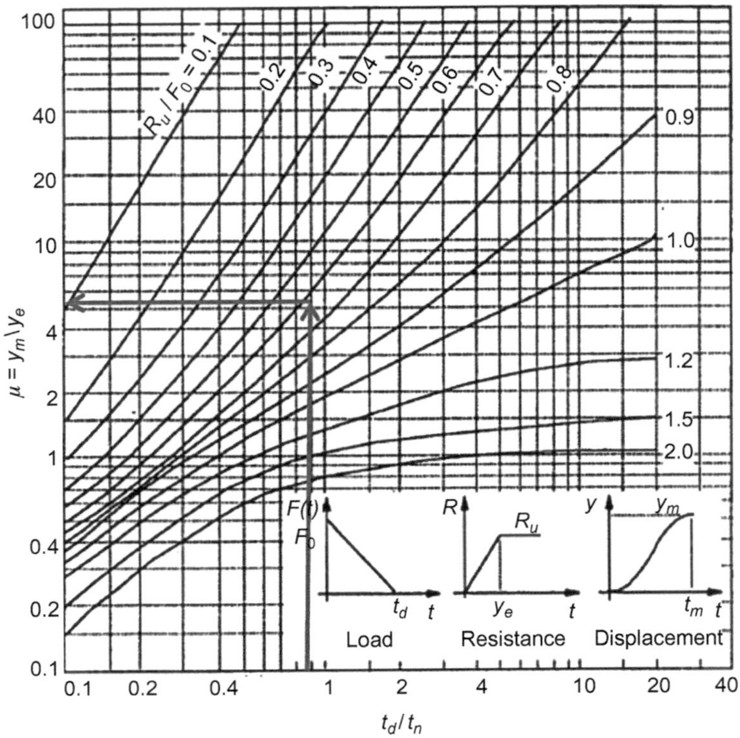

Figure 6.28 Graphic solution for SDOF system.

- period of vibration:
  $t_n = 2\ \pi(M_e/K)^{0.5} = 0.048$ s,
- time increment $= t_n/10 = 0.0048$ s,
- use 0.002 s to match for stiff supporting element,
- $V = 0.385R + 0.115F$.

The graphic solution presented by Biggs (1964) for elastic plastic SDOF system is depicted in the chart in Fig. 6.28.

$t_d/t_n = 0.042/0.048 = 0.88$

$R_u/P_o = 215/342 = 0.63$

$\mu_d = 5$

Max. deflection: $y_m = \mu_d \cdot Y_e = 17.3$ mm

Support rotation:

$\theta_a = \arctan\ (y_m/0.5l) = 0.68° < 2°$ OK

6.13.7 Numerical Integration Method

Time	F	u	v	a	Resistance	Reaction
(s)	(N)	(mm)	(mm/s)	(mm/s²)	(kN)	(kN)
0	342,000.000	0.000	0.000	220,645.161	0	39.33
0.002	325,714.286	0.429	423.291	202,645.576	11.2720649	37.46148
0.004	309,428.571	1.660	797.030	171,093.359	43.5906943	35.60107
0.006	293,142.857	3.568	1096.239	128,115.507	93.6792015	33.7475
0.008	276,857.143	5.982	1300.956	76,601.999	157.074225	31.89905
0.01	260,571.429	8.699	1397.570	20,011.475	228.42586	30.05366
0.012	244,285.714	11.496	1379.720	− 37,860.979	301.856853	28.20907
0.014	228,000.000	14.143	1248.721	− 93,138.332	371.356747	26.36297
0.016	211,714.286	16.421	1013.461	− 142,121.335	431.184533	24.51315
0.018	195,428.571	18.138	689.804	− 181,536.079	476.252851	22.65764
0.02	179,142.857	19.136	299.515	− 208,752.809	502.468015	20.79488
0.022	162,857.143	19.309	− 131.199	− 221,961.376	507.003148	18.92377
0.024	146,571.429	18.604	− 573.452	− 220,291.619	488.48619	17.04378
0.026	130,285.714	17.027	− 997.615	− 203,870.672	447.09029	15.15499
0.028	114,000.000	14.644	− 1375.299	− 173,813.425	384.520619	13.25804
0.03	97,714.286	11.574	− 1681.259	− 132,146.841	303.898598	11.35414
0.032	81,428.571	7.981	− 1895.079	− 81,673.275	209.551401	9.444963
0.034	65,142.857	4.064	− 2002.534	− 25,781.999	106.720921	7.532516
0.036	48,857.143	0.046	− 1996.538	31,778.378	1.21178349	5.619038
0.038	32,571.429	− 3.846	− 1877.608	87,151.326	− 100.99797	3.70683
0.04	16,285.714	− 7.394	− 1653.826	136,630.775	− 194.15742	1.798107
0.042	0.000	− 10.402	− 1340.286	176,909.156	− 273.12763	− 0.10515
0.044	0.000	− 12.703	− 947.694	215,682.731	− 333.54348	− 0.12841
0.046	0.000	− 14.151	− 492.040	239,972.154	− 371.55978	− 0.14305
0.048	0.000	− 14.649	− 3.902	248,165.669	− 384.65364	− 0.14809

The positive peak deflection is $y_m = 19.3$ mm at $t = 0.022$ s.
The plastic deformation: $y_p = y_m - y_e = 19.3 - 3.5 = 18.8$ mm.
The rebound elastic deformation $= y_m - y_p = 14.65 - 18.8 = -4.5$ mm.

- Support rotation
- $\theta_d = \arctan\ (y_m/0.5L) = \arctan[(19.3)/0.5(3600)] = 0.71^\circ < 2^\circ$ OK
- The peak reaction is (43.68kN) at $t = 0.012$.

6.13.8 Numerical Integration for Side Wall

Time	F	u	v	a	Resistance	Reaction
(s)	(N)	(mm)	(mm/s)	(mm/s²)	(kN)	(kN)
0	144,000.000	0.000	0.000	92,903.226	0	16.56
0.002	138,240.000	0.181	178.927	86,024.022	4.75837853	15.89943
0.004	132,480.000	0.703	338.342	73,390.878	18.4511113	15.2423
0.006	126,720.000	1.514	467.589	55,855.754	39.7662564	14.58811
0.008	120,960.000	2.547	558.043	34,598.021	66.8827501	13.93615
0.01	115,200.000	3.717	603.686	11,045.137	97.5928869	13.28557

(Continued)

Time	F	u	v	a	Resistance	Reaction
(s)	(N)	(mm)	(mm/s)	(mm/s^2)	(kN)	(kN)
0.012	109,440.000	4.930	601.508	− 13,223.200	129.450567	12.63544
0.014	103,680.000	6.091	551.703	− 36,581.006	159.935357	11.98478
0.016	97,920.000	7.107	457.657	− 57,464.950	186.621361	11.33265
0.018	92,160.000	7.896	325.714	− 74,478.980	207.339581	10.67823
0.02	86,400.000	8.391	164.747	− 86,487.651	220.322915	10.02082
0.022	80,640.000	8.543	− 14.433	− 92,691.895	224.324084	9.359965
0.024	74,880.000	8.329	− 199.807	− 92,682.187	218.698626	8.695399
0.026	69,120.000	7.748	− 378.954	− 86,465.573	203.447439	8.027127
0.028	63,360.000	6.825	− 539.885	− 74,464.765	179.216051	7.355398
0.03	57,600.000	5.608	− 671.839	− 57,489.409	147.25073	6.680692
0.032	51,840.000	4.163	− 766.010	− 36,681.462	109.314405	6.003686
0.034	46,080.000	2.573	− 816.130	− 13,438.374	67.5680632	5.325214
0.036	40,320.000	0.930	− 818.887	10,680.762	24.4256286	4.646204
0.038	34,560.000	− 0.671	− 774.147	34,059.787	− 17.607964	3.967621
0.04	28,800.000	− 2.137	− 684.953	55,133.773	− 56.104618	3.2904
0.042	23,040.000	− 3.385	− 557.326	72,493.769	− 88.875602	2.615383
0.044	17,280.000	− 4.346	− 399.851	84,980.999	− 114.11788	1.943265
0.046	11,520.000	− 4.971	− 223.106	91,764.210	− 130.53449	1.274544

$m = 8.5$ mm at 0.022 s

Support rotation:

$\theta_d = \arctan\,(y_m/0.5L) = \arctan[(8.5\text{ mm})/0.5(3600)] = 0.22° < 2°$ OK

6.13.9 Numerical Integration for Rear Wall

Real wall is not affect the design so it will be neglect which provide us a conservative design

6.13.9.1 Side Wall (In-Plane Loading)

- The side wall is a cantilever which transfers roof diaphragm reactions from the floor slab to foundations; the 5.2 m height is a bit conservative because some of the lateral load is removed at the floor slab.
- Height = 5.2 m, length = 20.2 m.
- Use the reaction from the roof diaphragm analysis.

6.13.9.2 Trial Size

- The same as the front wall.
- For bending moment, assume 12 steel bars at the corner provide the tension component for resisting in plane-building to consider the out-of-plane building will be complicated. Therefore, design for the response to be limited to the elastic range.

- $A_s = 2400\ mm^2$
- $b_w = 250$ mm
- $d = 20 - (0.25) + 0.035 + (1.6/2 \times 1000) = 19.8$ m
- $\rho = A_s/b \cdot d = 3.72/10 \times 792 = 0.0005 < 200/f_{dy}$ OK
- The cracks will be due to out-of-plane bending and by the construction joint at the base of the wall.
- $b_f = 0.6$ m for width of beam flange at the corner.
- $a = A_s\ f_{dy}/0.85\ (f_{dc})(b) = 375.2\ mm < 250$ mm (within the thickness of the wall)
- $M_p = A_s f_{dy}\ (d - a/2) = 25{,}650.37$ m-kN
- $R_b = M_p/L = 504$ t
- Calculate shear resistance from shear friction criteria
- Friction coefficient $\mu = 0.6(\lambda) = 0.6$
- $A_{vf} = 0.31/(6$ in. spacing 0 (800″ depth)(two faces) $= 51{,}668\ mm^2$
- $R_s = V_n = A_{vf} \cdot f_y \cdot \mu = 1633.4$ ton
- Compute SDOF equivalent system
- $R_u = = 1633$ ton
- For moment of inertia to approximate the effect of wall cracking due to out-of-plane loads use half wall thickness.
- Chord height $= 4b + t = 1250$ mm
- Include chord reinforcing, $nA_s = 8.04 \times 3.72 = 29.91$
- $I = \Sigma b\ (h_3)/12 + \Sigma A\ d^2 = 141.84\ m^4$
- Flexural deflection, $f = P(H)3/3\ (E)(I) = 0.002$
- Shear deflection, $S = 1.2\ HP/A \cdot G = 0.51$ mm
- Beam stiffness, $K = P/(f + s) = 7.9$ kN/mm
- Yield deflection, $Y_e = R_u/K = 0.625$ mm
- Tributary front and rear wall height,
  $W_1 = 2.5 \times 0.25 \times (28/2) \times (3.6/2)(2\ \text{walls}) = 31.5$ ton
- Tributary roof weight,
  $W_2 = 2.5\ (0.15)(28/2)(20.2) = 106$ ton
- Shear wall weight
  $W_3 = 2.5 \times 0.25 \times 20.2 \times 5.1/2 = 32.2$ ton
- Include 20% of the front wall and roof with the shear wall mass
- Beam mass, $M_e = [9w30 + 0.2(w2 + w1)]/\text{gravity} = 0.339\ k\text{-}s^2/in$
- Period of vibration, $t_n = 2\pi(M_e/K)0.5 = 0.017$ s
- Use time increment $= t_n/10 = 0.0017$ assume 0.002
- The side wall is an effectively SDOF the support reaction $V = R$

Time	Force	y	v	a	Resistance	Reaction
(s)	(kips)	(in.)	(in./s)	(in./s^2)	(kips)	(kips)
0.000	35.19	0.000	0.00	109.6	0.00	0.00
0.002	44.08	0.000	0.22	107.1	9.71	9.71
0.004	73.78	0.001	0.43	109.3	38.68	38.68
0.006	124.20	0.002	0.66	115.0	87.28	87.28
0.008	195.22	0.004	0.89	120.8	156.42	156.42
0.010	285.95	0.006	1.14	121.4	246.99	246.99
0.012	374.74	0.008	1.31	54.6	357.22	357.22
0.014	463.15	0.011	1.32	− 41.1	476.34	476.34
0.016	549.27	0.013	1.16	− 122.4	588.55	588.55
0.018	621.38	0.015	0.86	− 181.3	679.58	679.58
0.020	669.53	0.017	0.46	− 216.2	738.94	738.94
0.022	686.84	0.017	0.02	− 229.0	760.35	760.35
0.024	670.34	0.017	− 0.44	− 221.6	741.47	741.47
0.026	621.38	0.015	− 0.85	− 193.9	683.62	683.62
0.028	545.42	0.013	− 1.19	− 144.5	591.79	591.79
0.030	451.36	0.011	− 1.41	− 73.1	474.83	474.83
0.032	343.23	0.008	− 1.48	− 4.6	344.72	344.72
0.034	232.50	0.005	− 1.43	58.8	213.63	213.63
0.036	126.94	0.002	− 1.26	107.0	92.59	92.59
0.038	32.68	0.000	− 1.02	133.0	− 10.00	− 10.00
0.040	− 47.71	− 0.002	− 0.76	130.8	− 89.69	− 89.69

The positive peak deflection is $y_m = 0.017$ in. at $t = 0.024$ s.
The peak dynamic reaction = 345 ton.
Ductility ratio: $\mu_d = y_m/y_e = 0.68 < 1.0$ OK
Side wall interaction:

$$[\Delta_d/\Delta_a]_i^2 + [\Delta_d/\Delta_a]_o^2 \left[(0.68/1.0)^2 + (0.22/2)^2\right] = 0.47 < 1.0 \quad \text{OK}$$

6.13.10 Roof Slab (in Plane Loads)

- The roof diaphragm is designed to transfer wall loads to the shear walls.
- The diaphragm is fixed at both ends by a continuous attachment to the walls.
- The center of mass coincides with the center of rigidity to neglect any torsional effect.
- Span $L = 28 - 0.25 = 27.75$ m
- Depth $d = 20.2$ m
- Determine width of composite (wall) flange

- $A = L/12 = 2.31$ m
- $B - 6\ 6\ (b) = 6 \times 250 = 1500$ mm
- $C -$ (3.6 m wall span)/2 = 1.8 m
- Load case A (applied to long side of the building)
- Load = (front wall reaction − rear wall reaction delayed by 0.051 s).

  Slab concrete trial size:
- 5 in. (130 mm) thickness, plus 2 in. (50 mm) steel decking with an average depth of 6 in.
- Roof reinforcement used to resist shear 10 mm@ 200 mm, each face $A_s = 68.75$ mm^2
- Chord reinforcing used to resist bending: 250 mm concrete and 10 bars 25 mm, $A_s = 5100$ mm^2
- $d = 20 - 0.25/2 = 19.875$ m
- $R = A_s/b \cdot d = 0.0002 < 200/f_{dy}$
- The response will be limited to the elastic range even though cracking will probably be caused anyway by out-of-plane bending.
- $A = A_s f_{dy}/0.85 f'_{cd}\ b_f = 63.8$ mm within thickness of the wall
- $M_p = A_s f_{dy}(d - a/2) = 5{,}469{,}469$ cm-kN
- $R_b = 16\ M_p/L = 31{,}266$ kN.

  Compute shear resistance for deep beam criteria
- $V_c = (1/6)\ (f'_c)^{0.5}\ b \cdot d = 2237$ kN
- Because of out-of-plane bending, use only half the slab reinforcing for resisting in-plane shear.
- $A_v = A_{vh} = 69$ mm^2
- $L_n/d = = 1.39$
- $S = 200$ mm

6.13.11 Nominal Shear Capacity

- $V_s = [A_v\ (1 + l_n/d)/(12s) + A_{vh}(11 - l_n/d)/(12s)] f_{dy} \cdot d$
- $= 3670$ kN
- $V_n = V_c + V_s = 5907$ kN
- Critical section of shear is 0.15 L from support
- $R_{s1} = 2(V_n)/0.7 = 1721$ ton

  Compute shear resistance from shear friction criteria
- For shear friction $f_{dy} = 1.1 \times 1.1 \times 414 = 501$ MPa
- Friction coefficient $(\mu) = 1.0$
- $A_{vf} = 15{,}712.5$ mm^2
- $V_n = A_{vf} f_{dy}\ (\mu) = 827$ ton

- $R_{s2} = 2(V_n) = 1654$ ton

  Compute SDOF
- $R_{s2} < R_{s1} < R_b$ shear friction control

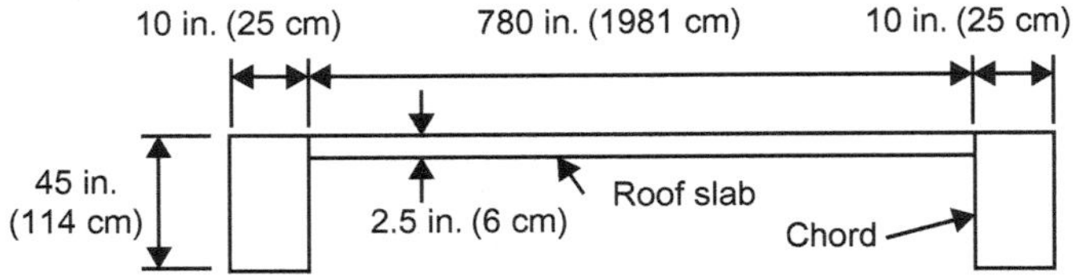

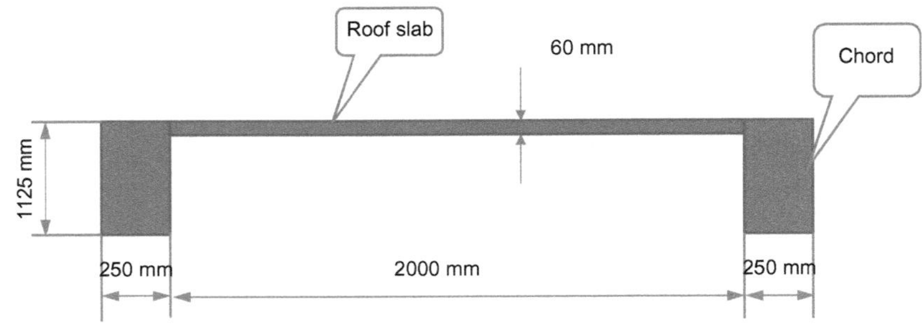

Chord height $= 4b + t = 4(250) + 125 = 1125$ mm

Because the roof diaphragm is a deep and relatively short beam, the stiffness must include shear deformation. Compute the mid-span deflection for a tribritary load of 180 N/mm.

- Use half the roof slab thickness for moment of inertia calculation to approximate the effect of roof cracking due to out-of-plane bending.
- $I = \Sigma b\,(h)^3/12 + \Sigma Ad^2 = 101.22\ \text{m}^4$

  Flexural deflection:
- $F = w\,(L)4/384\ E_c\,I = 0.103$ mm
- Shear modulus $= E_c/2(1 + \nu) = 10{,}356.497$ MPa

  Shear deflection:
- $S = 0.125\ F \cdot w \cdot (L)2/A(G) = 565.81$ kN/mm

  Effective stiffness:
- $K = w \cdot L/(f + S) = 5667.03$ kN/mm
- Yield deflection: $y_e = R_u/K = 2.75$ mm
- Include 20% of the tributary wall height with the beam mass
- Beam mass $= 224\ \text{N} \cdot \text{s/mm}$
- Use elastic value to compute K_{LM}

- $K_{LM} = 0.5/0.64 = 0.78$
- Equivalent mass $M_e = K_{LM}$ (beam mass) $= 0.174$ kN-s^2/mm
- Time period of vibration; $t_n = -0.035$ s
- By using numerical, the reaction column in the table for front wall will be multiplied by 28 m, which is the total length of the building. This reaction will represent the load to the side wall for in-plane.

Time	F	u	v	a	Resistance	Reaction
(s)	(N)	(mm)	(mm/s)	(mm/s^2)	(kN)	(kN)
0	1,101,240.000	0.000	0.000	4513.279	0	126.6426
0.002	1,048,921.513	0.009	8.604	4090.513	49.5565549	120.6451
0.004	996,829.908	0.034	15.989	3295.172	190.429753	114.7088
0.006	944,929.862	0.071	21.485	2200.277	404.866578	108.8228
0.008	893,173.260	0.118	24.591	906.102	668.42666	102.9723
0.01	841,502.431	0.168	25.028	− 469.020	952.220548	97.13938
0.012	789,854.017	0.216	22.760	− 1799.519	1225.55122	91.30505
0.014	738,163.226	0.257	17.996	− 2964.060	1458.71718	85.45038
0.016	686,368.169	0.287	11.176	− 3856.619	1625.72102	79.55824
0.018	634,414.006	0.301	2.923	− 4396.136	1706.63635	73.61467
0.02	582,256.605	0.298	− 6.007	− 4533.882	1689.41744	67.60994
0.022	529,865.494	0.277	− 14.799	− 4257.871	1570.98713	61.53936
0.024	477,225.881	0.240	− 22.651	− 3593.892	1357.50464	55.40362
0.026	424,339.633	0.188	− 28.848	− 2603.113	1063.79004	49.20862
0.028	371,225.132	0.126	− 32.827	− 1376.439	711.958979	42.96499
0.03	317,916.027	0.058	− 34.230	− 26.173	329.39357	36.68716
0.032	264,458.964	− 0.009	− 32.932	1324.269	− 53.764459	30.39208
0.034	210,910.452	− 0.072	− 29.056	2551.643	− 407.36874	24.09786
0.036	157,333.063	− 0.124	− 22.960	3544.130	− 704.01948	17.82225
0.038	103,791.242	− 0.163	− 15.204	4211.526	− 921.55965	11.58119
0.04	50,346.983	− 0.184	− 6.499	4493.453	− 1045.0887	5.387544
0.042	− 2944.316	− 0.189	2.359	4364.818	− 1068.3108	− 0.7499
0.044	− 3595.599	− 0.175	10.773	4049.581	− 993.29568	− 0.79591
0.046	− 4005.414	− 0.146	18.188	3365.439	− 827.87787	− 0.77936

- The positive peak deflection is $y_m = 0.3$ mm at 0.018 s.
- Ductility ration: $\mu_d = y_m/y_e = 0.3/2.75 = 0.11 < 1.0$ OK
- Slab is good until check out of plane.

6.13.12 Roof Slab (Out-of-Plane Bending)

- The roof panels are 5.56 m by 2.5 m at a 2.2:1 ratio, so they will be analyzed as one-way slabs.
- Span: $L = 2.5$ m design width is 1 m.

  Load case A (parallel to 2.5-m span)

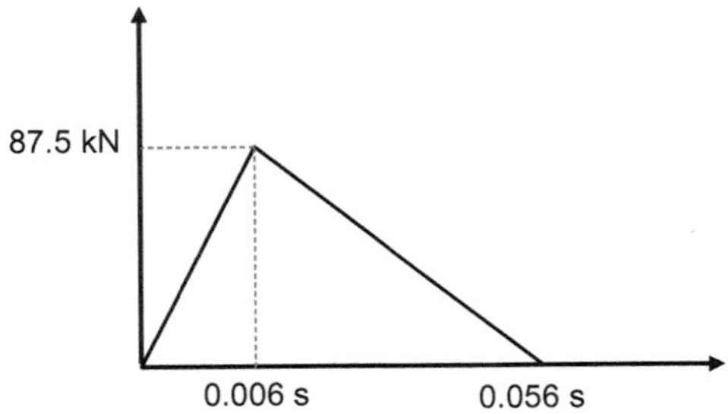

- Peak over pressure: $P_a = 35$ kN/m^2
- Rise time: $t_r = 0.006$ s
- Time of duration: 0.05 s.
- Peak load: $P_o = 35 \times 2.5 \times 1/1000 = 87.5$ kN

  Load case B (perpendicular to the span)

  $L_1 = 1$ m this leads to $C_e = 1.0$ and $t_r = 0.0$ s.

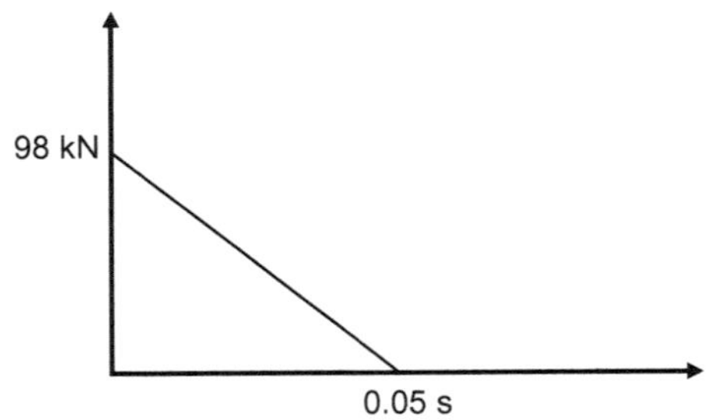

- Peak pressure: $P_a = C_e(P_{so}) + C_d(Q_o) = 39.2$ kN/m^2
- Peak load: $P_o = 39.2 \times 2.5 \times 1 = 98$ kN

  Compute bending strength

- Choose 5″ slab thickness plus metal deck (127 mm 10 mm @ 200 mm spacing each way top and bottom)
- $A_s = 78.5 \times 5/\text{m} = 471$ mm^2
- $d_c = 101.5$ mm
- $d_s = 126.5$ mm
- $\rho_c = A_s/b \cdot d_c = 0.0039$
- $\rho_s = A_s/b \cdot d_s = 0.0031 > 200/f_{dy}$ OK

  For dynamic bending:

- $f_{dy} = 1.1 \times 1.17 \times 414 = 532.8$ MPa
- $f'_{dc} = 1.0 \times 1.19 \times 30 = 35.7$ MPa
- $a = A_s \cdot f_{dy}/0.85 \cdot f'_{dc} \cdot b = 7.5$ mm
- $M_{pc} = A_s \cdot f_{dy}[dc - a/2] = 135$ m · kN

- $M_{ps} = 72.0$ k-in. $= 24.6$ m · kN
- $R_b = 8\ (M_{ps} + M_{pc})/L = 122$ kN

  Compute shear resistance:
- $f'_{dc} = 30$ MPa
- $V_n = (1/6)\ b \cdot d \cdot (f_{dc})^{0.5}$
- $R_s = V_n \cdot L/[0.5L - d] = 17.2$ kips $= 78$ kN

  Compute equivalent SDOF:
- Because $R_b < R_s$ bending control $R_u = R_b = 48.0$ kN
- Static load = 3.56 kN
- Positive $R_u = R_b -$ static load = 4.5 ton
- Negative $R_u = R_b +$ static load = −5.4 ton
- Average ρ = (0.0039 + 0.0031)/2 = 0.0035
- Gross moment of inertia:

  $I_g = b(h)^3/12 = 48.83 \times 10^6$
- Cracked moment of inertia:

  $nA_s = 956.25\ \text{mm}^2$

  $$c = \frac{-nA_s + \sqrt{nA_s(nA_s + 2b_d)}}{b} = 2.29\ \text{cm}$$

  $I_{cr} = bC^3/3 + nA_s(d - C)^2 = 7.03 \times 10^6\ \text{mm}^4$
- Average moment of inertia:

  $I_a = (I_g + I_{cr})/2 = 27.93 \times 10^6 \text{mm}^4$
- Effective bilinear stiffness:

  $K = 307EI/L^3 = 156.6$ kN/cm
- Positive yield deflection:

  $y_e =$ (positive R_u)$/K = = 2.75$ mm
- Negative yield deflection:

  $y_e =$ (negative R_u)$/K = -3.3$ mm
- Beam mass = beam weight/gravity

  $= 0.00028$ kN-s^2/mm
- Because of the expected response, use an average value for K_{LM}.
- Elastic $K_{LM} = 0.41/0.53 = 0.77$
- Plastic $K_{LM} = 0.33/0.5 = 0.66$
- Average $K_{LM} = (0.77 + 0.66)/2 = 0.715$
- Equivalent mass:

  $M_e = K_{LM}$ (beam mass) $= 0.00019$ kN-s^2/mm
- Period of vibration:

  $t_n = 2\pi\ (M_e/k)0.5 = 0.022$ s
- For support reaction:

  $V = 0.37R + 0.13F$

  $T_d/t_n = 0.05/0.0022 = 2.3$

  $R_u/P_o = 12.2$ ton$/9.8 = 1.2$

- Using the chart $\mu_d = 1.25$
- $\gamma_m = \mu_d \cdot \gamma_e = 1.25 \times 2.75 = 3.44$ mm
- $\theta_d = \arctan \gamma_m/(0.5L) = \arctan(3.44)/(0.5 \times 2500) = 0.17° < 2°$ OK
- Roof slab interaction:
  $[\Delta d/\Delta a]_i^2 + [\Delta d/\Delta a]_o^2 = [0.45/1.0]^2 + [0.14/2.0]^2 = 0.21 < 1.0$ OK

6.13.13 Roof Beam

- The roof beam is connected to the roof slab to prevent separation during rebound. The connection will be designed to prevent composite action between roof slab and roof beam because composite action increases bending capacity while not increasing the beam shear capacity; neglecting this effect could be very conservative.
- Beam span: $L = 5.4$ m pinned connection at the two end, tributary slab width, $B = 2.4$ m
  - Load case A (perpendicular to span of beam)
  - Travel time between slabs $= L_1/U = 2.5/400 = 0.006$ s
  - Load = 5.56[(slab reaction + (slab reaction with 0.006 s delay)]
  - Load case B (parallel to span of beam)
  - Travel time between slabs $= L_1/U = 5.56/400 = 0.014$ s
  - Load = (5.56 m) (average of delayed slab reactions to 0.014 s) 8(2 sides)

Beam section properties

- W14 × 38
- $d = 358$ mm flange width/thickness, $b_f/2t_f = 6.6$
- Web thickness:
  $t_w = 7.9$ mm web depth/thickness $h_c/t_w = 39.6$
- Radius of gyration:
  $r_y = 38.8$ mm
- Moment of inertia:
  $I = 541 \times 10^6$ mm^4
- Plastic modulus $Z = 960{,}938$ mm^3

Calculate bending resistance:

- $f_{dy} = (1.1)(1.29)36$ ksi = 51.1 ksi = 352 MPa

Check flange:

- $\lambda_p = 65/(f_{dy})^{0.5} = 9.1 > 6.6$ OK (AISC LRFD)
- Check web:
- $\lambda_p = 640/(Ff_{dy})^{0.5} = 89.5 > 39.6$ OK (AISC LRFD)

Unbraced length for plastic design:

- $L_{pd} = (3600 + 2200\ M_1/M_p)r_y/f_{dy} = 109$ in. = 2725 mm

- $M_p = Z(f_{dy}) = 355.11$ m-kN
- $R_b = 8MP/L = = 528$ kN

  Calculate shear resistance:
- $V_n = 0.6 f_{dy}$ (d) $(t_w) = 608$ kN
- $R_s = 2V_n = 1216$ kN

  Compute SDOF equivalent system:
- $R_b < R_s$ bending control, $R_u = R_b = 528$ kN
- Allowable response $\mu_a = 3.0$, $\theta_a = 2.0$ (low response range)
- Static load = beam load + slab load = 67.2 kN
- Positive $R_u = R_b$-static load = 451 kN
- Negative $R_u = -R_b$-static load = −585 kN
- Effective stiffness:

  $K = 384EI/5L^3 = 914$ kN/cm
- Positive yield deflection:

  $y_e = R_u/k = 29.8$ mm
- Negative yield deflection:

  $y_e = R_u/k = -38.8$ mm
- Include 20% of tributary slab weight
- Beam mass = [beam weight + 0.2(tributary slab weight)]/gravity = 0.00163 kN-s^2/mm
- Because of expected response, use an average value of $K_{1M} = 0.72$.
- Equivalent mass:

  $M_e = K_{LM}$ (beam mass) = 0.006 k-s^2/in.
- Period of vibration:

  $t_n = 2\pi(M_e/K)^{0.5} = 0.056$ s.
- The peak deflection:

  $y_m = 72.1$ mm at 0.042 s.
- $\mu_d = y_m/y_e = 72.1/29.8 = 2.42 < 3.0$ OK
- Support rotation:
- $\theta_d = \arctan(y_m)/0.5L = \arctan$ $72.1/0.5(5400) = 1.53 < 2$ OK
- It needs to provide lateral bracing at midspan.

  Roof girders
- The trial size is W21 × 111.
- Use the same procedure for the beam.
- $M_p = Zf_y = 161.082$ m-kN
- $R_b = 8M_p/L = = 134.7$ ton
- The bending control, so $R_u = R_b = 134.7$ ton
- Static load = girder weight + beam weight + slab weight = (282 kN)
- R_u positive = R_b-static load = 1040 kN

- R_u negative $= -161.5$ kN
- $K = 384EI/5L^3 = 18.68$ kN/mm
- $Y_e = R_u/K = 1040/18.68 = 55.7$ mm
- $M_e = 0.72$ (0.007 kN-s^2/mm) $= 0.029$ kN-s^2/mm
- Period of vibration:
  $t_n = 0.104$ s
- By numerical solution, the peak positive deflection is $y_m = 146.6$ mm at $t = 0.072$ s.
- $\mu_d = y_m/y_e = 146.6/55.7 = 2.63 < 3.0$ OK
- $\theta_d = \arctan(y_m)/0.5L = \arctan 146.6/0.5(9600) = 1.7° < 2°$ OK
- It is recommended to add lateral bracing at 8-ft spacing at beam connections.

  Column design:
- The check for section W10 × 45
- $A = 8312.5$ mm^2, $r_y = 50.25$ mm
- $f_{dy} = 1.1 \times 1.19 \times 240$ ksi $= 321$ MPa

  Calculate compression resistance:
- $\lambda_c = K\,L(f_y/E)^{0.5}/\pi_r = 0.92K < 1.5K$ OK (AISC LRFD)
- $F_{cr} = (0.658)^{(\lambda c)0.5}\,(f_y) = 228$ MPa (AISC LRFD)
- Compression capacity:
- $P_n = A_g f_{cr} = 199.6$ ton

  Calculate tension resistance:
- $P_n = A_g \times f_{dy} = 283.94$ ton

  Compute SDOF equivalent system:
- $R_u = 199.6$ ton (compression)
- $R_u = 283.94$ ton (tension)
- Because the column supports major structural roof components, the allowable response will be limited to elastic range, thus $\mu_a = 1°$.
- Effective stiffness:
  $K = A(E)/L = 469$ kN/mm
- As the column is already a SDOF, so there are no transformation factors.
- Roof slab = 261.3 kN
- Beam weight = 9.5 kN
- Girder weight = 14.5 kN
- Column weight = 1.22 kN
- Total static load = 281 kN
- Positive $R_u = 1710$ kN
- Negative $R_u = -3125$ kN
- Positive yield deflection $= R_u/K = 3.65$ mm

- Negative yield deflection $= R_u/K = 6.5$ mm
- Include 20% of weight of supported components
- Column mass: $M = 0.006$ kN-s^2/mm
- $t_n = 0.022$ s for support reaction $V = R$
- The peak positive deflection:
  $y_m = 3.1$ mm (load case A)
- The peak positive reaction = 1459 kN
- The peak rebound reaction = −107 kN
- Ductility:
  $\mu_c = y_m/y_e = 0.130/0.14 = 0.93 < 1.0$ OK

6.13.14 Foundation

In this example, precast concrete piles will be used with an allowable compressive force of 356 kN and an allowable tension force of 222 kN, both with $F_S = 3$ against ultimate capacity; because battered piles will resist all lateral forces without need for passive soil pressure, S_F of 1.2 may be used.

Permissible compression:
$P_c = (356)(3/1.2) = 890$ kN
Permissible tension:
$P_t = (222)(3/1.2) = 556$ kN
Pile batter will be 3 units horizontal to 12 units vertical.
Resultant axial dimension $= [(3)^2 + (12)^2]^{0.5} = 12.4$
From front-wall analysis:
$V = 143$ kN/m blast direction, −99 kN/m rebound direction
From side-wall in-plane analysis:
$V = -3305$ kN
From rear-wall analysis:
$V = 91.5$ kN/m′ to −33.9 kN/m
For column analysis:
$P = 1459$ kN compression, −107 kN tension.
Total lateral force on building in direction of blast was determined from a combination of front and rear peak reactions and wall peak reactions, neglecting any difference in time phasing.
$V = 2\ (3382) + (28)\ (143\ \text{kN/m} + 33.9\ \text{kN/m}') = 11717$ kN
Total lateral force on building in opposite direction of blast:
$V = 2\ (3305) + (28)(99 + 91.5) = 11{,}944$ kN

- Lateral load on battered piles: For blast on long side of building, there are 48 pairs of battered piles resisting blast loads, in addition to normal vertical piles in the corners and underneath the internal columns.

- Lateral component of battered pile:
  $R_h = 11{,}944/(2 \times 48) = 125$ kN
- Axial component of battered pile:
  $R_1 = 125\ (12.4/3) = 516.7$ kN
- Front wall foundation
- There is a combination between piles for columns and walls
- Static load for wall:
  P_1 = wall weight = pile cap weight + soil weight + slab weight = 239.3 kN
- Static load from steel column:
  P_2 = pier weight + side column roof load
  $= 4 + (281.1)/2 = 145$ kN
- Maximum blast load = 1459/2 = 730 kN
- Minimum blast load = −107/2 = −53.5 kN
- $P_{2max} = (145) + (730) = 874$ kN (down)
- $P_{2min} = (145\text{ kN}) - (53.5\text{ kN}) = 91$ kN (up)
- V_1 + (8 battered piles) (125 kN) = 1000 kN
- Maximum axial pile compression:
  $R_{max} = (R_1) + (12.4/12)[p_1 = p_{2max}]/(\text{piles}) \leq 200$ kips OK
  $= 516 + (12.4/12)[(239.3 + 874)/4] = 805.1$ kN
- Maximum axial pile tension:
  $R_{min} = -R_1 + (12.4/12)[(239.3 - 91)/4\text{ piles}] = 476\text{ kN} < 889.6$ kN

Side wall foundation

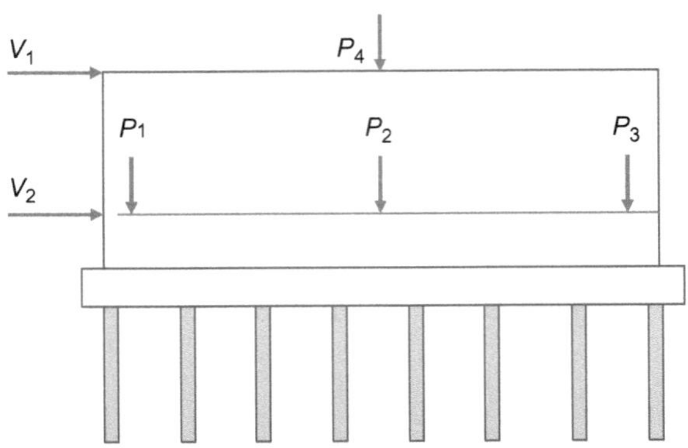

N (Number of piles) = 38

Moment of inertia: $I = \Sigma ad^2$ = (4 piles) $[(1.2)^2 + (2.4)^2 + (3.6)^2 + (4.8)^2 + (6.0)^2 + (7.2)^2 + (8.4)^2] + (8\text{ piles})(9.8)^2 = 1586.7\text{ m}^4$

For corner columns:

Static load = = 70.3 kN

Max. blast load = 1459 × 0.25 = 364.8 kN

Min. blast load = = 47.6 kN

Pier weight = = 4 kN

P_{1max} = 70.3 + 364.8 + 4 = 439.1 kN

P_{1min} = = 70.3 − 26.7 + 4 = 44.6 kN

For side column (use results from P_2, front wall calc):

P_{2max} = (874 kN) (down)

P_{2min} = (91 kN) (up)

Weight of wall and pile cap: P4 = 2126 kN

- From roof diaphragm calculation:

  V_1 = 2981.8 kN

  $V_2 = V_1$ − (4 battered piles) (28.1 kips) = 2482.1 kN

- Total vertical downward load:

  $P = P_{1max} = P_{2max} + P_{1min} + P_4$ = = 3486.96 kN

- Overturning moment at grade:

  $M = V_1$ (roof height) + $[P_{1max} - P_{1min}]$ (column spacing)

  = 2981.8 (3.6) + [(439.1) − (44.6)](9.6) = 14522 kN · m

- Maximum axial compression:

  $R_{max} = R_1 + [(P/N) + M_c/I](12.4/12)$ = 516 + [(3486.96/38 e_a) + (14,522)(10)/1586.7](12.4/12) = 702.82 kN < 900 kN OK

- Total vertical upward load:

  $P = P_{1max} + p_{2min} + p_{1min} + p_4$ = 2522 kN

- Maximum axial tension:

  $R_{min} = R_1[(P/N) + (M)(c)/I](12.4/12)$ = 542.7 kN < 556 kN OK

Column foundation

Static load from column = roof load + weight of pier + pile cap + soil = = 250.88 kN

P_{2max} = 1459.33 kN

P_{2min} = −107.4 kN

Max. pile compression:

$R_{max} = [P_{2max} + p_1]$/piles = [(1459.33) + (250.88)]/4e_a = 427.6 kN < 900 kN.

REFERENCES

ACI 318, 1989. Building Code Requirements for Reinforced Concrete (ACI 318-89) and Commentary (ACI 318R-89), *ACI* Committee 318. American Concrete Institute Detroit, MI.

AISC LRFD, 1993. Load and Resistance Factor Design Specification for Structural Steel Buildings. American Institute of Steel Construction, Chicago, IL.

ASCE, "Design of Blast Resistant Buildings in Petrochemical Facilities", 1997.

ASCE, "Structural Design for Physical Security, State of the Practice", 1999.

ASCE, "Design of Blast Resistant Buildings in Petro-Chemical Facilities", 2010.

ASCE Manual 41, 1971. Plastic Design in Steel: A Guide and Commentary, *ASCE Manual Number 41*, Second ed American Society of Civil Engineers, New York, NY.

Baker, W.E., Cox, P.A., Westine, P.S., Kulesz, J.J., Strehlow, R.A., 1983. Explosion Hazards and Evaluation, Fundamental Studies in Engineering, Vol. 5. Elsevier, Amsterdam.

Biggs, J.M., 1964. Introduction to Structural Dynamics. McGraw Hill, New York.

Newmark, N.M., Hansen, R.J., 1961. Design of blast resistant structures. In: Harris, Crede (Eds.), Shock and Vibration Handbook, Vol.3. McGraw-Hill, New York.

Norris, C.H., et al., 1959. Structural Design for Dynamic Loads. McGraw-Hill, New York.

Roark, 1989. In: Roark, Raymond J., Young, Warren C. (Eds.), Formulas for Stress and Strain, Sixth ed. McGraw-Hill Book Company, New York.

TM 5-1300, 1300. Structures to Resist the Effects of Accidental Explosions, Technical Manual TM 5-1300. Department of the Army, Navy, and Air Force, Washington, DC.

TNO Green Book, 1992. Method for the Determination of Possible Damage to People and Objects Resulting from Releases of Hazardous Materials (CPR 16E). Committee for the Prevention of Disasters Due to Dangerous Substances, The Director-General of Labour, The Hague, The Netherlands.

Unified Facilities Criteria (UFC), 2008. UFC 3-340-02 .Structures to Resist the Effects of Accidental Explosions. , U. S. Army Corps of Engineers, Naval Facilities Engineering Command, Air Force Civil Engineer Support Agency.

CHAPTER 7

Design of a Foundation Under Vibrating Equipment

7.1 INTRODUCTION

The analysis and design of foundations and structures subjected to dynamic loads for vibrating rotating machines is considered a very complex problem because of the interaction of structural engineering, geotechnical engineering, and the theory of vibration.

The structure design of machine foundation depends mainly on the information provided by the machine manufacturer and the geotechnical investigation report from the geotechnical consultant. Before starting the geotechnical investigation, the geotechnical consultant should refer to the plot plan to find the location of the machine's foundation, in order to define any building near to it to find out which required soil investigation data should be included in the report. During the design phase, however, it is necessary to adjust the dimensions of the foundation to meet the design criteria or to avoid interface or clash with other fixed facilities such as the pipelines, other foundations, or the neighboring buildings.

The category of the foundation will be either a shallow foundation or a deep foundation; the most famous type is the pile, which is used in case of weak soil. The other structure configuration of the foundation will be:

1. A block-type foundation, which consists of a thick slab of concrete directly fixed to the machine and other fixed auxiliary equipment. This type is the most traditional in oil and gas process plants and petrochemical plants.
2. An elevated pedestal foundation, consisting of a base-slab and vertical columns supporting a grid of beams at the top in which rest skid mounted machinery; this type of support is most traditionally in electrical power station projects.

7.2 MACHINE REQUIREMENTS

There are two large groups of machines, which are the centrifugal or reciprocating machines.

Onshore Structural Design Calculations.
DOI: http://dx.doi.org/10.1016/B978-0-08-101944-3.00007-3

In order to design the foundation under a machine, it is essential to know the machine technical data that shall be available by the manufacturer.

The required machine data, properties, and parameters include the following items that should be considered as a checklist for the required data:

- Outline drawings of machine assembly
- Function of the machine
- Weight of the machine and the weight of the rotor
- Location of the center of gravity in the space: x, y, and z
- Speed ranges of the machine or free frequency of unbalanced primary and secondary forces
- Magnitude and direction of unbalanced forces both vertically and horizontally, and their points of application
- Limits imposed on the foundation with respect to differential deflection between points on the plan area of foundation
- Foundation requirements from the soil investigation report

The magnitude and direction of the unbalanced forces are often not available from the machine manufacturer. Their justification is that their centrifugal machines are perfectly balanced. For reciprocating machines, the unbalanced forces, which are generally of considerable magnitude, are provided by the machine manufacturer.

The machines should be connected by pipes; therefore, the settlement should have a limit to avoid any damage to the pipes. Limits of differential settlement in case of high pressure gas 50,000 psi piping should be less than 0.0025 mm (0.0001 in.).

7.3 FOUNDATION DESIGN GUIDELINE

The main physical concept in designing the foundation under the vibrating equipment is based on the logic that a lightweight body cannot vibrate a heavyweight body. In other words, no body can vibrate another body heavier than itself. So the main concept is to choose a machine foundation that has a weight higher than the weight of the machine. The design guide here details how you can design the foundation by selecting a foundation weight higher than the machine weight, with reasonable dimensions.

The following guidelines are from Ary et al. (1984), which is very useful to choose the dimensions of the foundation and then check with the analysis, which in most cases will be acceptable.

- A rigid block type foundation resting on soil should have a weight of two to three times the weight of the machine for centrifugal type.
- For a reciprocating machine type, the foundation weight should be from five to seven times the weight of the machine.
- The top of the foundation is usually 300 mm above the finished floor or pavement elevation to avoid damage from surface water runoff in case of rain or a cleaning of the facilities.
- The depth of the foundation is governed by the length of the anchor bolt, plus a minimum of 100 mm. The depth of the foundation should not be less than 600 mm under any condition. By defining the concrete depth you can calculate the foundation weight.
- The foundation should be wide to increase damping in rocking mode. Therefore, the width should be at least 1 to 1.5 times the vertical dimension from the machine centerline to the foundation base.
- After knowing the physical dimension of the machine from the vendor we can define the foundation length and width by adding clearance for about 300 mm from the machine edges to the foundation edge for maintenance purposes.
- The length and width of the foundation should be adjusted to verify that the center of gravity of the machine plus the equipment coincides with the center of gravity of the foundation.

In case of a block foundation resting on piles the following precaution should be considered:

- The pile cap weight should be 1.5 to 2.5 times the machine weight for centrifugal machines and from 2.5 to 4 times for reciprocating machines.
- Arrange the pile to ensure that the centroid of the pile group coincides with the center of gravity of the combined structure and machine load (Table 7.1).

Fig. 7.1 presents a diesel generator package, which is considered to be a type of reciprocating machine.

Most industries use a centrifugal machine in different applications, such as for a compressor, a pump, or as shown in Fig. 7.2, which presents the gas turbine generator package.

For the gas turbine generator, the vibration is usually low compared to applications. Fig. 7.3 presents a photo of a gas turbine with a centrifugal gas compressor package. Fig. 7.4 shows that there are two bolts; one of them is the anchor bolt, and the other bolt will have a steel plate underneath which is fixed during the pouring of the concrete. So when

Table 7.1 Summary of derived expressions for a single-degree-of-freedom system

Expression	Constant force excitation, F_o constant	Rotating mass type excitation, $F_o = m\,e\omega^2$
Magnification factor	$M = \dfrac{1}{\sqrt{(1-r^2)^2 + (2Dr)^2}}$	$M = \dfrac{r^2}{\sqrt{(1-r^2)^2 + (2Dr)^2}}$
Amplitude at frequency, f	$Y = M(F_o/k)$	$Y = M_r\,(m_1 e/m)$
Resonant frequency, f_{mr}	$f_{mr} = f_n\sqrt{1-2D^2}$	$f_{mr} = \dfrac{f_n}{\sqrt{1-2D^2}}$
Amplitude at resonant frequency, f_r	$Y_{mnr} = \dfrac{(F_o/k)}{2D\sqrt{1-D^2}}$	$Y_{mnr} = \dfrac{(m_1 e/m)}{2D\sqrt{1-D^2}}$
Transmissibility factor	$T = \dfrac{\sqrt{1+(2Dr)^2}}{\sqrt{(1-r^2)^2 + (2Dr)^2}}$	$T = \dfrac{r^2\sqrt{1+(2Dr)^2}}{\sqrt{(1-r^2)^2 + (2Dr)^2}}$

Where:
$r = \omega/\omega_n$,
ω_n (undamped natural circular frequency) $= k/m^{0.5}$,
D (damping ratio) $= C/C_c$,
C_c (critical damping) $= 2(km)^{0.5}$,
T_r = force transmitted/F_o,
T_r = force transmitted/$m_1 e\omega_n^2$.

Figure 7.1 Diesel generator.

Figure 7.2 Gas turbine generator.

Figure 7.3 Gas compressor package.

Figure 7.4 Fixing the skid to concrete.

rotate the bolt which is resting on this steel plate and from the moving of these bolts can adjust the machine leveling during installation.

7.3.1 Foundation Dynamic Analysis

Machinery foundations shall be designed based on the equipment manufacturer's recommendations, and published design procedures and criteria for dynamic analysis.

If equipment manufacturer's vibration criteria are not available, the maximum velocity of movement during normal steady-state operation shall be limited to 0.12 in. (3.0 mm) per second for centrifugal machines and to 0.15 in. (3.8 mm) per second for reciprocating machines.

Any support structures or foundations for centrifugal machinery greater than 500 horsepower shall be designed for the expected dynamic forces using dynamic analysis procedures.

For centrifugal machinery less than 500 horsepower, in the absence of a detailed dynamic analysis, the foundation weight shall be designed to be at least three times the total machinery weight, unless specified otherwise by the equipment manufacturer.

In case of using reciprocating machinery less than 200 horsepower, in the absence of a detailed dynamic analysis, the foundation weight shall be designed to be at least five times the total machinery weight, unless specified otherwise by the manufacturer.

7.3.2 Soil Parameter

The allowable soil-bearing or allowable pile capacity for foundations for equipment designed for dynamic loads shall be a maximum of half of the normal allowable for static loads. Noting that, the allowable soil bearing will be provided from the geotechnical consultant office, which will perform a soil investigation test.

The maximum eccentricity between the center of gravity of the combined weight of the foundation and machinery and the bearing surface shall be 5% in each direction.

Structures and foundations that support vibrating equipment shall have a natural frequency that is outside the range of 0.80 to 1.20 times the exciting frequency.

For this type of foundation, there are some soil parameters that are required in case of vibrating equipment.

- The allowable bearing capacity
- The soil density: ρ
- The shear modulus: G
- The shear wave velocity shall be calculated from the following formula:

$$V_s = \sqrt{\frac{G}{\rho}}$$

- Dynamic coefficient of subgrade reaction for vertical vibration (K_z).

  For a rigid rectangular footing of plan area A on a semi-infinite elastic half-space (Barkan, 1962; Richart et al., 1970):

$$K_z = \frac{B_z}{\sqrt{A}}\left(\frac{G}{(1-\nu)}\right)$$

  where:

  V = Poisson's ratio and equal to 0.3 as a typical value for the sand,

  G = shear modulus,

  B_z = the ratio of the length to breadth of footing of about 1.40.

  Barkan suggests that for allowable static bearing capacity of about 0.15 N/mm^2, K_z should not be less than 0.03 N/mm^3.

- Coefficient of subgrade reaction for horizontal vibrations: K_x

$$K_x = 0.5\ K_z$$

- Coefficient of subgrade reaction for rocking vibrations: K_ϕ

$$K_\phi = 2\ K_z$$

- Coefficient of subgrade reaction for torsion vibrations: K_θ

$$K_\theta = 0.75\ K_z$$

Table 7.2 provides the equations to calculate the effective radius of footing in case of horizontal rocking, vertical rocking, and torsional vibration (Table 7.3).

Table 7.2 Embedment coefficient for spring

Mode of vibration	Effective radius of footing, r_o	Embedment coefficient
Horizontal vibration	$\sqrt{BL/\pi}$	$\eta_x = 1 + 0.6(1-\nu)(h/r_o)$
Vertical vibration	$\sqrt{BL/\pi}$	$\eta_z = 1 + 0.55(2-\nu)(h/r_o)$
Rocking	$\sqrt[4]{(BL^3)/3\pi}$	$\eta_\phi = 1 + 1.2(1-\nu)(h/r_o) + 0.2(2-\nu)(h/r_o)^3$
Torsional	$\sqrt[4]{\dfrac{BL(B^2+L^2)}{6\pi}}$	–

B and *L* are the width and length of the foundation, respectively. *h* is the foundation depth embedment below the grade.

Table 7.3 Effect of depth of embedment on damping ratio

Mode of vibration	Damping ratio embedment factor
Vertical	$\alpha_\varepsilon = \dfrac{1 + 1.9(1-\nu)(h/\gamma_0)}{\sqrt{\eta_\varepsilon}}$
Horizontal	$\alpha_\gamma = \dfrac{1 + 1.9(2-\nu)(h/\gamma_0)}{\sqrt{\eta_\gamma}}$
Rocking	$\alpha_\phi = \dfrac{1 + 0.7(1-\nu)(h/r_0) + 0.6(2-\nu)(h/r_0)^3}{\sqrt{\eta_\phi}}$

Source: Ary et al., 1984.

$$\eta_s = 1 + 0.6(1-\nu)\left(\frac{h}{r_0}\right)$$

$$\eta_x = 1 + 0.55(2-\nu)\left(\frac{h}{r_0}\right)$$

$$\eta_\phi = 1 + 1.2(1-\nu)\left(\frac{h}{r_0}\right) + 0.2(2-\nu)\left(\frac{h}{r_0}\right)^3$$

Table 7.4 provides a summary of the calculation of the mass ratio, the damping ratio, the spring constant, and the natural frequency for different modes of vibration based on the foundation dimensions and the geotechnical data from the soil investigation tests.

7.3.3 Modes of Motion

There are different modes of motion for the machine which is presented in Fig. 7.5.

There will be a combination between two modes of oscillation presented in Fig. 7.6.

Table 7.4 Main parameters

Vibration mode	Mass ratio (*B*)	Damping parameters	Damping ratio	Spring constant (*K*)	Natural frequency, f_n
Vertical	$B_z = \frac{(1-\nu)m}{4\rho r_o^3}$	$c_z = \frac{3.4 r_o \sqrt{\rho G}}{1-\nu}$	$D_z = \frac{0.425}{\sqrt{B_z}}$	$k_z = \frac{4Gr_o}{1-\nu}$	$f_{nz} = \frac{1}{2\pi}\sqrt{\frac{k_z}{m}}$
Horizontal	$B_x = \frac{(7-8\nu)m}{32(1-\nu)\rho r_o^3}$	$c_x = \frac{18.4(1-\nu)\sqrt{\rho G}}{7-8\nu}$	$D_x = \frac{0.288}{\sqrt{B_z}}$	$k_x = \frac{32(1-\nu)Gr_o}{7-8\upsilon}$	$f_{nx} = \frac{1}{2\pi}\sqrt{\frac{k_x}{m}}$
Rocking	$B_\psi = \frac{3(1-\nu)m}{8\rho r_o^3}$	$c_\psi = \frac{0.8 r_o^4 \sqrt{\rho G}}{(1-\nu)(1+B_\psi)}$	$D_\psi = \frac{0.15}{(1+B_\psi)\sqrt{B_\psi}}$	$k_\psi = \frac{8Gr_o^3}{3(1-\nu)}$	$f_{n\psi} = \frac{1}{2\pi}\sqrt{\frac{k_\psi}{I_\psi}}$
Torsional	$B_\theta = \frac{I_\theta}{\rho r_o^5}$	$c_\theta = \frac{4\sqrt{B_\theta \rho G}}{1+2B_\theta}$	$D_\theta = \frac{0.50}{1+2B_\theta}$	$k_\theta = \frac{16}{3} Gr_o^3$	$f_{n\theta} = \frac{1}{2\pi}\sqrt{\frac{k_\theta}{I_\theta}}$

7.3.4 Mass Moment of Inertia

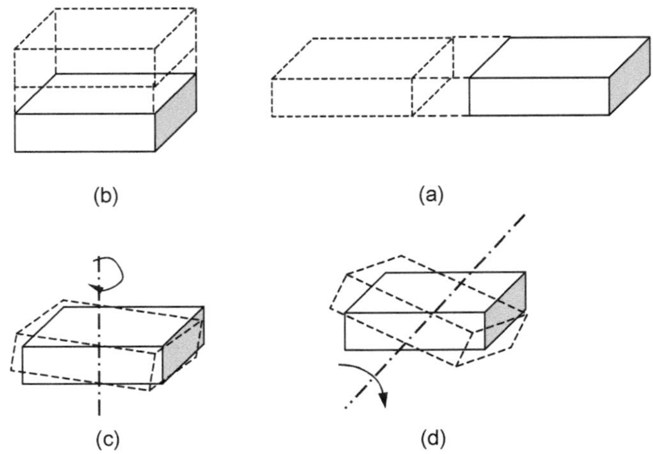

Figure 7.5 Modes of excitation: (a) horizontal excitation, (b) vertical excitation, (c) torsional excitation, and (d) rocking excitation.

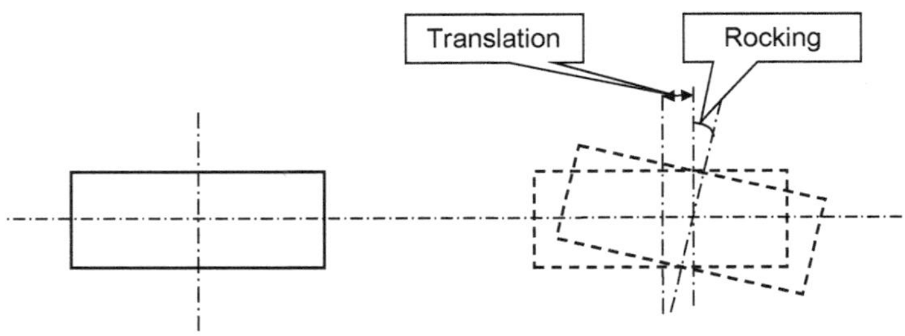

Figure 7.6 Combination mode of oscillation.

The mass moment of inertia for a prism can be calculated from the following equation as shown in Fig. 7.7:

$$I_x = \frac{1}{12} m(a^2 + b^2)$$

$$I_y = \frac{1}{12} m(a^2 + L^2)$$

$$I_z = \frac{1}{12} m(b^2 + L^2)$$

$$I_{z'} = I_z + mL^2/4$$

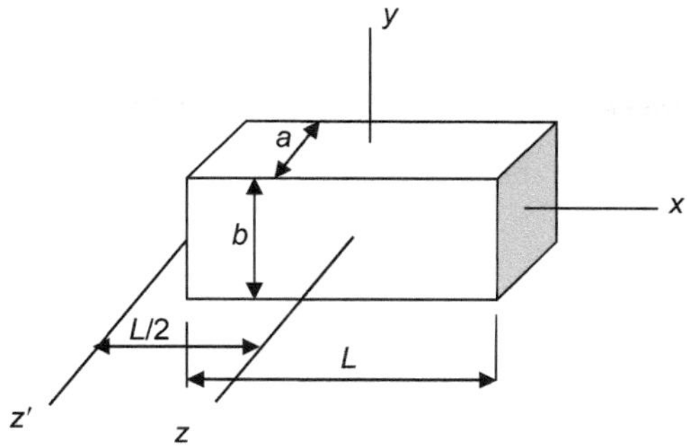

Figure 7.7 Moment of inertia calculation in three directions.

Table 7.5 Design eccentricity for centrifugal machines

Operating speed (rpm)	Eccentricity (mm)
750	0.356–0.812
1500	0.203
3000	0.051

Example 1: Centrifugal machine

The machine data will be as follows:

1. *Machine data*

Compressor:

Weight (W_c) = 35,270 Ibs = 16 ton

Rotor weight (W_r) = 2100 Ibs = 0.952 ton

Operating speed (ω) = 6949 rpm

(f) = 727.7 rad/s

Critical speed (ω_c) = 3400 rpm and at 9000 rpm

Eccentricity of unbalanced mass: (e') = 0.0015 in (0.0381 mm).

This value is provided by the manufacturer for the static condition.

The value of eccentricity due to an imbalance of mass is usually not available from the manufacturer. There were, however, some studies performed in 1962 which presented some values that can assist us in predicting this value as shown in Table 7.5.

The following is the equation delivered by API for centrifugal compressors:

$e(\text{mil}) = \alpha\ (12{,}000/\omega)0.5 < 1.0$ mil,

where $\alpha = 0.5$ at installation time

$= 1.0$ after several years of operation

1 mil = 0.001 in. = 0.0254 mm
The dynamic eccentricity $(e) = e'/(1-(\omega/\omega_c)^2) = 0.000472$ in. = 0.0118 mm
Centrifugal force $F_o = (W_r/g)ef^2 = 0.616$ ton
Turbine data:
Weight $(W_t) = 7.26$ ton
Rotor weight $(W_r) = 0.247$ ton
Operating speed $(\omega) = 6949$ rpm
Critical speed ω_c at 2000 and 9000 rpm
Dynamic eccentricity at operating speed:
$e = 0.0381/[1-(6949/2000)^2] = 0.00343$ mm
Centrifugal force $F_o = 0.0458$ ton
Total centrifugal force = 0.616 + 0.0458 = 0.662 ton
Skid weight $(W_s) = 2.27$ ton
Total machine weight $= W_c + W_t + W_s = 16 + 7.26 + 2.27 = 25.53$ ton.

2. *Soil data*
Soil density $(\gamma) = 20\ \text{kN/m}^3$
Shear modulus $(G) = 44.82\ \text{N/mm}^2$
Poisson's ratio $(\nu) = 0.45$
Soil internal damping ratio $(D) = 0.05$
Static allowable bearing capacity $(q) = 72\ \text{kN/m}^2$

Solution:

3. *Selection of foundation criteria*
The foundation trial data are as shown in Fig. 7.8.

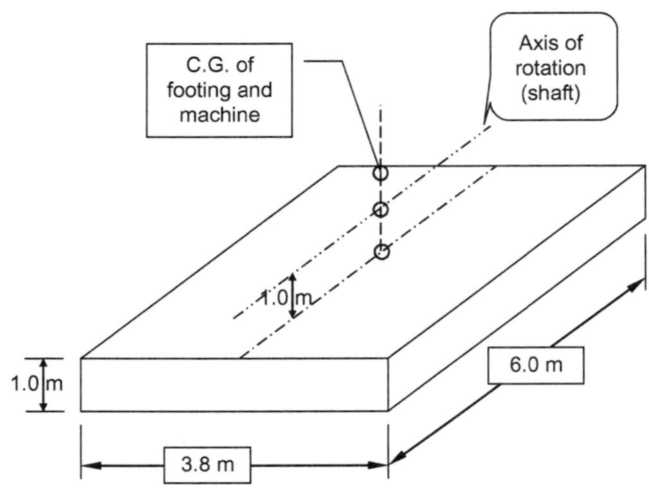

Figure 7.8 Centrifugal machine foundation dimensions.

Weight of the footing (W_f) = 45.58 ton

Total static weight = machine weight + footing weight = 25.53 + 45.58 = 71.11 ton

Actual soil pressure = 71.11/3.8(6.0) = 3.11 t/m^2

$< 0.5\ q_{\text{all}}$ OK

$W_f/W_m = 1.78$ which is close to 2 OK

4. *Dynamic analysis*

   The axis of rotation of the shaft is located 6 ft (1.8 m) above the foundation bottom.

   The dynamic force will excite the structure in three different modes: vertical, horizontal, and rocking.

5. *Static condition check*
   - **a.** Static bearing capacity is 50% of the allowable bearing capacity.
   - **b.** Static settlement must be uniform therefore the Center of Gravity (C.G.) of the footing and machine is coincide.
   - **c.** Bearing capacity for static plus dynamic loads will be less than 0.75 of the allowable bearing capacity.
   - **d.** The magnification factor should be less than 1.5.

   Table 7.6 has the dynamic analysis parameter for three modes of failure.

7.3.5 Limiting Dynamic Condition

The area zones in Fig. 7.9 are classified as follows:

- Zone (E) is dangerous, so it needs to be shut down immediately.
- Zone (D) is near failure, and it needs 2 days maximum to perform the correction before failure.
- Zone (C) is faulty, and it needs a correction within 10 days to save a lot of money.
- Zone (B) is a minor fault, and a correction would be a waste of money.
- Zone (A) has no fault; it is like new equipment.

1. By knowing the vibration amplitude and operating frequency and from Fig. 7.9 presents the allowable safe limit for the amplitude and the operating frequency (rpm) to define the location of the graph if it is safe within the limits or not. In our case the vertical vibrating amplitude = 0.000019 in at 6949 rpm. From the figure it is locate in safe limit.
2. The velocity = $2\pi f$ (CPS) × displacement amplitude as calculated above. Compare with the limiting value in Table 7.7 so the velocity is equal to 0.0138 in./s (0.345 mm/s) and this velocity falls in smooth operation.

Table 7.6 Dynamic analysis

Item	Parameters	Source	Vertical	Horizontal	Rocking oscillation
1	r_o	Table 7.2	$r_o = 2.71$ m	$r_o = 2.71$ m	$r_o = 2.44$ m
2	Mass	$M = W/g$	72.5 kN · s^2/m	72.5 kN · s^2/m	$I_\psi = 105{,}128$ Ib · s^2-ft
3	Mass ratio	Table 7.3	0.24	0.35	0.166
4	Geometric damping ratio	Table 7.3	0.868	0.494	0.361
5	Spring coefficient	Table 7.3	2.15	0.95	0.46
6	Equivalent spring constant	Table 7.3	8.4 kN/mm	595 kN/mm	3.32×10^6 kN · m/rad
7	ω_n	Table 7.3	1040.9 rpm	873.9 rpm	1449.8 rpm
8	Resonance frequency, f_{mr}	Table 7.1	Not possible	1221.4 rpm	1686.1
9	Magnification factor, M	Table 7.1	0.022	0.016	0.045
10	Dynamic force		6.62 kN	6.62 kN	13 m · kN
11	Vibration amplitude	Table 7.1	1.42×10^{-5} mm	1.45×10^{-5} mm	0.1611×10^{-6} rad
12	Component of rocking oscillation		2.558×10^{-5} mm	$I_\psi \cdot R_v = 0.1611 \times 10^{-6} \times 1.8$ 2.47×10^{-5} mm	
13	Resultant vibration amplitude		(0.000019 in.) 0.000483 mm	(0.000018 in.) (0.000457 mm)	
14	Transmissibility factor, T_r		0.256	0.127	0.162

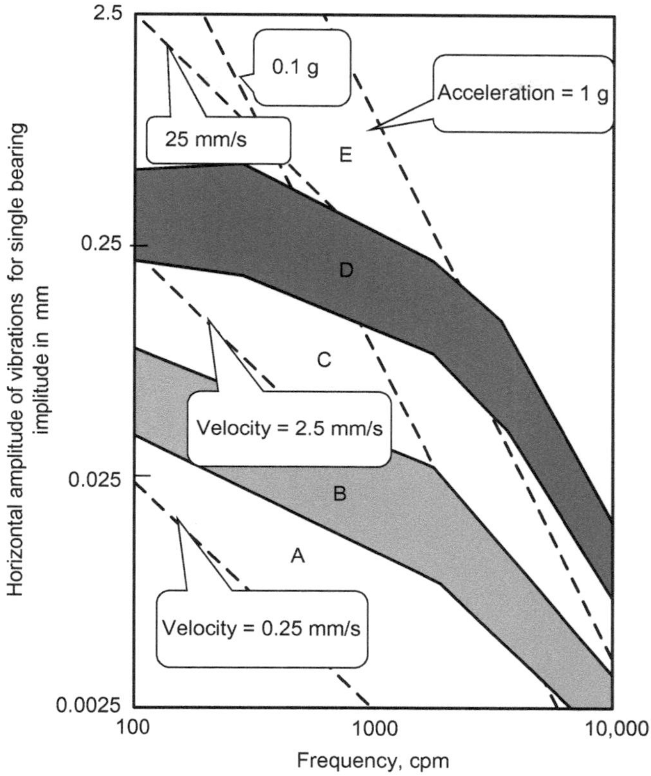

Figure 7.9 Dynamic limits condition.

Table 7.7 General machinery vibration severity data

Horizontal peak velocity, in./s (mm/s)	Machine operation
<0.005 (<0.127)	Extremely smooth
0.005–0.010 (0.127–0.245)	Very smooth
0.010–0.020 (0.254–0.508)	Smooth
0.020–0.040 (0.508–1.016)	Very good
0.04–0.080 (1.016–2.032)	Good
0.080–0.160 (2.032–4.064)	Fair
0.160–0.315 (4.046–8.00)	Slightly rough
0.315–0.630 (8.00–16.00)	Rough
>0.630 (> 16.00)	Very rough

Source: After Baxter, R.L., Bernhard, D.L., 1967. Vibration Tolerances for industry. ASME paper 67-PEM-14, plant engineering and maintenance conference, Detroit, Michigan.

3. The magnification factor should be less than 1.5, which is applicable in this example.
4. The machine frequency should have at least a difference of ±20% of the resonance frequency.

$$0.8\omega_{mr} \geq \omega \geq 1.2\omega_{mr}$$

$$\omega/\omega_{mr} = 6949/1221.4 = 5.69 > 1.2 \quad \text{OK}$$

A transmissibility factor is usually applied only to high-frequency spring mounted machine. The transmissibility factor should normally be less than 1 for spring mounted machines having an inertia block. In the example, T_r is less than 1, indicating that dynamic forces are not amplified.

Example 2: Reciprocating machine

The following data are for the machine, and the auxiliary equipment should be supplied by the machine vendor.

1. *Machine data*

   Compressor = 12.75 ton
   Gas coolers = 1.972 ton
   Scrubbers = 3.179 ton
   Motor = 8.164 ton
   Rotor weight = 2.721 ton (included in compressor weight)
   Total machine load = 26.07 ton

2. *Dynamic forces*
   a. Compressor speed primary (operating): 585 rpm, secondary = 1170 rpm
   b. Vertical primary force: $F_z = 0.603$ ton
   c. Vertical secondary force: $f_z = 0.251$ ton
   d. Horizontal primary force: $F_x = 0.329$ ton
   e. Rocking primary moment: $T_\psi = 1.563$ mt
   f. Pitching primary moment: $T_\phi = 4.70$ mt
   g. Pitching secondary moment: $T_\phi = 1.71$ mt

3. *Soil and foundation parameters*

   Soil density: 18.7 kN/m^3
   Shear modulus: $G = 96.53$ N/mm^2
   Poisson's ratio: $\nu = 0.35$
   Soil internal damping ratio: $D = 0.05$
   Static allowable bearing capacity: $q_{all} = 120$ kN/m^2
   Permanent settlement of soil: 5 mm at 120 kN/m^2

Solution:

4. *Trial sizing for the foundation*

The footing trial size based on the previous trial as a guide, as illustrated in Section 7.3 will be as shown in Fig. 7.10.

Weight of the footing, $W_F = 147.35$ ton

Total static load = machine weight + footing weight = 173.41 ton

Footing weight/machine weight = 5.65 > 5 OK

Actual soil pressure = 173.41/8.5(4.8) = 4.3 t/m^2

Less than $0.5q_{all}$ OK

This trial dimension is OK.

1. Mass and mass moment of inertia

$M = \frac{W}{g}$

$$I_{\psi+\varphi} = \sum_{1}^{n} \left[\frac{m_t}{12}(a_1^2 + b_1^2) + m_1 k_1^2\right]$$

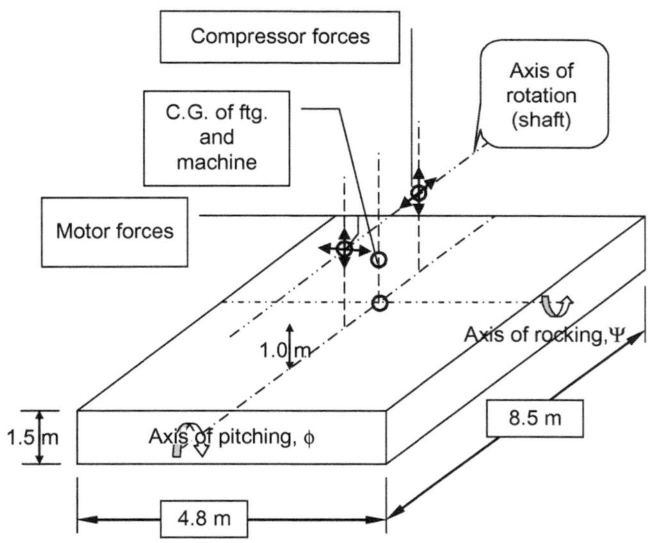

Figure 7.10 Sketch for reciprocating machine.

Vertical excitation (*z*-direction)	Horizontal direction (*x*-direction)	Rocking excitation (ψ-direction)	Pitching excitation (ϕ-direction)
$m = 173.41$ kN·s^2/m	$m = 173.41$ kN·s^2/m	I_ψ (machine) = 1399.51 I_ψ (footing) = 10,513.4 Total = 11,912.91 kN·s^2/m	I_ϕ (machine) = 1.4 I_ϕ (footing) = 4.27 Total = 5.67 kN·s^2/m

2. Spring constant

These data will be calculated from the following table.

In the rock direction: $L = 8.5$ m and $B = 4.8$ m.

In the pitching direction: $L = 4.8$ m and $B = 8.5$ m.

Parameter	Vertical excitation (z-direction)	Horizontal direction (x-direction)	Rocking excitation (ψ-direction)	Pitching excitation (ϕ-direction)
Equivalent radius (r_o)	$r_o = 3.58$ m	$r_o = 3.58$ m	$r_o = 4.16$ m	3.15 m
Embedment factor (η)	1.1	1.232	1.175	1.235
Spring constant coefficient (B)	2.15	0.95	0.58	0.45
Equivalent spring constant (k)	227,167 t/m'	197,302 t/m'	3481 mt/rad	1626 mt/rad

3. Damping ratio

Parameter	Vertical excitation (z-direction)	Horizontal direction (x-direction)	Rocking excitation (ψ-direction)	Pitching excitation (ϕ-direction)
Embedment factor (α)	1.254	1.623	1.024	1.041
Mass ratio (B)	0.328	0.408	0.115	0.222
Effective damping coefficient			1.6	1.58
Geometrical damping ratio (D)	0.931	0.732	0.302	0.195
Internal damping (D)	0.05	0.05	0.05	0.05
Total damping	0.981	0.782	0.352	0.245

4. Natural frequency

$$f_n = \frac{60}{2\pi}\sqrt{\frac{k}{m}}$$

5. Machine parameter

$F_{zo}(P) = 0.603$ ton

$F_{zo}(S) = 0.25$ ton

$F_{xo}(p) = 0.329$ ton

$T_{\psi o}(P) = 1.563 + 0.329 \times 2.44 = 2.366$ mt

$T_{\phi o}(P) = 4.7$ mt

$T_{\phi o}(S) = 1.71$ mt

Displacement response:

The displacement response will be calculated from the following equation:

In vertical direction, z: $Z = \Sigma\, M_z F_{zo}/K_z$

In horizontal direction, x: $X = \Sigma\, M_x F_{xo}/K_x$

In rocking excitation: $\Psi = \Sigma\, M_\psi F_{\psi o}/K_\psi$

In pitching excitation: $\phi = \Sigma\, M_\phi F_{\phi o}/K_\phi$

Total displacement in z-direction $= Z + \psi\, L/2 + \phi B/2$.

Parameter	**Vertical excitation (z-direction)**	**Horizontal direction (x-direction)**	**Rocking excitation (ψ-direction)**	**Pitching excitation (ϕ-direction)**
Natural frequency	1082.8	1009.1	1677.6	1611.2
Resonance frequency	$2D^2 > 1$	$2D^2 > 1$	$f_{m\psi} = 1934.3$	$f_{m\phi} = 1717.6$
Vibrating force	$F_{zo}(P) = 0.603$ ton $F_{zo}(S) = 0.25$ ton	$F_{xo}(p) = 0.329$ ton	$T_{\psi o}(P) = 2.366$ mt	$T_{\phi o}(P) = 4.7$ mt $T_{\phi o}(S) = 1.71$ mt
Magnification factor	M_z max not possible	M_x max not possible	$M_y(P) = 1.111$@ 585RPM $M_y(S) = 1.454$@ 1170 RPM	$M_f(P) = 1.128$@ 585RPM $M_f(S) = 1.690$@ 1170 RPM
Displacement response	0.10233×10^{-2} in. 0.0026 mm	0.058308×10^{-2} in. 0.0015 mm	0.8128×10^{-6} rad	5.0369×10^{-6} rad
Total displacement	0.717×10^{-3} in. *0.018 mm	0.4432×10^{-3} in. 0.011 mm		

$$^*Z_t = Z + \psi R_\psi + \phi R_\phi$$

$$Z_t = 0.0026 + 0.828 \times 10^{-6} \times 4200 + 5.6369 \times 10^{-6} \times 2400$$

Parameter	**Vertical excitation (z-direction)**	**Horizontal direction (x-direction)**	**Rocking excitation (ψ-direction)**	**Pitching excitation (ϕ-direction)**
Transmissibility factor	$T_r(P) = 1.142$ $T_r(S) = 1.102$	$T_r(P) = 1.201$	$T_r(P) = 1.144$ $T_r(S) = 1.620$	$T_r(P) = 1.146$ $T_r(S) = 1.793$
Transmitted force	$P_v = 1.142 \times 0.603$ $+ 1.102 \times 0.25$ $= 0.96$ ton	$P_h = 1.201 \times 0.329$ $= 0.395$ ton	$PT_\psi = 1.144 \times 2.366$ $+ 1.620 \times 0.2095$ $= 3.05$ mt	$PT_\phi = 1.146 \times 4.7$ $+ 1.793 \times 1.7$ $= 8.43$ mt

Transmittal dynamic vertical force = 0.96 ton

Transmittal moment $T_\psi = 3.05 + 0.395 \times 2.44 + 0.96 \times 0.84 = 4.82$ mt

$T_\phi = 8.43$ mt

$q_{dyn} = 0.96/(8.5 \times 4.8) + 4.82 \times (6)/(4.8)(8.5)^2 + 8.43 \times (6)/(8.5) \times (4.8)^2 = 0.37\ t/m^2$

Total static plus dynamic bearing capacity $= 4.3 + 0.37 = 4.67\ t/m^2 = 0.467 < 0.75\ q_{all}$ OK

Limiting the dynamic condition:

1. The maximum vibrations are 0.000717 in. (0.0182 mm) and 0.000443 in. (0.01125 mm) in the vertical and horizontal dimensions, respectively.

   Fig. 7.11 shows that it falls in the safe zone at the operating frequency of 585 rpm.
2. The velocity is equal to $2\pi f/(\text{cps}) \times \text{displacement}$ as calculated above, then compared to the limiting values in Table 7.7.
3. Acceleration is calculated from the following equation: $4\pi^2 f^2/(\text{cps}) \times \text{displacement}$.

   This check will consider if the above checks are not satisfied and compare acceleration with the same figure.
4. The magnification factor should be less than 1.5, which is accepted in this example.
5. The acting frequency of the machine should have at least a difference of ±20 with the resonance frequency. For horizontal and vertical direction there is no resonance.

   For rocking mode: $0.8f_m = 0.8 \times 1934.4 = 1547.4$ rpm
   $1.2f_m = 1.2 \times 1934.4 = 2321.2$ rpm

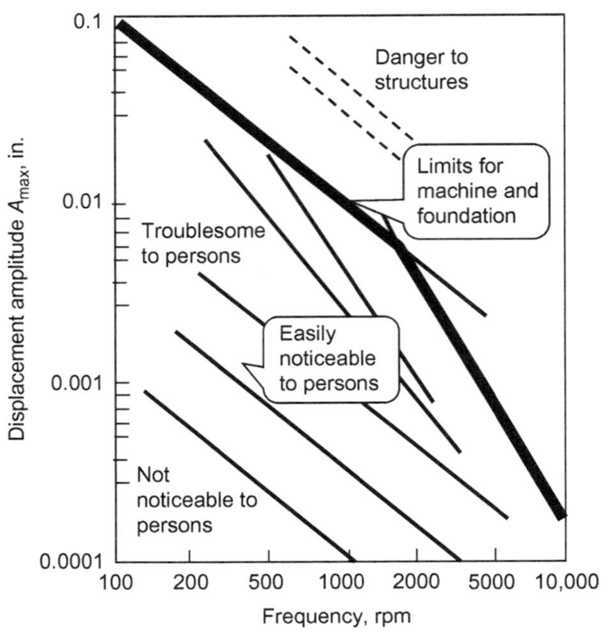

Figure 7.11 Frequency limits.

As the primary and secondary are 585 and 1170 rpm, it will be safe, and no resonance will occur.

For pitching mode: $0.8f_m = 0.8 \times 1717.6 = 1374.1$ rpm

$1.2f_m = 1.2 \times 1717.6 = 2061.1$ rpm

As the primary and secondary are 585 and 1170 rpm, it will be safe, and no resonance will occur.

6. The transmissibility factor is usually considered for high frequency machines mounted on springs. The transmissibility factor should be less than 1.

Example 3: Design of Steel Structure Support Under Machine Skid

Check steel frame support gas generator package with the following data as shown in Fig. 7.12.

Dry weight = 14.0 ton

Wet weight = 15.5 ton

Skid dimension: 5.8 m × 1.0 m

Machine has 1000 rpm

Beam section: W10 × 22 with steel ASTM A36 with F_y equal to 240 MPa

Solution:

Figure 7.12 presents the structure system with the applied load.

Maximum beam design moment = 4.07 mt

Maximum beam stress = $4.07/3.8 \times 10^{-4} = 10705.5$ t/m^2 = 105 MPa

Allowable beam stress = $0.66 \times 240 = 158$ MPa

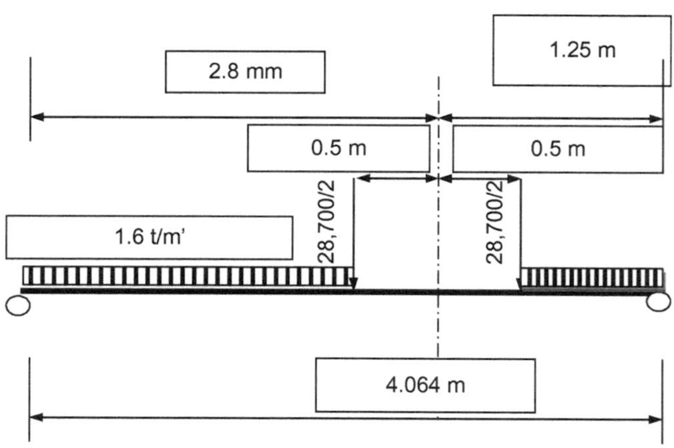

Figure 7.12 Structure system and loads.

Unity check ratio (UC) = 105/158 = 0.66 < 1.0 OK

A local check has been performed to ensure no resonance will occur due to seating the new gas generator package over the deck.

$$\Delta = \frac{Pa^2b^2}{3EIL}$$

Beam stiffness (K) = $P/\Delta = 3EIL/a^2b^2$
$= 3 \times 20 \times 10^7 \times 0.00184 \times 4.0/(2.8)^2$
$(1.25)^2$
$= 1{,}261{,}710$ kN/m

Mass (M) = $W/g = 15.5$ kN · s^2/m

$$T_n = 2\pi\sqrt{\frac{M}{K}} = 0.103 \text{ s} = 0.26 \text{ s}$$

$f_n = 1/T_n = 3.84$ Hz
$\omega_n = 2\pi f_n = 27.32$ rad/s = 1640 rpm
$\omega_n = 2\pi f_n = 24.13$ rad/s = 230.4 rpm
The new generator frequency: $\omega = 900$ rpm
The frequencies ratio is:
$\omega/\omega_n = 1000/230.4 = 4.34 > 1.2$ OK

7.4 VIBRATION ISOLATION

Vibrating, rotating, reciprocating, and impacting equipment creates machine-induced vibration and/or shock, which is transmitted into their support systems. Rotating machines and equipment that are not properly balanced produce centrifugal forces, creating steady state and random vibration.

Machines generating pulses or impacts, such as forging presses, injection molding, impact testers, hammers, centrifugal pumps, and compressors are the most predominate sources of vibration and shock.

The inertia block or pad of isolation is an important factor in isolating engine vibration from the surrounding structure. However, the proper design of the foundation is not enough. There are several additional techniques that can be used to isolate the vibration.

7.4.1 Isolating Liners

A liner can be fabricated and used to line the pit into which the concrete inertia block is poured, as shown in Fig. 7.13. A number of suitable liners are available commercially.

Figure 7.13 Liner isolation before concrete pouring.

It is important to consult the liner manufacturer for specific information. The principle for all liners is the same; line the bottom and sides of the pit and pour the concrete inertia block inside the isolator lining. The engine and/or common mounting skid will still vibrate, but the vibration is dampened and largely confined within the liner.

Be sure to construct the liner so that no liquid concrete can flow into gaps between the liner slabs. If concrete seeps between the inertia block and the pit, the vibration absorption value of the liner will be greatly reduced.

Other materials, such as sand or gravel, may be used as isolating mediums. One foot of well-tamped, settled gravel under the inertia block will be satisfactory.

Do not bridge the gap between the inertia block and the surrounding floor with concrete or a similar solid material. If for reasons of neatness or appearance it is necessary to close this gap, use an expansion joint or a similar resilient material.

7.4.2 Spring and Rubber Mounts

Spring and rubber mounts of various sizes and resiliencies are available for installation purposes. These mounts can be positioned between the common skid and the inertia block or pad or between the inertia block and bottom of the pit. As with the isolating liners, we recommend contacting the manufacturer of the mounts for specific instructions.

Isolation systems have natural frequencies as low as 6.0 Hz and can be designed to exhibit low or high internal damping. These solutions are used under large concrete foundations supporting heavy machinery, buildings, measuring machines, roller mills, and similar equipment. Isolators are available in two types to suit a wide range of design options when considering the formwork required to provide proper support during the foundation construction process. Fig. 7.14 presents a type of isolation made of a molded elastomer, which has been compounded and designed to offer low-frequency isolation and high-load capacity.

For units installed in basements or on ground floors (floors with no other floors beneath), neoprene waffle-type pads which provide 50% vibration reduction or sandwich type pads of rubber and cork which provide 75% vibration reduction can be used.

Where engine-generator sets are to be installed above the ground floor, the more critical type of isolators should be used. For units up to 200 kW, the type of isolator made of rubber bonded to metal can be used and will provide about 90% isolation. Larger units should use spring type vibration isolators that provide about 95% isolation. All percentages are approximate, and exact information for your particular application should be discussed with the machine manufacturer to ensure that the right type of isolator is selected.

Figure 7.14 Isolations from mould elastomer under foundation.

7.4.3 Inertia Block Bolt or Pad Mounting Bolt Installation

The bolt diameter will be determined by the hole diameter in the engine mounting base or common skid frame. The bolts should be long enough to provide a minimum embedded length of 30 times the bolt diameter plus 3–4 in. (76–102 mm) for a hook. (The bolt should have a "J" or "L" shaped hook on the nonthreaded end to increase its holding power.) Approximately 7 in. (178 mm) needs to protrude above the top surface of the inertia block or pad. These 7 in. (178 mm) will provide the length needed for:

- The grout (if used): 2 in. (51 mm)
- Sole plate (if used): 3/4 in. (19 mm)
- Chock: 1/2 in. (13 mm)
- Shims and engine base: 1-3/4 in. (44.5 mm)
- Washer, nut, and small variations in levelness: 7/8 in. (22 mm)

For a common skid-mounted engine, only 5-1/2 in. (140 mm) of bolt need to protrude above the inertia block or pad surface.

Bolt placement in the inertia block or pad can be determined by making a template from 1 × 6 in. (25 × 152 mm) boards. Suspend the template over the inertia block or pad and hang bolts and sleeves through the template holes. A total of 7 in. (178 mm) of bolt must extend from the top surface of the inertia block or pad.

A sleeve of convoluted plastic tubing 2–3 in. (51–76 mm) in diameter should be placed around the bolts before they are embedded in the concrete.

This will allow the bolts to bend and conform to the dimensions of the sole plate (if used) if the template was not exact. The sleeve may be 10–12 in. (254–305 mm) long. The top end of the sleeve should be slightly above the top level of the inertia block or pad so that the concrete will not spill into the sleeve and interfere with bolt adjustments.

7.4.4 Grouting

Grouting can be done only after the installation of the inertia block or pad has fully cured and the sole plates have been positioned and leveled. On sole plate installations, grouting is important, as it anchors the sole plates in place. Since the sole plates support the engine, it is important that the grout be installed properly to hold the plates level.

Engines and common skids can be mounted directly to the grout without the use of sole plates. When this is done, the engine must be mounted and leveled before the grout is poured. After installing the machine the mechanical department will install shims to adjust the leveling by doing alignment and balancing to the machine and after that the civil department is responsible for pouring the grouting under the engine skid to have a strong bond between skid and the foundation.

Grouting procedure is critical to follow under good supervision and a competent applicator.

The first step is to make a form around the inertia block or pad. If possible, pour the grout from one point on the inertia block or pad only, and allow the grout to flow under the machine skid. This pouring procedure will help decrease the chances of air pockets being trapped between the engine and the inertia block or pad. Air pockets will decrease the contact area between the grouting and machine skid. Also, a metallic-based grout will expand into these spaces and force the engine out of alignment. If the pour point on the engine or common skid is slightly higher than the rest of the inertia block or pad, the grout will flow more easily under the machine skid.

The best way to install a concrete, metallic-based grout is to form wedge-shaped grout pads. These pads should run the length of the engine or common skid. Slope the grout outward in a wedge shape towards the inertia block or pad to provide better support. Sole plates can be embedded in this run of grout, or the engine base can be installed directly on it.

The advantage of this grouting technique is that it will keep grout out from under the engine. The grout will not be able to expand up into the hollow area under the engine base and force the engine out of alignment.

Grouting should be worked into place using rods or chain lengths. Work the material gently to avoid air entrapment.

When using sole plates, pour in enough grout to embed the plates 1/2 in. (13 mm) into the grout. When sole plates are not used, never allow the grout to come up over the engine base or common skid, to allow for future adjustments.

Follow the grout manufacturer's instructions for applying the grout and recommendations for curing times. Concrete grouts must be sealed after curing. All metallic based grouts should be sealed to prevent rust from destroying the grout.

If the grout is allowed to settle at a slight outward slope, oil and water will be able to run off the inertia block or pad.

After the grout has cured, remove the leveling screws and remove any accumulation from the common skid or engine base. Save enough grout to pour into the inertia block bolt sleeves after the engine has been aligned.

Many epoxy grouts are also available, which provide superior performance for these applications.

7.5 DESIGN CHECKLIST

The checklist as presents in Table 7.8 is very important to follow if you review the calculation for the foundation design under vibrating equipment or you perform the design by yourself, so it will be your final step of the checking process.

Table 7.8 Foundation under machines design checklist

Item	Design condition	Check point
1	Static bearing capacity	50% of the allowable soil pressure
2	Settlement	The eccentricity of the machine center of gravity and foundation center of gravity is within 5% of any linear dimension.
3	Static plus dynamic loads	75% of the allowable soil pressure
4	Settlement of static plus dynamic load	Satisfy 2 and in case of rocking motion, the axis of rocking should coincide with the principal axis of the footing.
5	Magnification factor (applied for machine-generating unbalanced forces)	The calculated values of M should be less than 1.5 at resonance frequency.
6	Resonance	The acting frequencies of the machine should not be within + 20% of the resonance frequency.
7	Transmissibility factor	Transmissibility factor should be less than 3%.
8	Resonance of individual structure component	The resonance condition with the lowest natural frequency shall be avoided by maintaining the frequency ratio <0.5 and >1.5.

REFERENCES

API standard for centrifugal compressor, Section 2.18.4.
Ary, S.C., O'Neil, M.W., Pincus, G., 1984. Design of Structures and Foundation for Vibrating Machine. Gulf of Publishing, USA.
Barkan, D.D., 1962. Dynamics of bases and foundation. McGrow Hill, NY.
Baxter, R.L., Bernhard, D.L., "Vibration Tolerances for industry", ASME paper 67-PEM-14, plant engineering and maintenance conference, Detroit, Michigan, April 1967.
FBREEKA, vibration and shock control manual, 2009.
Major, A., 1962. Vibration Analysis and Design of Foundation for Machines and Yurbines. Akademiai Kiado, Budapest, Collet's Holding Limited, London.
Richart, F.E., Hall, J.R., Woods, R.D., 1970. Vibrations of soils and Foundations. Prentice Hall, Englewood Cliffs.
Waukesha manual, installation of Waukesha engines and Eginator system, 2004.

CHAPTER 8

Soil Investigation and Pile Design

8.1 INTRODUCTION

Site investigation is considered the first essential operation to be performed in the analysis, design, and choice of a foundation. Such investigation provides the necessary information about geological, physical, and geotechnical properties of soil, which are prerequisites to a safe and economical construction of all civil engineering works.

Insufficient or inadequate information with respect to the character and bearing capacity of the underlying soil beneath an engineering structure may result in serious damage or collapse of the structure.

In general, the purposes of making a site investigation are:

1. To provide the civil engineer with reliable and detailed information about soil and groundwater condition for the selection of the most efficient and economical type of foundation.
2. To plan for a suitable method of construction and to decide on the efficient equipment used.
3. To investigate whether any aggressive subsoil water or organic impurities exist in the soil.
4. To study any further conditions that may affect the structure (groundwater level, future excavation, etc.).

Because soil is a very important construction material in which, on which, or by means of which civil engineers build structures, it is doubtful if any major civil engineering works are constructed at present without site exploration being undertaken. Soil investigation should provide data on the following items:

1. location of the groundwater level,
2. bearing capacity of soil,
3. settlement predictions,
4. selection of alternative depths of foundation,
5. data of earth pressures on retaining structures and excavation supports,
6. data necessary for construction purposes (permeability, stratification, etc.).

Onshore Structural Design Calculations.
DOI: http://dx.doi.org/10.1016/B978-0-08-101944-3.00008-5

8.2 SOIL EXPLORATION METHODS

Various methods of subsurface investigation are known and widely used to obtain a considerable amount of information about the subsoil. Among these methods: test pits and open cuts, borings (wash boring, auger boring, rotary and percussion drilling), soundings, load tests, and geophysical methods.

Some methods will obviously yield more information than others, and soil engineers should constantly keep in mind the purpose of the project and the relative costs involved when planning the exploration program. Two main points are to be considered in soil investigation:

1. The size of the project. If it is a large project, geological studies will be necessary to determine the type and the extent of other methods of investigation. If it is a small project, some borings and the study of the neighboring foundation may be sufficient.
2. The type of structure and its sensitivity to future settlement. This study in many cases requires the complete change of the static system of the superstructure.

The choice of the exploration method is also determined by the location of the site. For compacted sites in towns where local information is available, borings and soundings will provide the required data. For new sites, site surveys and geological studies are necessary to determine the program for soil investigation.

8.2.1 Planning the Program

The actual planning of a subsurface exploration program includes some or all of the following steps:

1. Assembly of all valuable information: dimensions, column spacing, type and use of structure, basement requirements, bridge span and pier loadings, heights of retaining structure, etc.
2. Reconnaissance of the area: field trips to the site to examine topography, the type and behavior of adjacent structures (such as cracks or noticeable settlement), erosion in existing cuts, and soil stratification. The reconnaissance may also be in the form of a study of geological maps or aerial photographs.
3. Preliminary site investigation: a few borings or a test pit to establish types of soil, stratification, and possibly location of the ground water table (G.W.T).
4. Detailed site investigation: for complex projects and/or erratic soil conditions, samples are collected for shear strength determination and settlement analysis.

8.2.2 Organization of Field Work

After the study of the geological data, maps, and photographs, the engineer can consider the number, type, and depth of exploratory borings required. The engineers in charge of soil investigation have to inquire about the results as soon as any samples are extracted, and do not need to wait until all borings have been completed, otherwise there will be a considerable waste of money and efforts (the chances are high that repetitive information will have been collected while some data will still be lacking).

To obtain the most useful information with minimal cost and effort, the soil engineer in charge of an investigation should follow a number of sample rules as follows:

1. The engineer must know exactly what kind of information the investigation is expected to provide. This means that he must be completely familiar with the type of construction planned on the site. If the soil investigation focuses on a certain building location, the geotechnical engineer must know the exact value and kind of the loads on that building foundation. If future extensions or additional stories are planned, he should know their locations and loads. He should know if there are social requirements regarding allowable total or differential settlements.
2. The engineer must write down minimum requirements for bore hole spacing and depth, while taking into consideration the type of structure and estimated soil condition. The borings should provide sufficient data to enable reliable soil profile along two principal axis to be drawn. The minimum depth of boring should be related to the extent to which the subsoil will be affected by the proposed construction. This can be evaluated by using the concepts of stress distribution in soils. A useful approximate rule to determine the depth of investigation is to find the depth at which the increase in vertical stress caused by the purposed structure will be less than 10% of the originally existing vertical stress.

   For cuttings, retaining walls, and other structures where the lateral stability is the main concern, the investigation should extend below the depth of the deepest possible surface of failure. In soils, the sampling is usually governed by a maximum allowable spacing of 1−1.5 m, plus the condition that the sample must be taking each change in soil conditions; investigation for dams, reservoirs, and landslides; demand continuous sampling is obtained by a special apparatus.

3. The engineer must select the proper drilling and sampling equipment with attention to local preference for certain types of equipment which exists throughout the world.
4. The engineer should select the field supervisor and must be able to layout boring locations with reasonable accuracy.

8.2.3 Soil Boring Methods

The main purpose of drilling a test hole or boring is to obtain samples of the various strata and to acquire an accurate picture of the subsoil profile at one location. The soil sample as shown in Fig. 8.1. Different methods of soil boring are available as follows:

1. Test pits and open cuts:

   This method is ideal for soils which required little support for excavation. This test hole is made large enough to permit an observer to descend to log the soil profile visually, and the succession of strata can be examined in the wall of the pit. Block samples may be taken from the excavation sites in cohesive soils for laboratory testing.

   The trial pits are generally used for exploration of shallow foundations (a depth up to 4 m). as depths of pits increase, the cost increases very rapidly, particularly in water-bearing strata. Trial pits are also used to explore depth, cables, and other underground public works and/or old structures.
2. Hand and powered augers:

   Auger drilling is becoming the most common method of soil exploration to depths of up to 60 m. The speed of operation in most

Figure 8.1 Rig for the SPT.

types of soil is greater than that of any other drilling method. Distributed samples may be obtained from the soil brought to the surface, and undistributed samples may be taken from the bottom of the hole at any required depth. A powered auger with large machines is capable of making borings up to 1.0 m in diameter. In suitable soil where no casing is required, the size of the hole can be examined directly as in trial pits.

8.2.3.1 Wash Borings

In this method, the soil is loosened by a high-pressure water jet from a pipe passing down the borehole. The washings are brought to the surface in the water which passes back up the outside of the jet pipe. This method is extremely cheap and rapid; on the other hand, the washings are usually so disturbed that the sample is used only for identification of soil layers.

8.2.3.2 Sampling Methods

There are two methods of sampling, described in the following sections.

1. *Disturbed Samples*

   There are some instances when sample disturbance is of little importance since only the soil type and stratification needs to be determined. Disturbed samples are more satisfactory where classification of the soil is the prime objective since they are cheaper and faster to obtain (wash boring, auger boring). Disturbed samples are adequate for the investigation of borrow material to be used in fills, highway work, and other grading work.

2. *Undisturbed Samples*

   Undisturbed samples are necessary to study soil behavior and to perform laboratory tests under conditions as similar as possible to those in the original environment. Shear strength, consolidation, and permeability tests must be carried out in the laboratory using undisturbed samples. The major problems associated with undisturbed samples are:

   a. The samples are unloaded of their in-situ overburden pressure.
   b. Friction of the sites of samples, which creates some disturbance.
   c. Samples from levels below Ground Water table (G.W.T) may drain during the recovery process.
   d. Changes in bore pressure will disturb the sample.

   Undisturbed soil samples are obtained by forcing a thin-walled seamless stainless steel sampling cylinder into the soil at the bottom of a bore, or at the bottom and/or in the wall of a test pit. The forcing is accomplished by jacking or a continuous push.

In its simplest and most frequently used form, a thin-walled sampler consists of a 60- to 90-cm-long pipe with a wall thickness of about 1.5 mm and a diameter of 5 to 8 cm. At one end, a cutting edge is provided, while at the other end, the type is attached to an adapter filled with a ball value and vents. After samples have been taken, the sampling tube can be detached from the adapter and the ends sealed off with wax for transport to the laboratory. There, it may be stored, or the wax may be removed, and the sample extruded for immediate testing.

3. *Spacing of Borings*

The spacing of borings cannot be determined with absolute exactness. They depend upon many factors such as the nature and condition of soil, the shape and extent of the structure, etc. Furthermore, the spacing of borings should conform to the importance, size, and system of the structure. The following spacings are often used in planning boring work (Table 8.1).

In some technical practice which is specific for oil and gas define the spacing between bore hole depends on the type of the facilities and make a ranking if the structure is sensitive to soil or not and its criticality with the expected in deviation on soil profile. Table 8.2 presents the minimum and maximum bore holes for easy structure with no expect in

Table 8.1 Soil boring spacing

Structure or project	Spacing of borings, m
Highway (subgrade survey)	300–600
Earth dams, dikes	30–60
Borrow pits	30–120
Multistory buildings	15–30
One-story manufacturing plants	30–90

Table 8.2 Soil-boring space in oil and gas plants

Facility type	Bore hole	
	Minimum	Maximum
Process area	20-m grid	100-m grid
Roads	50 m	200 m
Pipe rack	50 m	200 m
Pipe bridges	1 at each end	At each end
Building	25-m grid	100-m grid
Pipeline routes	250 m	2000 m

Table 8.3 The number of bore holes for different tank diameter[a]

Tank diameter	Bore holes in the perimeter		Bole holes evenly spaced at tank half-radius
	Minimum	Maximum	
<10[b]	0	0	0
10–24	3	4	0
25–49	4	8	0
50–59	6	12	0
60–90	4	12	4
>90	6	16	4–6

[a]For any tank shall be soil boring in the center.
[b]For a diameter of 10 m or less, there will be only one soil boring at the center.

soil strata and the maximum bore hole number in case of critical facilities and there is an expectation to have a change in soil profile. In between it will be the judgment by the consultant engineer.

In the case of tanks, the number of bore holes depends on the tank diameter; in any case, it will be a soil boring in the tank center, but the minimum and maximum number of bore holes that are located on the perimeter are depicted in Table 8.3.

4. *Boring Depth*

The depth explored is generally the depth beyond which the effect of structure load is negligible. It should include all stressed zones of soil involved in the foundation system and should cover all layers of soil which affect the settlement of the structure. The depth of boring may be taken as 1.5 times the breadth of the raft or between 3 and 5 times the breadth of footing. If a sand layer is reached, it must be penetrated sufficiently to ensure its continuity.

A better simple rule for structures such as hospital or office buildings is to relate the estimated boring depth d to the number of stories:

$$\text{Light steel or narrow concrete,}\ d = 3(S)^{0.7}\ \text{m}$$

$$\text{Heavy steel or wide concrete,}\ d = 6(S)^{0.7}\ \text{m}$$

where S is the number of stories.

5. *Boring Report*

A boring report should contain:

1. A situation plan of the construction site drawn to scale and oriented with respect to north.

2. Identification, data, place, borehole locations and their coordinates from a reference axis, and drilling equipment used.
3. Description of soil profiles over the full depth of the borehole with a classification of the soil encountered in each boring. The boring log should be drawn to a stated scale and should contain information of soil types and thickness encountered.
4. Description of techniques for advancing and stabilizing the borehole.
5. The location of the groundwater table, the groundwater condition, and/or surface drainage observation.
6. Information on any difficulties or obstructions encountered in boring operations (boulders, roots, drains, telephone and electric cables, pipes, old foundations, etc.).
7. A listing of samples obtained and a report on any testing of soil in place.

8.2.4 Standard Penetration Test

The standard penetration test (SPT) is a well-established and unsophisticated method which was developed in the United States c.1925. It has since undergone refinements with respect to equipment and testing procedure. The testing procedure varies in different parts of the world.

Therefore, standardization of SPT was essential in order to facilitate the comparison of results from different investigations. The equipment is simple, rugged, and relatively inexpensive. Another advantage is that representative soil samples are obtained, though they are disturbed.

The reliability of the method and the accuracy of the result depend largely on the experience and care of the engineer on site.

A split-barrel sampler is driven from the bottom of a prebored hole into the soil by means of a 63.5-kg hammer, dropped freely from a height of 0.76 m. The diameter of the prebored hole normally varies between 60 and 200 mm. If the hole does not stay open by itself, casing or drilling mud should be used. The sampler is first driven to a depth of 15 cm below the bottom of the prebored hole, then the number of blows required to drive the sampler another 30 cm into the soil, the so-called N30 count, is recorded. The rods used for driving the sampler should have sufficient stiffness. Normally, when sampling is carried out to depths greater than around 15 m, 54 mm rods are used.

The quality of test results depends on several factors, including the actual energy delivered to the head of the drill rod, the dynamic properties

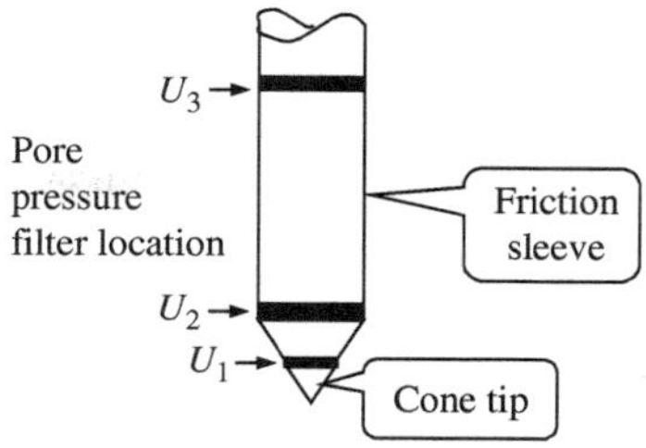

Figure 8.2 Sketch for CPT.

(impedance) of the drill rod, the method of drilling, and borehole stabilization. The actual energy delivered can vary between 50% and 80% of the theoretical free-fall energy. Therefore, correction factors for rod energy (60%) are commonly used, according to Seed and De Alba (1986). The SPT can be difficult to perform in loose sands and silts below the groundwater level (typical for land reclamation projects), as the borehole can collapse and disturb the soil to be tested. The following factors can affect the test results: nature of the drilling fluid in the borehole, diameter of the borehole, the configuration of the sampling spoon, and the frequency of delivery of the hammer blows.

Therefore, it should be noted that drilling and stabilization of the borehole must be carried out with care. The equipment of the test is shown in Fig. 8.2. The measured N-value (blows/0.3 m) is the so-called standard penetration resistance of the soil. The penetration resistance is influenced by the stress conditions at the depth of the test. Peck et al. (1996) proposed, based on settlement observations of footings, the following relationship for correction of confinement pressure: the measured N-value is to be multiplied by a correction factor CN to obtain the reference value N_1, corresponding to an effective overburden stress of 1 t/ft^2 (approximately 107 kPa).

$$N_1 = N \cdot \mathrm{CN} \tag{8.1}$$

where CN is a stress correction factor and p' is the effective vertical overburden pressure.

$$\mathrm{CN} = 0.77 \log_{10}\left(\frac{20}{p'}\right) \tag{8.2}$$

8.2.5 Cone Penetration Test

The cone penetration test (CPT) was invented and developed in Europe but has gained increasing importance in other parts of the world,

Figure 8.3 Cone penetration test.

especially in connection with soil compaction projects. Different types of mechanical and electric cone penetrometers exist, but the electric cone is most widely used. A steel rod with a conical tip (an apex angle of 60° and a diameter of 35.7 mm) is pushed at a rate of 2 cm/s into the soil. The steel rod has the same diameter as the cone.

The penetration resistance at the tip and along a section of the shaft (friction sleeve) is measured. The friction sleeve is located immediately above the cone and has a surface area of 15,000 mm^2. The electric CPT is provided with transducers to record the cone resistance and the local friction sleeve.

A CPT probe equipped with a pore-water pressure sensor is called a CPTU. It is important to assure complete saturation of the filter ring of the pore-water (piezo) element as shown in Fig. 8.3. Otherwise, the response of the piezo-transducer, which registers the variation of pore water pressure during penetration, will be slow and may give erroneous results. The CPTU offers the possibility to determine hydraulic soil

Figure 8.4 Vane test.

properties (such as hydraulic conductivity—permeability), but it is most widely used for identification of soil type and soil stratification.

The CPT can also be equipped with other types of sensors, including vibration sensors (an accelerometer or geophone) for determination of vibration acceleration or velocity. The "seismic cone" is not yet used on a routine basis, but because of the relative simplicity of the test, it has potential for wider application especially on soil compaction projects.

The CPT is standardized, and the measurements are less operator-dependent than the SPT, thus giving more reproducible results. The implementation of the CPT is presented in Fig. 8.4.

The CPT measures the cone resistance (q_c) and the sleeve friction (f_s) from which the friction ratio FR can be determined. FR is the ratio between the local sleeve friction and the cone resistance, expressed as a percentage (f_s/q_c). In spite of the limited accuracy of sleeve friction measurements, the valuable information which can be obtained in connection with compaction projects has not yet been fully appreciated. As will be discussed later, the sleeve friction measurement reflects the variation of lateral earth pressure in the ground, and can be used to investigate the effect of soil compaction on the state of stress. Cone and sleeve friction measurements are also strongly affected by the effective overburden pressure. It is necessary to take this effect into account, similar to the SPT.

One important objective of the CPT investigations in connection with soil compaction is to obtain information concerning soil stratification and variation in soil properties both in horizontal and vertical directions. The friction ratio is often used as an indicator of soil type (grain size) and can provide valuable information when evaluating alternative compaction methods.

Measurement of the excess pore water pressure with the CPTU can detect layers and seams of fine-grained material (silt and clay). It is also possible to obtain more detailed data concerning soil permeability and thus soil stratification.

8.2.6 Vane Test

Vane shear testing is one of the most common in situ methods for the estimation of the undrained shear strength of the soil as presented in Figs. 8.5 and 8.6.

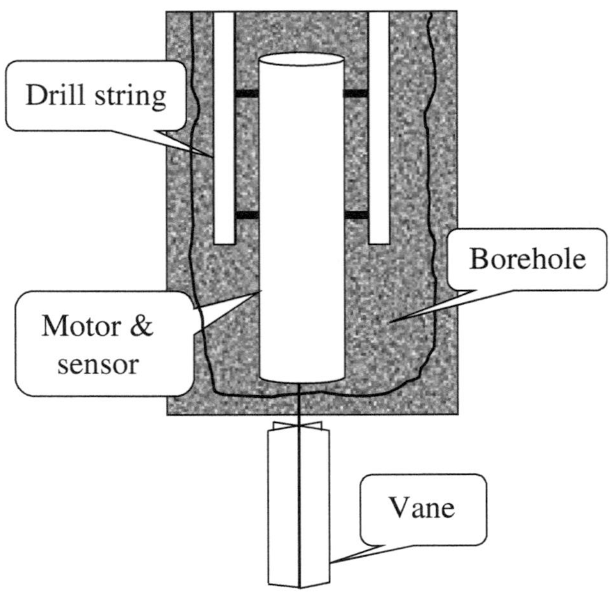

Figure 8.5 Sketch for vane test.

The Fig. 8.5 illustrates the vane shear tip used in the tests. The tip has a diameter of 75 mm and a height of 112 mm. Typically the ratio of height to diameter should be equal to 2.

The vane is introduced into the borehole to the depth where the measurement of the undrained shear strength is required. Then it is rotated, and the torsion force required to cause shearing is calculated. In the following figure, you can see a manual vane shear. The blade is rotated at a specified rate that should not exceed 0.1° per second (practically 1° every 10 s). The procedure and the equipment should typically follow the procedures suggested by ASTM D2573-72.

The shear strength of the material is calculated from the torque by dividing by the constant *K*, which depends on the dimensions and the shape of the vane. More can be found in ASTM D 2573-72.

The results of the corrected vane shear strength measured in-situ are shown in the following plot. The plot is shear strength (corrected) versus time, and the test was performed at a rate of 1° per 10 s. You can see in the plot the peak shear strength and the residual. After the completion of the test, the vane was rotated twice; then the residual strength was measured.

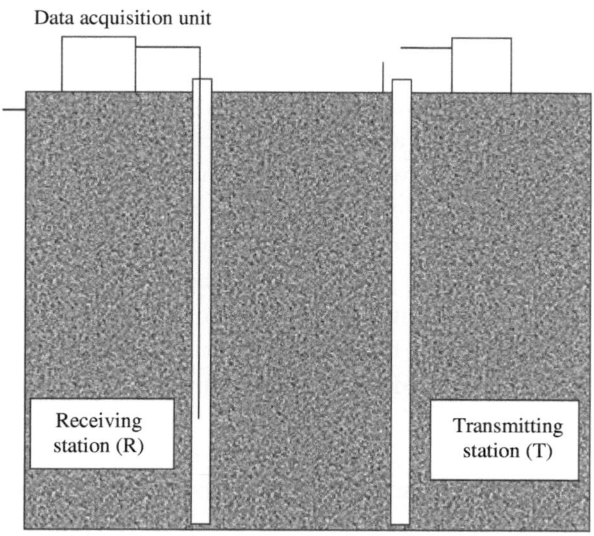

Figure 8.6 Schematic of cross-hole test setup.

8.2.7 Cross-Hole Test

In case of very critical projects, such as for new facilities with large storage tanks, it is important to take into account the phenomena of the soil through the long depth for engineering decisions.

This test is important for defining the wave propagation through the soil depth, which is very critical information, especially in earthquake regions.

The target from this test is to estimate the profile of the seismic velocity of longitudinal and transversal waves along boreholes at the selected location using the cross-hole seismic technique. This test consists of two boreholes 95 m deep drilled using a rotary drilling. The borehole will be cased with 97.6 mm diameter PVC pipes to accommodate the source and the receiver probes. The boreholes were spaced around 5 m from center to center at the ground surface.

The inclination of the two boreholes will be measured by using an inclinometer setup. The coordinates obtained from the inclinometer readings were then used to accurately calculate the distance between the two boreholes at 1.0-m depth intervals starting at the ground surface and ending at the end of the boreholes.

The raw data which will be obtained from the cross-hole survey is the travel times of the compression (*P*), shear (*S*), waves from the source hole (*T*), and the receiver hole (*R*), as shown in Fig. 8.6. The travel times of the seismic waves are measured and calculated for each *T* and *R* position with the known distance to calculate the apparent velocities *P* and *S* for each depth interval.

The instrumentation and data acquisition consists of a probe, a control unit, and a cable with a cable reel. The probe is characterized by a reliable clamping system, which his obtained through the progressive bending of a harmonic steel spring beside the probe. There will be a continuous current motor controlled by the surface electronics which moves the piston inside the probe, controlling the bending of the spring and providing for the clamping and unclamping operations. Inside the probe, there are three 10 Hz geophones oriented along the x-y-z axis to determine the arrival time from the seismic waves of type *S* to the geophone.

A specially designed down-hole hammer that fits through the 10-mm casing will be used as the transmitting source. The hammer has gripper plates that expand out to make firm contact with the inside of the casing. The hammer operation is pneumatic and electronic.

The equipment for the cross-hole test has many design features which facilitate accurate shear wave velocity measurements in accordance with the procedures described in ASTM D4428/D4428M-91 for cross-hole seismic testing.

The velocity of the propagation of stress waves through media is a function of the material properties of the media and depends upon the elastic modulus of the material, Poisson's ratio, and the material density.

8.2.7.1 Body Waves

These waves can be either primary or *P*-wave. This is a longitudinal wave in which the directions of motion of the particles are in the direction of propagation. This motion is nonrotational, and the wave is one of dilation-propagated speed.

Body waves can also be secondary, shear, or *S*-waves. This is a transverse wave in which the direction of motion of the particles is perpendicular to the direction of propagation.

Compressive waves are the fastest, followed by the shear waves, which are followed by other various types of waves. A fluid such as pore water does not transmit shear waves. Accordingly, in soil, shear waves are conducted through the soil skeleton only.

8.2.7.2 Surface Waves

In a uniform, infinite medium only *P*- and *S*-waves appear. If the medium is bounded or nonuniform, such as surface soils, other simple types of waves may appear. The most important of these are the surface waves that are propagated near the surface of a solid. Surface waves have differing depths of penetration depending on their wavelength. In nonuniform media, surface waves travel at a velocity that depends on their frequency.

8.3 SOIL INVESTIGATION REPORT

The following check can assist you in providing the scope of work for geotechnical soil investigation or reviewing the outcomes from the report.

The final report of the geotechnical consultant shall include information and recommendations on items marked in the following list, as a minimum (owner to indicate project/site specific requirements).

1. Introduction
   - ☐ Description of proposed construction
   - ☐ Purpose and scope of investigation
   - ☐ Abstract of findings and recommendations
2. Site Conditions
   - ☐ General description of site geology
   - ☐ Potential geologic hazards

- ☐ Site surface description
- ☐ General description of site topography
- ☐ Description of aboveground obstructions

3. Subsurface Conditions
   - ☐ Stratigraphy
   - ☐ General description of subsurface material properties
   - ☐ Groundwater elevations and expected variations
   - ☐ Description of underground obstructions encountered or otherwise identified
4. Field Investigation
   - ☐ Summary of operations
   - ☐ Description of sampling procedures
   - ☐ Description of field tests
   - ☐ Logs of borings, soundings, pits, wells, etc. in accordance with *ASTM D5434* and containing the following:
     - ☐ Complete descriptions and thicknesses of all strata, including near-surface materials such as paving, base course, topsoil, fill, etc.
     - ☐ Locations referenced to plant coordinate system
     - ☐ Ground surface elevations referenced to plant datum, if available; if not, then referenced to mean sea level
     - ☐ SPT values in blows per 150 mm increments
     - ☐ Results of all field tests
   - ☐ Location plan, containing as a minimum the following:
     - ☐ Scale plan with locations of borings, soundings, pits, wells, etc.
     - ☐ Plant coordinate system
5. Laboratory Tests
   - ☐ Description of tests
   - ☐ Test results
6. Hydrology
   - ☐ Erosion potential
   - ☐ Surface run-off coefficients
   - ☐ Percolation
7. Seismic Analysis
   - ☐ Seismicity based on seismic risk map
   - ☐ Soil profile type and site coefficient(s)
   - ☐ Site specific seismic risk study
8. Foundation Recommendations
   - ☐ Type(s) of foundation(s) recommended
   - ☐ Basis for selecting recommended foundation type(s)

- ☐ Recommendations for foundation type(s) selected
- ☐ Recommendations for deep foundations regardless of foundation type selected
- ☐ Recommendations for shallow foundations regardless of foundation type selected
- ☐ Soil strength parameters used in determining design capacities

9. Shallow Foundation Recommendations

- ☐ Spread footings: Depth below grade, size, and shape restrictions
- ☐ Mat foundations: Depth below grade and modulus of subgrade reaction
- ☐ Tank foundations: Recommendations and restrictions, excavation and backfill, ringwall or mat considerations, extended water tests
- ☐ Vibratory equipment foundations: Dynamic shear modulus, Poisson's ratio, and other considerations, based on the following:
  - ☐ Correlations from published literature
  - ☐ In-situ testing
- ☐ Ultimate and allowable net soil-bearing capacity
  - ☐ As a function of the shape and size of foundation, depth of embedment, and soil strength
  - ☐ Any increase in net allowable bearing capacity for hydrotest loads and short-term loads such as wind or an earthquake
- ☐ Foundation settlement
  - ☐ As a function of loading, shape and size of foundations, and compressibility of subsoils
  - ☐ Immediate settlement during construction
  - ☐ Long-term settlement
  - ☐ Time rate of settlement
  - ☐ Adjacent foundation settlement
  - ☐ Differential settlement for tanks
    - ☐ Along the perimeter
    - ☐ Center of tank to perimeter
    - ☐ Slope of tank bottom after anticipated settlement
    - ☐ Limitations or recommendations for hydrotest procedures to minimize differential settlement
    - ☐ Anticipated settlement and rebound during hydrotest and specific measurements during hydrotest

10. Deep Foundation Recommendations

- ☐ Type of pile or drilled pier and basis for recommendation
- ☐ Ultimate and allowable axial compression capacity through end bearing and skin friction

- ☐ Capacity vs length
- ☐ Any increase in capacity for hydrotest loads, or for short term loads such as wind or an earthquake

☐ Minimum and maximum tip elevations, when applicable

☐ Ultimate and allowable axial uplift capacity
- ☐ Uplift capacity vs length
- ☐ Any increase in capacity for hydrotest loads, or for short term loads such as wind or an earthquake

☐ Allowable lateral capacity
- ☐ Applied lateral loading vs deflection of pile head
- ☐ Pile moment vs depth
- ☐ *P-Y* curves
- ☐ Recommendations for the generation of *P-Y* curves and required parameters (cohesion, friction angle, E_{50})

☐ Down-drag considerations

☐ Spacing, group action, and use of batter piles

☐ Settlement considerations

☐ Vibratory equipment foundations, spring constants in each direction for recommended pile type

☐ Driven pile installation considerations
- ☐ Driving criteria, including refusal criteria
- ☐ Wave equation analysis
- ☐ Predrilling requirements/restrictions
- ☐ Potential problems and recommended solutions
- ☐ Pile installation near existing facilities

☐ (Nondriven) drilled pile and pier installation considerations
- ☐ Installation equipment requirements
- ☐ Casing/slurry considerations
- ☐ Installation criteria and recommendations
- ☐ Potential problems and recommended solutions

☐ Load test requirements, procedures, and acceptance criteria

11. Earth Pressures

☐ Active earth pressure and at-rest earth pressure

☐ Ultimate and allowable passive soil resistance for on-site soils, and recommended fill and backfill material

☐ Groundwater considerations

☐ Drainage requirements

12. Soil Properties
- ☐ Coefficient of friction or adhesion values between soil and concrete
- ☐ Unit weight of soil
- ☐ Cohesion and angle of internal friction
- ☐ Chemical analysis and other properties of soil and groundwater at depths of proposed structural elements and utilities, as follows:
    - ☐ pH value
    - ☐ Electrical conductivity (laboratory determination)
    - ☐ Chloride ion (Cl) concentration
    - ☐ Sulfate ion (SO_4) concentration
    - ☐ Electrical resistivity of soil (field determination)
    - ☐ Yearly average moisture content of soil
    - ☐ Thermal resistivity of soil
- ☐ Permeability
    - ☐ Laboratory determination
    - ☐ In-situ determination

13. Slabs and Pavements
- ☐ Natural soil and fill, subgrade suitability
- ☐ Recommended California bearing ratio value for pavement design
    - ☐ Based on correlations from published literature
    - ☐ Based on laboratory testing
    - ☐ Based on in-situ testing
- ☐ Recommended modulus of subgrade reaction for slab design
- ☐ Treatment for improving subgrade, if required
- ☐ Base course, subbase course, and shoulder recommendations
- ☐ Surfacing recommendations
- ☐ Base, subbase, and subgrade drainage recommendations
- ☐ Complete pavement system design

14. Other Considerations; Discuss and Provide Recommendations for the Following:
- ☐ Frost susceptibility of soils, frost depth
- ☐ Liquefaction potential of soils
- ☐ Swelling potential of soils, including the depth of the zone of soil moisture content fluctuation
- ☐ Collapsible or dispersive soils
- ☐ Effects of proposed construction on existing facilities or adjacent property
- ☐ Geologic or other potential hazards

15. Excavation Considerations
- ☐ Allowable excavation slope inclinations, temporary and permanent
- ☐ Groundwater control
  - ☐ Recommended dewatering method
  - ☐ Temporary and permanent groundwater control
  - ☐ Flow quantities
- ☐ Foundation subgrades
  - ☐ Heave control
  - ☐ Protecting/preserving the integrity of subgrade
- ☐ Effects on existing facilities
- ☐ Potential excavation problems
- ☐ Rock excavation
  - ☐ Rippability of rock
  - ☐ Definition of rock for contract documents
  - ☐ Rock quantity estimate guidance
- ☐ Pressure diagrams for shoring design
- ☐ Applicability of specialized shoring/stabilization procedures
- ☐ Classification of soil types per OSHA regulations (Types A, B, C)

16. Dikes and Embankments
- ☐ Recommended slope inclination
- ☐ Slope stability analysis
- ☐ Settlement
- ☐ Seepage analysis
- ☐ Erosion protection of slopes
- ☐ Foundation and subgrade preparation
- ☐ Fill material: Type, compaction, and moisture content control

17. Railroads
- ☐ Natural soil and fill subgrade preparation
- ☐ Ballasting, with consideration to the availability of local materials

18. Earthwork
- ☐ Topsoil: Thickness for stripping and definition for contract documents
- ☐ Suitability of on-site material for structural and nonstructural fill
- ☐ Special preparations or other requirements for use of on-site material
- ☐ Availability of imported fill
- ☐ Subgrade preparation
- ☐ Recommended compaction criteria and moisture content control
- ☐ Potential compaction difficulties and recommended solutions

8.4 DEEP FOUNDATION

Pile foundation is the most common type of deep foundation used to transmit structural loads into the deeper layers of firm soil in such a way that the layers of soil or rock can sustain the loads. A pile foundation is generally more expensive than an ordinary shallow foundation, and is used where soil at or near the surface is of poor bearing capacity or settlement problems are anticipated.

The main functions of piles are:

1. To carry more load from the superstructure to the lower, more resistant soil strata, thus increasing the load capacity of the site.
2. To reduce the settlements to the minimum value, and consequently the differential settlements. The settlement is very critical in case of sensitive structures which, by virtue of their sensitive structural statically system, cannot undergo appreciable differential settlements.
3. To avoid excavation underwater for sites where G.W.T. is high. This may represent an expensive item in the cost of the foundation and may also cause reduction of the strength of certain soils.
4. To densify the soil by driving compaction piles in loose cohesionless soil deposits.

Generally, piles are made of timber, steel, concrete, or some combination of those materials.

Technically, from a design point of view, there are two types of piles: end bearing piles and friction piles. The two options have advantages and disadvantages.

Fig. 8.7 presents the end bearing pile; in this case, most of the load transfers to the rock layer by the end bearing of the pile. Most loads will be taken by its base, and the rest of the load is taken by the section friction. The skin friction along the system of the pile could be neglected, and the bearing capacity of the pile is derived only from the point bearing resistance of the soil under pile tip.

In order to obtain the full benefit of the ultimate strength of the firm layer under the pile tip, the pile should penetrate the bearing stratum to a depth at least three times the pile diameter.

Fig. 8.8 presents the friction piles that transfer their loads to the surrounding soil by friction developed along their sides.

If the pile penetrates a clay layer, the skin friction is equal to the cohesion (C) of that layer. In the case of granular material, the skin

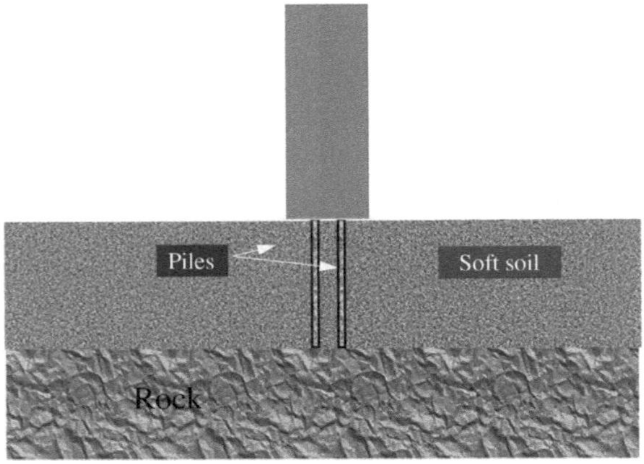

Figure 8.7 End bearing pile.

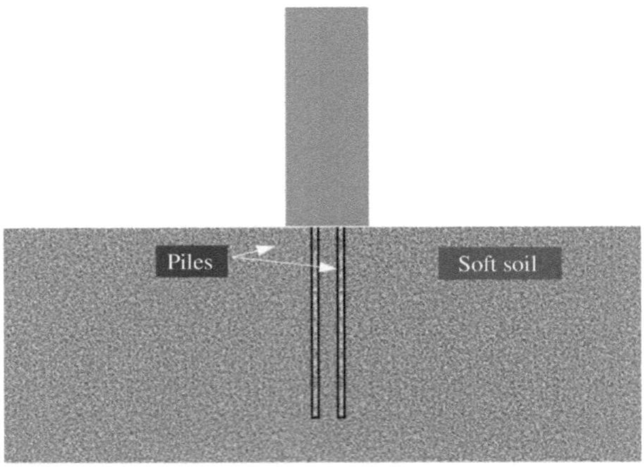

Figure 8.8 Friction pile.

friction is proportional to the intensity of the earth pressure and can be considered varying linearly with enough accuracy.

There are different types of piles used, and the differences between them depend on their material and manner of execution.

8.4.1 Steel Piles

These are usually rolled sections of (*H*) shapes or steel pipes. Wide-flange and *I*-beams may also be used.

Splices in steel piles are made in the same manner as in steel columns, including welding, riveting, or bolting. Pipe piles are either a welded or seamless-steel pipe which may be driven either open-ended or closed-ended. Pipe piles are usually used on offshore structure platforms.

8.4.2 Concrete Piles

Concrete piles have two methods of construction: the cast-in-place piles and the precast concrete piles.

Cast-in-place piles are formed by making a hole in the ground and filling it with concrete. They may be drilled or formed by driving a shell. The steel piles shown in Figs. 8.9 and 8.10 are driven or to make the hole and fill it with concrete.

Figure 8.9 Steel pile with cone tip.

Figure 8.10 Raymond shell piles.

The steel shell is usually withdrawn during or after pouring concrete and is sometimes left to protect the concrete from mixing with the mud or cement washed away by the groundwater.

8.4.3 Prestressed Pile

Piles are formed to the specified length, cured, and shipped to the construction site. A primary consideration with precast piles is the handling stress. To take care of handling stresses, some of which are tensile, the piles are reinforced and in some cases prestressed.

Precast reinforced concrete piles may have square or octagonal cross sections. They should be adequately reinforced to withstand driving and handling stresses. Long precast piles should be driven with careful guides to prevent buckling during driving in the part of the pile in the driving rig. To overcome the driving stresses, the lateral steel reinforcements should become closely spaced at the top and bottom of the pile to resist the stress wave concentration at the ends of the driven pile.

There will be a bending moment affect the pile due to handling for relatively short piles (less than 12.0 m), pile is usually lifted from one end and the pile is treated as a beam carrying its own weight. For long piles, the piles should be lifted from two, three, or four points at the specified distances indicated in Fig. 8.11 in order to reduce the bending stresses to a minimum. Lifting points should be marked by hooks or bolts that will be removed later (Fig. 8.12).

Special care should be given to the material of concrete piles to remain intact and prevent aggressive soil from attacking the pile. It is necessary to use siliceous aggregates and rich cement content (350 kg of cement per cubic meter of finished concrete). In addition, concrete piles must be protected from the dissolved sulfates or chemicals present in the underground water. If the ratio of sulfur trioxide

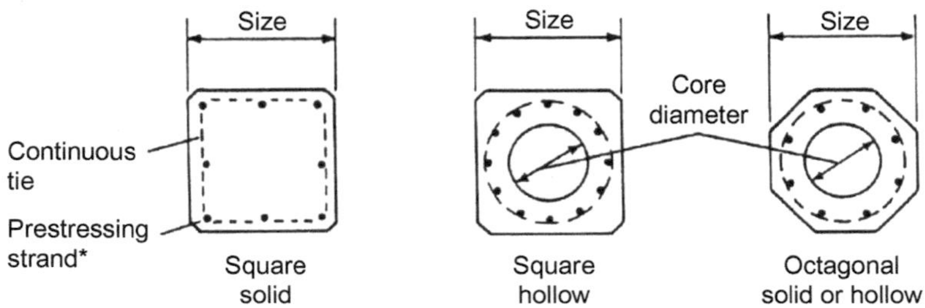

Figure 8.11 Prestressed pile sections.

Figure 8.12 Lifting of the concrete pile.

(SO_3) in the soil water increases over 0.03% (300 mg/L) in stagnant water, or 0.015 in running water, and if in addition the ratio of SO_3 in the soil itself is 0.2%, ordinary Portland cement cannot be used. In this case, special sulfate-resisting cements should be used. In all cement used, the presence of free lime or calcium traces should be minimized as possible.

Table 8.4 summarizes the comparison between different types of piles.

8.4.3.1 Pile Calculation

There are many methods and theories for calculating the pile capacity, but the basic concept is shown in Figs. 8.8 and 8.9, which is based on the fact that the bearing capacity (Q_d) of a pile may be assessed using soil mechanics. The capacity is assumed to be the sum of skin friction and end-bearing resistance; that is,

$$Q_d = Q_f + Q_p = f A_s + q A_p \tag{8.3}$$

where:

Q_f: skin friction resistance,
Q_p: total end bearing resistance,
F: unit skin friction capacity,
A_s: side surface area of pile,
q: unit end bearing capacity,
A_p: gross end area of pile.

Table 8.4 Typical pile characteristics and uses

Characteristic	Pile type			
	Concrete filled steel pipe piles	**Composite piles**	**Precast concrete (including prestress)**	**Cast in place (thin shell driven with mandrel)**
Maximum length	Practically unlimited	55 m	30 m for precast, 60 m for prestress	30 m for straight section, 12 m for tapered section
Optimum length	12–36 m	18–36 m	12–15 m precast 18–30 m prestress	12–18 m for straight 5–12 m for tapered
Application material specifications	ASTM A36 for core ASTM A252 for pipe ACI 318 for concrete	ASTM A36 for core ASTM A252 for pipe ACI 318 for concrete ASTM D25 for timber	ASTM A15 reinforcing steel ASTM A82 cold drawn wire ACI 318 for concrete	ACI
Recommended maximum stress	0.4 f_y reinforcement, 0.5 f_y or core 0.33 fc' for concrete	Same	0.33 fc' 0.4 f_y for reinforcement unless prestress	0.4 f_y if steel gauge ≤ 14, 0.35 f_y if shell thickness ≤ 3 mm
Maximum load for usual condition	1800 kN with cores 18,000 kN for large section with steel core	1800 kN	8500 kN for prestress 900 kN for precast	675 kN
Optimum load range	700–1100 with cores 4500–14,000 kN with cores	250–725 kN	350–3500 kN	250–550 kN

There are many methods for calculating the pile capacity for cohesive and noncohesive soil. One of these methods from API is presented below as a guideline for calculating the pile capacity.

1. For cohesive soil

a. Skin friction

$$f = \alpha S_u \tag{8.4}$$

α = a dimensionless factor,
S_u = undrained shear strength of the soil at the point in question.

$$\alpha = 1.0(S_u \leq 25 \text{ kPa}) \tag{8.5}$$

$$\alpha = 0.5(S_u \geq 70 \text{ kPa}) \tag{8.6}$$

$$\alpha = 1 - \left(S_u - \frac{25}{90}\right) \tag{8.7}$$

where P_o is the effective overburden pressure at the point in question.

$$N_c = 9 \text{ for } \frac{L}{D} \geq 3; \quad (s_{u\in})_b > 25 \text{ kPa} \tag{8.8}$$

b. End bearing

$$N_c = 6 \text{ for } (s_u)_b \leq 25 \text{ kPa} \tag{8.9}$$

$$q = 9\ S_u \tag{8.10}$$

2. Noncohesive soil

a. Skin friction

$$f = kp_o \tan \phi \tag{8.11}$$

k = coefficient of lateral earth pressure (ratio of horizontal to vertical normal effective stress),
ϕ = friction angle between the soil and pile wall.

b. End bearing

$$q = p_o N_q \tag{8.12}$$

N_q = dimensionless bearing capacity factor.

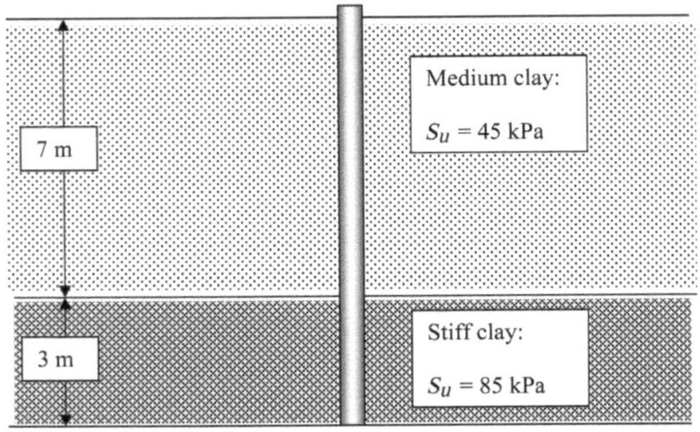

Figure 8.13 Example calculation.

Example:

Calculate the pile capacity for a precast reinforced concrete pile 0.25 × 0.25 m with a length of 10 m that consists of two layers, a medium and a stiff clay (Fig. 8.13).

Solution:

For medium clay layer: $S_u = 45$ kPa, $\alpha = 1 - (45 - 25/90) = 0.78$

For stiff clay layer: $S_u = 85$ kPa, $\alpha = 0.5$

Friction capacity: $Q_{fs} = 0.78 \times 45 \times (4 \times 0.25) \times 7 + 0.5 \times 85 \times (4 \times 0.25) \times 3 = 373.2$ kN

End bearing capacity: $Q_b = 9\ S_u\ A_b = 9 \times 85 \times (0.25 \times 0.25) = 47.82$ kN

The pile capacity: $Q_t/F_S = (373.2 + 47.82)/3 = 140$ kN

8.4.4 Pile Caps

Pile caps are the concrete foundations, which affects by the column loads from up and Piles reaction at the point of contact piles to the caps. In this type of foundation, ignore the impact of the soil where the soils are not contact with the caps in rigid or flexible manner to allow to carry any part of the column load and the stiffness of the piles is great, so the piles carry all the load.

Often, the column load cannot be carried by one pile only, so the column will need more than one pile to carry the load; it needs a piles cap to distribute the column load to these piles equally.

In this phase of design, care should be taken to distribute the load to the piles equally; this can happen by putting the center of gravity of the column to coincide with the center of gravity of the pile cap.

To ensure the transfer of the load from the column to the pile, the pile steel reinforcement should extend inside the pile cap with at least 600 mm to guarantee transfer of this load by the bond between concrete and steel.

The pile caps will be design as a rigid foundation and assume the piles carrying equal loads from the column load, so the pile cape thickness will be designed to resist the punching stresses and the tension in top and bottom.

8.5 RETAINING WALL

A retaining wall is used to confine water or soil. In industrial projects, it is usually used around the tanks to confine the oil from seepage in case of tank failure or leaking from any of the tanks.

In general, the following are the types of retaining walls:

1. Gravity-retaining walls
2. Cantilever walls
3. Counter fort walls
4. Buttress walls
5. Bridge abutments
6. Basement walls
7. Semigravity walls
8. Walls with relieving slabs
9. Precast and prestressed walls

In process plants, it is traditional to use cantilever walls. The forces which affect the retaining walls after defining the preliminary dimensions to the walls will be as follows:

1. The wall height
2. The dead and live load on the wall
3. The lateral pressure that affects the wall due to the soil weight behind the wall
4. The lateral pressure due to a live load or moving load on the soil that exists behind the wall
5. The wave effect in the case of a wall adjacent to the sea
6. Earthquake load
7. The water or oil lateral pressure in case of a bundle wall
8. Any other load that affects the structure during the execution or operation phase

8.5.1 Preliminary Retaining Wall Dimensions

After defining the preliminary dimensions, the design of the retaining wall will start by checking the stability of the retaining wall against the following three criteria:

1. Stability against overturning
2. Stability against sliding
3. Stability against bearing capacity.

After checking the stability of the retaining wall, calculate the stress on the wall to design the reinforced concrete section of the wall, then check stability again for a second time until the compatibility check of stability with safe reinforced concrete design sections dimensions.

8.5.1.1 Check Stability Against Overturning

Checking stability of the retaining wall against overturning will be done by applying the following equation, which ensures that the sum of moments due to forces tending to resist overturning is higher than the sum of the moments due to forces tending to overturn the walls. The forces on the retaining wall are shown in Fig. 8.14.

$$F_S = \frac{M_R}{M_O} \tag{8.13}$$

where:

M_R = Sum of the moments due to forces tending to resist overturning about point (*C*).

M_O = Sum of the moments due to forces tending to overturning the wall about point (*C*).

F_S = factor of safety about 1.5–2.

The following Fig. 8.14 presents the moments that affect the wall overturning and resistance against overturning.

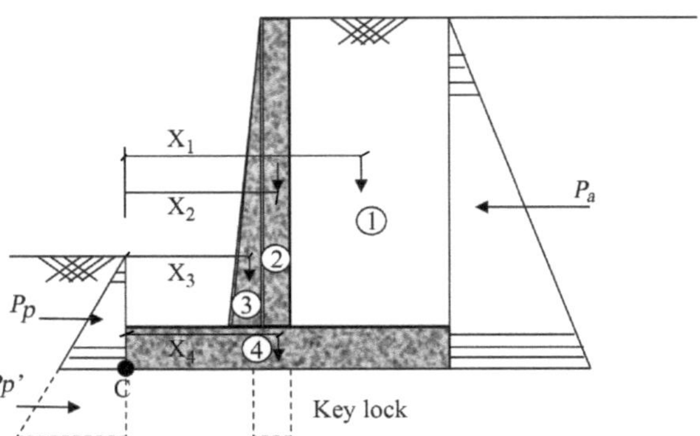

Figure 8.14 Forces drive and resist sliding.

Table 8.5 Resisting overturning moment

Section	Area	Weight	Moment arm measure from (C)	Moment about (C)
1	A_1	$W_1 = A_1 \times \gamma_s$	X_1	M_1
2	A_2	$W_2 = A_2 \times \gamma_c$	X_2	M_2
3	A_3	$W_3 = A_3 \times \gamma_c$	X_3	M_3
4	A_4	$W_4 = A_4 \times \gamma_c$	X_4	M_4
		ΣV		ΣM_R

Where:
γ_c = concrete density,
γ_s = soil density.

$$P_a = K_a \gamma H \tag{8.14}$$

where P_a is the active force and $K_a = 0.3$.

M_R can be calculated from Table 8.5.

Calculate F_S. If it is within 1.5–2.0, then accept it; if it is less, then the retaining wall dimensions need to be changed in order to achieve the required overturning stability factor of safety.

8.5.1.2 Check Stability Against Sliding

The factor of safety for stability will be calculated from the following equation:

$$F_S = \frac{\Sigma F_R}{\Sigma F_d} \tag{8.15}$$

where:

ΣF_R = Sum of the forces that resist the retaining wall from sliding.
ΣF_d = Sum of the horizontal driving forces to slide the retaining wall.
$\Sigma F_d = P_a$

$$\begin{aligned} \Sigma F_R &= S_V \times \tan \phi^2 + P_P \\ &= S_V \tan(k_1 \phi^2) + k_2 C_2 \end{aligned} \tag{8.16}$$

where $k_1 = k_2 = 2/3$.

P_P is the passive force, but from a practical point of view, it should not be taken into calculation to increase the factor of safety. However, usually during construction filling in front of the wall will be the last stage.

If the factor of safety is less than the limit, you can increase the stability against sliding by using a key lock, as shown in Fig. 8.14.

8.5.1.3 Check Stability Against Bearing Capacity

The soil bearing capacity is obtained from the soil report. The retaining wall should check that the load which is transferred from the wall to the soil is less than the soil bearing capacity (Fig. 8.15).

$$F_S = \frac{q_{\text{max}}}{q_{\text{all}}} \tag{8.17}$$

where:

q_{all} = maximum allowable bearing capacity,
q_{max} = maximum applied load to the soil,
F_S = factor of safety between 2.5 and 3.

$$M_n = \Sigma M_R - \Sigma M_o \tag{8.18}$$

$$X = \frac{M_n}{\Sigma V} \tag{8.19}$$

$$e = \frac{B}{2} - X \tag{8.20}$$

Fig. 8.16, The second step is to start designing the concrete sections. In the main forces that affect each element are presented.

$$M_1 = M_2 + M_3 \tag{8.21}$$

The shape of the steel reinforcement detail will be as shown in the Fig. 8.16.

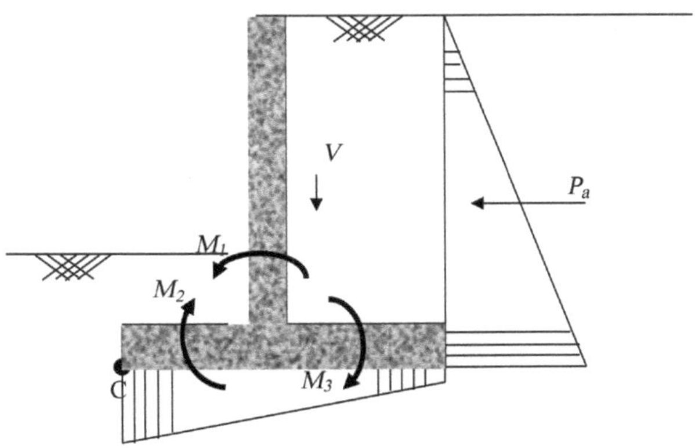

Figure 8.15 Stresses check against bearing capacity.

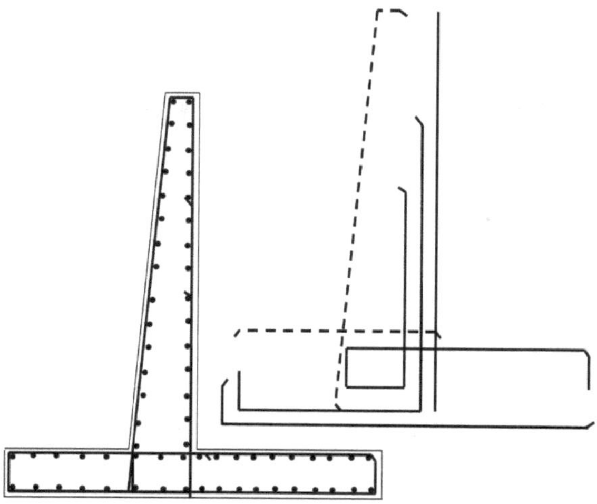

Figure 8.16 Retaining wall steel detailing.

Example:

Retaining wall as in the following drawing. The height of the retaining wall is 4.0 m, and the surcharge load is 0.3 t/m^2.

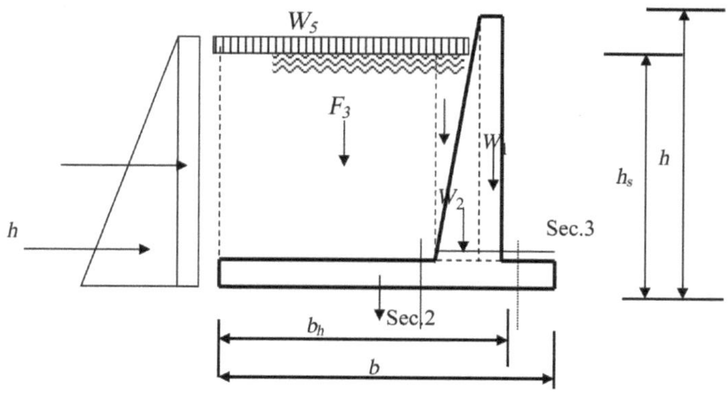

Solution:

The solution is laid out in a Microsoft Excel worksheet, which may be viewed at the following website: www.elreedyma.comli.com or www.elreedyman.tk.

Input data for retaining wall design

Wall dimensions

Height of wall (h) = 4 m
Height of soil (h_s) = 4 m
Thickness of the wall top (t_t) = 0.25 m
Thickness of the wall bottom (t_b) = 0.4 m
Width of base (b) = 3 m
Width of toe = 0.75 m
Width of heel (b_h) = 2.25 m
Base thickness (t) = 0.5 m

Properties of material

$\gamma_c = 2.5$ t/m^3
$\gamma_s = 1.8$ t/m^3
$k_a = 0.333$
Friction factor, $\mu = 0.42$

Applied loads

Intensity of surcharge load = 0.3 t/m^2

Results of stability analysis, stresses, and straining actions

Calculation of weights

$W_1 = 2.1875$ t/m'
$W_2 = 0.66$ t/m'
$W_3 = 3.75$ t/m'
$W_4 = 0.47$ t/m'
$W_5 = 11.66$ t/m'
Net weight = 18.72 t/m'
Stabilizing moment = 32.93 mt/m'

Earth and water pressure at top of foundation level

Force for backfill (F_1) = 3.671325 t/m'
Surcharge load (F_2) = 0.34965 t/m'

Earth and water pressures at foundation level

Force due to backfill (F_1) = 4.7952 t/m'
Force due to surcharge (F_6) = 0.3996 t/m'
Sum of sliding forces = 5.2 t/m'
Overturning moment = 7.2 mt/m'

Safety factors

Overturning safety factor = 4.6 > 2 OK
Sliding safety factor = 1.51 > 1.5 OK

Stresses on soil

Eccentricity(e) = 0.125 m
Stress at toe = 7.8 t/m^2
Effective width of base = 3 m

(*Continued*)

Straining actions at critical sections
Stress at Section 1 = 7.02 t/m^2
Stress at Section 2 = 6.6 t/m^2
Moment at Section 1 = 1.77 mt/m'
Moment at Section 2 = −3.81 mt/m'
Moment at Section 3 = 4.9 mt/m'

Example:

This is the same as the previous example but considers a water table 3 m above the foundation.

Solution:

Input data for retaining wall design

Wall dimensions
Height of wall (h) = 4 m
Height of soil (h_s) = 4 m
Thickness of the wall top (t_t) = 0.25 m
Thickness of the wall bottom (t_b) = 0.4 m
Width of base (b) = 3 m
Width of toe (b_t) = 0.75 m
Width of heel (b_h) = 2.25 m
Base thickness (t) = 0.5 m

Properties of material
$\gamma_c = 2.5\ t/m^3$
$\gamma_s = 1.8\ t/m^3$
Submerged soil density, $\gamma_{ss} = 0.8\ t/m^3$
$k_a = 0.333$
Friction factor, $\mu = 0.42$

Applied loads
Intensity of surcharge load = 0.3 t/m^2

Water heights above foundation level
Water height heel side = 3 m
Water height toe side = 0 m

Results of stability analysis, stresses, and straining actions

Calculation of weights
$W_1 = 2.1875$ t/m'
$W_2 = 0.65625$ t/m'

$W_3 = 3.75$ t/m'
$W_4 = 0.4725$ t/m'

(Continued)

$W_5 = 11.655$ t/m'
$W_6 = 0$ t/m'
Uplift = 0 t/m'
Net weight = 18.72125 t/m'
Stabilizing moment = 32.932 mt/m'

Earth and water pressure at top of foundation level
Force due to backfill (F_1) = 0.2997 t/m'
Force due to backfill (F_2) = 1.4985 t/m'
Force due to backfill (F_3) = 0.8325 t/m'
Force due to water pressure at heel side (F_4) = 3.125 t/m'
Force due to surcharge (F_6) = 0.34965 t/m'

Earth and water pressures at foundation level
Force due to backfill (F_1) = 0.2997 t/m'
Force due to backfill (F_2) = 1.7982 t/m'
Force due to backfill (F_3) = 1.1988 t/m'
Force due to water pressure at heel side (F_4) = 4.5 t/m'
Force due to surcharge (F_5) = 0.3996 t/m'
Sum of sliding forces = 8.1963 t/m'
Overturning moment = 10.1943 mt/m'

Safety factors
Overturning safety factor = 3.230433 >2 OK
Sliding safety factor = 0.959326 <1.5 Unsafe

Stresses on soil
Eccentricity (e) = 0.28546 m
Stress at toe = 9.8 t/m^2
Stress at heel = 2.77 t/m^2
Effective width of base = 3 m

Straining actions at critical sections
Stress at Section 1 = 8.0 t/m^2
Stress at Section 2 = 7.07 t/m^2
Moment at Section 1 = 2.23858 mt/m'
Moment at Section 2 = −5.83137 mt/m'
Moment at Section 3 = 6.63208 mt/m'

In this case, the sliding mode of failure is unsafe; the two solutions are to either increase the heel to 4.5 m or to construct a stem to contact with the footing to have a passive earth pressure.

8.6 SOIL WITH PROBLEMS

Most of the industrial plants, specifically oil and gas production plant, are to be in remote areas. In some cases, the location of the plant has weak soil, but there are many methods of solving this problem depending on the soil type and the soil investigation result.

8.6.1 Dynamic Compaction

Dynamic compaction (DC), also known as *dynamic deep compaction*, was advanced in the mid-1960s by Louis Menard. The process involves dropping a heavy weight on the surface of the ground to compact soils to depths as great as 12.5 m.

The purpose of this method is to reduce foundation settlements, reduce seismic subsidence and liquefaction potential, permit construction on fills, densify garbage dumps, and reduce settlements in collapsible soils.

This method is most effective in permeable, granular soils, as cohesive soils tend to absorb the energy and limit the technique's effectiveness.

The groundwater table should be at least 1.8 m (6 ft) below the working surface for the process to be effective. In organic soils, DC has been used to construct sand or stone columns by repeatedly filling the crater with sand or stone and driving the column through the organic layer.

This method is done by using a cycle duty crane to drop the weight, although specially built rigs have been constructed. The crane is typically rigged with a sufficient boom to drop the weight from heights of 15.4 to 30.8 m (50 to 100 ft), with a single line to allow the weight to nearly "free fall," maximizing the energy of the weight striking the ground. The weight to be dropped must be below the safe single line capacity of the crane and cable. Typically, weights range from 10 to 30 tons and are constructed of steel to withstand the repetitive dynamic forces.

The compaction procedure involves repetitively lifting and dropping a weight on the ground surface. The layout of the primary drop locations is typically on a 3.0 to 6.0 m grid with a secondary pass located at the midpoints of the primary pass. Once the crater depth has reached about 1 m, the crater is filled with granular material before additional drops are performed at that location.

This process produces large vibrations in the soil, which can have adverse effects on nearby existing structures. It is important to review the nearby adjacent facilities for vibration sensitivity and to document their preexisting condition, especially structures within 150 m of planned drop locations.

Vibration monitoring during DC is also prudent. Extreme care and careful monitoring should be used if treatment is planned within 200 ft (61.5 m) of an existing structure.

The craters resulting from the procedure are typically filled with a clean, free-draining granular soil. A sand backfill can be used when treating sandy soils. A crushed stone backfill is typically used when treating finer-grained soils or landfills.

The depth of influence is related to the square root of the energy from a single drop (weight times the height of the drop) applied to the ground surface. The following correlation was developed by Dr. Robert Lucas based on field data:

$$D = k(W \cdot H)^{0.5}$$

where D is the maximum influence depth in meters beneath the ground surface, W is the weight in metric tons (9 kN) of the object being dropped, and H is the drop height in meters above the ground surface. The constant k varies with soil type and is between 0.3 and 0.7, with lower values for finer-grained soils.

8.6.2 Vibro Compaction

Vibro compaction was developed in the 1930s in Europe. The process involves the use of a down-hole vibrator (vibroflot), which is used to increase bearing capacity, reduce foundation settlements, reduce seismic subsidence and liquefaction potential, and permit construction on loose granular fills.

Applicable soil types: The VC process is most effective in free draining granular soils. The spacing is based on a 165-horsepower (HP) (124 kW) vibrator. Although most effective below the groundwater table, VC is also effective above (Table 8.6).

Table 8.6 Expected improvement and typical probe spacing with vibro compaction

Soil description	Expected improvement	Typical probe spacing (m)[a]
Well-graded sand <5% silt, no clay	Excellent	27–33
Uniform fine to medium sand with <5% silt, no clay	Good	22.5–27
Silty sand with 5–15% silt, no clay	Moderate	18–22.5
Sand/silt >15% silt	Not applicable[b]	–
Clays and garbage	Not applicable	–

[a]Probe spacing to achieve 70% relative density with 165 HP vibroflot; higher densities require closer spacing.
[b]Limited improvement in silts can be achieved with large displacement and stone backfill.

The vibrator is lowered into the ground, assisted by its weight, vibration, and typically water jets in its tip. If difficult penetration is encountered, predrilling through the firm soils may also be performed. The compaction starts at the bottom of the treatment depth.

The vibrator is then either raised at a certain rate to achieve relative densities of 70% to 85%. Treatment as deep as 37 m (120 ft) has been performed.

8.6.3 Compaction Grouting

Compaction grouting, one of the few ground improvement techniques born in the United States, was developed by Ed Graf and Jim Warner in California in the 1950s. This technique densifies soils by the injection of a low mobility, low slump mortar grout. The grout bulb expands as additional grout is injected, compacting the surrounding soils through compression.

Besides the improvement in the surrounding soils, the soil mass is reinforced by the resulting grout column, further reducing settlement and increasing shear strength. The method is used to reduce foundation settlements, reduce seismic subsidence and liquefaction potential, permit construction on loose granular fills, reduce settlements in collapsible soils, and reduce sinkhole potential or stabilize existing sinkholes in karst regions.

Three main pieces of equipment are required to perform compaction grouting: one to batch the grout, one to pump the grout, and one to install the injection pipe. In some applications, ready-mix grout is used, eliminating the need for on-site batching.

The injection pipe is typically installed with a drill rig or is driven into the ground. It is important that the injection pipe is in tight contact with the surrounding soils. Otherwise, the grout might either flow around the pipe to the ground surface, or the grout pressure might jack the pipe out of the ground.

Augering or excessive flushing could result in a loose fit. The pump must be capable of injecting a low slump mortar grout under high pressure. A piston pump capable of achieving a pumping pressure of up to 1000 psi (6.9 MPa) is often required.

Compaction grouting is typically started at the bottom of the zone to be treated at ground surface and can be terminated at any depth. The technique is very effective in targeting isolated zones at depth. It is generally difficult to achieve significant improvement within about 2.5 m of the ground surface. Some shallow improvement can be accomplished using the slower and more costly top-down procedure. In this procedure, grout is first pumped at the top of the treatment zone. After the grout sets up, the pipe is drilled to the underside of the grout and additional grout is injected. This procedure is repeated until the bottom of the treatment zoné is grouted. The grout injection rate is generally in the range of 0.087 to 0.175 m^3/min, depending on the soils being treated. If the injection rate is too fast, excess pore pressures or fracturing of the soil can occur, reducing the effectiveness of the process.

8.6.4 Wick Drain

The purpose of wick drains is to accelerate the consolidation settlement of soft, saturated clays by reducing the drainage path. A wick drain is a prefabricated drainage strip that consists of a plastic core surrounded by a nonwoven polypropylene geotextile jacket. The geotextile (a filter fabric) allows passage of water into the core that is then pumped out.

A wick drain is installed by enclosing the drain in a tubular steel mandrel and supporting it at the base by an anchor plate. The mandrel is then driven into the soil by a vibratory rig or pushed by a hydraulic rig. At the desired depth, the mandrel is removed, the drain and anchor plate remain in place, and the drain is cut off with a tail about 300 mm long.

Generally, a surcharge program is considered when the site is underlain by soft fine-grained soils which will experience excessive settlement under the load of the planned structure. Using consolidation test data, a surcharge load and duration is selected to preconsolidate the soils sufficiently such that when the surcharge load is removed and the planned structure is constructed, the remaining settlement is acceptable.

The time required for the surcharge settlement to occur depends on the time it takes for the excess pore water pressure to dissipate. This is dictated by the soil's permeability and the square of the distance the water has to travel to get to a permeable layer. The Prefabricated Vertical Drain (PVD) accelerate the drainage by shortening the drainage distance. The spacing is designed to reduce the consolidation time to an acceptable duration. The closer the drains are installed (typically 3–6 ft on center) the shorter the surcharge program is in duration.

Procedure: Fill soil is typically delivered to the area to be surcharged with dump trucks. Dozers are then used to push the soil into a mound. The height of the mound depends on the required pressure to achieve the required improvement.

The wick drains are typically in 300-m (1000-ft) rolls and are fed into a steel rectangular tube (the mandrel) from the top. The mandrel is pushed, vibrated, driven, or jetted vertically into the ground with a mast mounted on a backhoe or crane. An anchor plate or bar attached to the bottom of the wick drain holds it in place in the soil as the mandrel is extracted. The wick drain will be cut off slightly above the ground surface, and another anchor is attached. The mandrel is moved to the next location, and the process is repeated. If obstructions are encountered during installation, the wick drain location can be slightly offset.

REFERENCES

Peck, R.B., Hanson, W.E., Thornburn, T.H., 1996. Foundation Engineering, second ed. Wiley, New York.

Seed, H.B., De Alba, P., 1986. Use of SPT and CPT tests for evaluating the liquefaction resistance of sands,. Proc. In Situ '86. ASCE, USA, pp. 281−302.

CHAPTER 9

Assessment for Structures in Industrial Plants

9.1 INTRODUCTION

Normally, the main cause of the deterioration in the reinforced concrete structure in oil and gas facilities is corrosion. Most of the oil and gas is delivered from offshore to facilities that are onshore, so these facilities are usually near the sea, and they experience the effect of chloride from the seawater.

When we see a concrete structure has signs of corrosion, we want to repair it; but before we do that, we must evaluate the building with high accuracy and identify the environmental conditions that surround the building.

After performing a risk assessment to the structure, we can identify the parts of the concrete elements that will be repaired and determine the method of repair.

The assessment method should be correct and technically accurate enough to determine the degree of building risk, then the repair process will be technically accurate and adequate, and more precise.

On the other hand, if the diagnosis of disease is correct, the selected medicine also will be correct. The preliminary steps of a building assessment are the diagnosis of the causes of defects in the structure as a result of corrosion, and the identification of the causes that led to the corrosion.

Therefore, we will think in two directions. One of them is to determine the method of repair, and the other is to define the reasonable protection method to protect steel bars from future corrosion.

The process of assessing the structure is often in two stages. The first stage is the initial assessment of the building; at this stage, we will define the problem and develop a plan for a detailed evaluation of the structure.

In a detailed assessment of the building, we will carefully define the problem and the reasons for them; at this stage the detailed inspection of the whole building will be carried out. A basis has been established for

Onshore Structural Design Calculations.
DOI: http://dx.doi.org/10.1016/B978-0-08-101944-3.00009-7

the inspection of buildings for evaluation in the Technical Report 26 (Society Concrete, 1984), as well as the American Concrete Institute.

There are many reasons for the deterioration of the reinforced concrete structures varies from the presence of the cracks until falling the concrete cover among those reasons are increasing in stresses, plastic shrinkage, as a result of frost or cracked concrete for plastic concrete, move the wood form during construction, the presence of aggregate has the capability to shrink or deteriorate as a result of the aggregate contaminated with the alkalinity.

There are also some structures such as reinforced concrete pipelines in sanitary projects and underground concrete structures like tunnels that come in direct contact with the soil; in this case, the carbonate present in the water will damage the concrete. This is a type of deterioration that is important to consider, but besides that, we will focus here on the corrosion result of the exposure of concrete structure to the atmosphere.

As we have stated previously, there are several types of cracks. Therefore the engineer who is responsible for performing the evaluation must be highly experienced in that area so that he can specify precisely why the incidence of deterioration in concrete was observed with the wrong diagnosis, which will be result in a wrong repair, which will cause a heavy financial loss, and which will affect structure safety.

The evaluation of a structure is not only a structural assessment, but an assessment of the state of concrete in terms of the presence of corrosion and the rate of corrosion and collapse of concrete. From the previous information, we will have the ability to decide if the concrete member can withstand loads. In addition, a case study will be performed to assess the member to see if an excessive deflection has happened more than the allowable limit that will cause cracks, or in case of reduction, in cross-section area or in falling concrete cover.

9.2 PRELIMINARY VISUAL INSPECTION

This inspection process start by identify the geographical location of the structure, the nature and circumstances of weather conditions surrounding the building, the method of constructing the building, its structural system, and the method of loading.

This assessment is often performed preliminary by visual inspection, concentrated on the cracks and the fall of concrete cover and also collecting data about the thickness of the concrete cover and the nature of structure in terms of the quality of concrete and quality of construction and structure system whether it is consists of beam and column system or slab on load bearing masonry or it is prestressed or precast concrete.

All such information is preliminary, and it is necessary to perform some simple measurements, for example to determine the depth of the chloride in the concrete. Take a sample from concrete that has fallen and perform a laboratory test.

The safety of the structure must be calculated precisely, especially after reducing the cross-sectional area of steel due to corrosion, as well as the fall of the concrete cover, which reduces the total area of the concrete member, as these two areas are the main coefficients that have a direct effect on the capacity of the concrete member to carry the loads.

Visual inspection is the first step in any process of technical diagnosis, and it starts with a general view of the structure as a whole and a concentration on the general defects. After this, one defines precisely the corrosion deterioration and the extent of corrosion on the steel reinforcement.

As we have stated before, the process of assessing the building must be performed by an expert, as the cracks in the concrete structure may not be caused by corrosion, as corrosion is not the only factor that causes cracks. However, it is the main factor for majorly deteriorated structures. Fig. 9.1A–C illustrates the various forms and causes of cracks in beams, slabs, and columns.

In general, cracks in concrete have many causes. They may affect only appearance, or they may indicate significant structural distress or a lack of durability. Cracks may represent the total extent of the damage, or they may point to problems of greater magnitude. Their significance depends on the type of structure, as well as the nature of the cracking. For example, cracks that are acceptable for building structures may not be acceptable in water retaining wall structures.

The proper repair of cracks depends upon knowing the causes of the cracks and selecting the repair procedures that take these causes into account; otherwise, the repair may only be temporary. Successful long-term repair procedures must prevent the causes of the cracks as well as the cracks themselves.

The cracks may be happening in plastic concrete or in hardened concrete. The sample of the plastic concrete cracks is plastic shrinkage cracking, settlement cracking and after hardening there are a dry shrinkage cracking will be illustrated as follow based on ACI code:

Concrete Society Technical Report 54 mentioned that different types of cracks occur at different times in the life of a concrete element, as

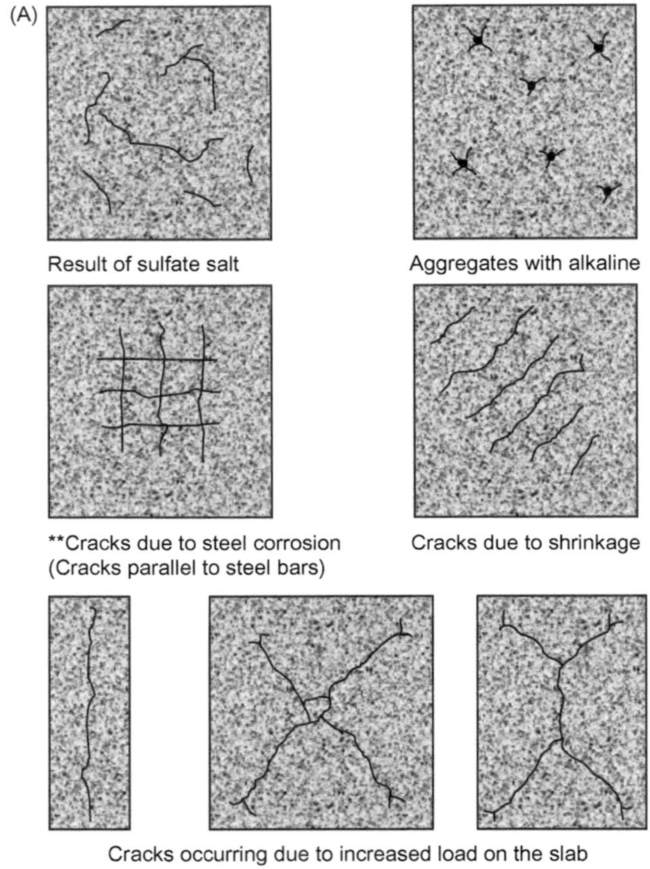

Figure 9.1 (A) Types of cracks in reinforced concrete slabs; (B) types of cracks in reinforced concrete beams and (C) types of cracks in reinforced concrete columns.

shown in Table 9.1. Therefore, as well as a recognition of a crack pattern, a knowledge of the time of the first appearance of cracks is helpful in diagnosing the underlying cause.

9.2.1 Plastic Shrinkage Cracking

Plastic shrinkage cracking in concrete occurs most commonly on the exposed surfaces of freshly placed floors and slabs or other elements with large surface areas when subjected to a very rapid loss of moisture caused by low humidity and wind, high temperature, or both.

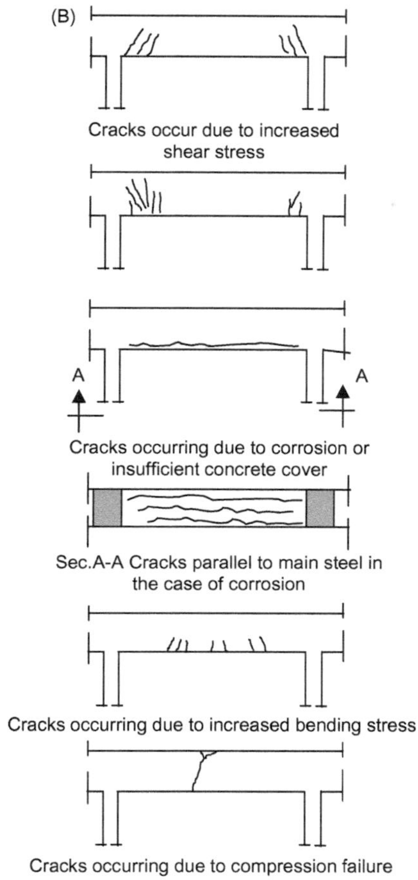

Figure 9.1 (Continued)

Plastic shrinkage usually occurs prior to final finishing, before curing starts.

When moisture evaporates from the surface of freshly placed concrete faster than it is placed by bleed water, the surface concrete shrinks.

Due to the restraint provided by the drying surface layer of the concrete, tensile stresses develop in the weak, stiffening plastic concrete, resulting in shallow cracks that are usually not short and run in all directions.

In most cases, these cracks are wide at the surface. They range from a few millimeters to many meters in length and are spaced from a few centimeters to as much as 3 m apart. Plastic shrinkage cracks may extend the full depth of elevated structural slabs.

(C)

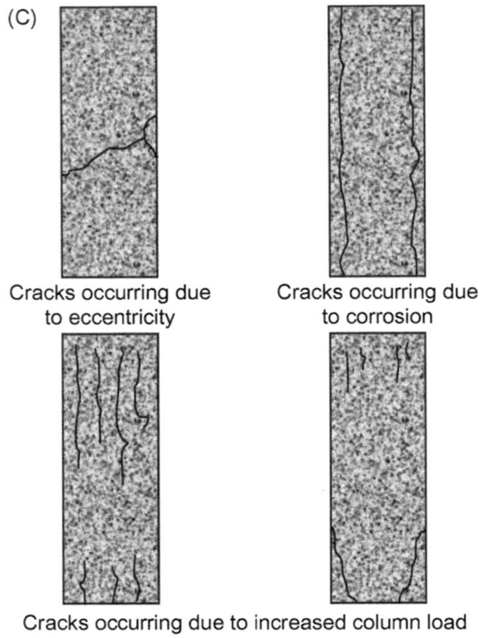

Figure 9.1 (Continued)

Table 9.1 Typical times for appearance of defects

Type of defect	Typical time of appearance
Plastic settlement cracks	10 min to 3 h
Plastic shrinkage cracks	30 min to 6 h
Crazing	1–7 days, sometimes much longer
Early thermal contraction cracks	1–2 days until 3 weeks
Long-term drying shrinkage cracks	Several weeks or months

From Concrete Society Technical Report 54.

9.2.2 Settlement Cracking

After initial placement, vibration, and finishing, concrete has a tendency to continue to consolidate. During this period, the plastic concrete may be locally restrained by reinforcing steel, a prior concrete placement, or formwork. This local restraint may result in voids, cracks, or both adjacent to restraining the element. When associated with reinforcing steel, settlement cracking increases with increasing bar size, increasing slump, and decreasing cover. The degree of settlement cracking will be magnified by insufficient vibration or the use of leaking or highly flexible forms.

Proper form design and adequate vibration, provision of sufficient time interval between the placement of concrete in slabs and beam, the use of the lowest possible slump, and an increase in concrete cover will reduce settlement cracking.

9.2.3 Drying Shrinkage

The common cause of cracking in concrete is drying shrinkage. This type of shrinkage is caused by the loss of moisture from the cement paste constituent, which can shrink by as much as 1% per unit length. Unfortunately, aggregate provides internal restraint that reduces the magnitude of this volume change to about 0.05%. On wetting, concrete tends to expand.

These moisture-induced volume changes are a characteristic of concrete. If the shrinkage of concrete could take place without any restraint, the concrete would not crack. It is the combination of shrinkage and restraint, which is usually provided by another part of the structure or by the subgrade, that causes tensile stresses to develop. When the tensile stress of concrete is exceeded, it will crack. Cracks may propagate at much lower stresses than are required to cause crack initiation.

In massive concrete structure elements, as in cases of a foundation under a heavy machine, tensile stresses are used by differential shrinkage between the surface and the interior concrete. The larger shrinkage at the surface causes cracks to develop; these may, with time, penetrate deeper into the concrete.

The magnitude of the tensile stresses is influenced by a combination of factors which will be as follows:

1. the amount of shrinkage,
2. the degree of restraint,
3. the modulus of elasticity,
4. the amount of creep.

The amount of drying shrinkage is influenced mainly by the amount and type of aggregate and the water content of the mix. The greater the amount of aggregate, the smaller the amount of shrinkage. The higher the stiffness of the aggregate, the more effective it is in reducing the shrinkage of the concrete, meaning that the concrete containing sandstone aggregate has a about twice the shrinkage of concrete containing granite, basalt, or limestone. The higher the water content, the greater the amount of drying shrinkage.

Surface crazing on walls and slabs is an excellent example of drying shrinkage on a small scale. Crazing usually occurs when the surface layer of the concrete has higher water content than the interior concrete. The result is a series of shallow, closely spaced, fine cracks.

Shrinkage cracking can be controlled by using properly spaced contraction joints and proper steel detailing. Shrinkage cracking may also be controlled using shrinkage-compensating cement.

9.2.4 Thermal Stresses

The temperature differences within a concrete structure may be due to cement hydration or changes in ambient temperature conditions or both. These temperature differences result in differential volume changes. The concrete will crack when the tensile strains due to the differential volume changes exceed their tensile strain capacity.

The effects of temperature differentials due to the hydration of cement are normally associated with mass concrete such as large columns, piers, beams, footing, retaining walls, and dams, while temperature differentials due to changes in the ambient temperature can affect any structure.

The concrete rapidly gains both strength and stiffness as cooling begins. Any restraint of the free contraction during cooling will result in tensile stress. Tensile stresses developed during the cooling stage are proportional to the temperature change, the coefficient of thermal expansion, the effective modulus of elasticity, and the degree of restraint. The more massive the structure, the greater the potential for temperature differentials and degrees of restraint.

Procedures to help reduce thermally induced cracking include reducing the maximum internal temperature, controlling the rate at which the concrete cools, and increasing the tensile strain capacity of the concrete.

Hardened concrete has a coefficient of thermal expansion that may range from 7 to $11 \times 10^{-6}/°C$ with an average of $10 \times 10^{-6}/°C$. When one portion of a structure is subjected to a temperature-induced volume change, the potential for thermally induced cracking exists.

The designers must give special consideration to structures in which some portions are exposed to temperature changes, while other portions of the structure are either partially or completely protected. A drop in temperature may result in cracking in the exposed element, while increases in temperature may cause cracking in the protected portion of the structure. So the designer must consider the structure movement by recommend using contraction joint and provide its detailed drawing.

These structures, which have a high difference in temperature, are usually concrete structures built in the area near the desert where there is a greater difference in the temperature between afternoon and midnight. Moreover, in cases of countries with high temperatures, they usually use air conditioners inside the building, so there is a high probability of cracks occurring due to the difference in temperature inside and outside the building. The designer should take these stresses into consideration.

9.2.5 Chemical Reaction

As shown in Fig. 9.1A, starlike cracks in the concrete surface indicate a chemical reaction. This reaction occurs if there is some aggregate containing active silica and alkalis derived from cement hydration, admixtures, or external sources such as curing water, groundwater, or alkaline solutions stored or used in the finished structure.

The alkali-silica reaction results in the formation of a swelling gel, which tends to draw water from other portions of the concrete. This causes local expansion and accompanying tensile stresses, and may eventually result in the complete deterioration of the structure.

Groundwater with sulfate presents a special durability problem for concrete. Due to the sulfate penetrates so it comes in contact with hydrated calcium aluminates and calcium sulfoaluminate will form, accompanied by an increase in volume, resulting in high tensile stresses which cause cracking. Therefore, it is recommended to use Portland Cement Type II and Type V, which contain low tricalcium aluminates and will reduce the severity of the cracks.

As noted in Fig. 9.1, in cases of corrosion in the steel reinforcement the cracks you will see are cracks parallel to the steel bars for any concrete members such as beams, slabs, and columns.

Moreover, you will see spots of brown color on the concrete surface, which is an indication of the corrosion in the steel reinforcement. In this stage, we use only the experience of the engineers who do the inspection.

The disadvantage of the visual inspection is that it completely depends on the experience of the engineer who does the inspection, as there are some cracks that have more than one reason with or without the corrosion effect.

9.3 COLLECTING DATA

The following data must be collected before performing the visual inspection and must be stated in the structure assessment report:

- construction year,
- the contractor and engineering office data,
- the structure system,
- the drawings and project specifications,
- the construction method,
- the concrete tests, if available,
- the environmental condition (near the sea or a chemical factory).

All the above data must be provided to you from the owner if you are a third party expert. In most cases, all these data cannot be collected. If you inspect an old structure over 20 or 30 years old, the drawing and specifications will not be available as there were no computers or software drawing programs from 30 years ago. All the hard-copy drawings in most cases are depleted or in bad condition. In cases where the owner has a big organization with an engineering department, they may have a document control system so you can find the drawing. Usually these buildings are office and industrial buildings, but for residential buildings you will be very lucky if the owner has these drawings.

The construction year is the critical information, as from this you can define the codes and specifications that were applied in this year, and this depends on your experience.

From knowing the construction year, the contractor who constructed the building, and the engineering office that provided the engineering design, one can imagine the condition of the building based on the reputation of the contractor and the engineering office.

The construction system can be deduced by visual inspection, the drawing, or by talking with the users of the building and the old engineers who shared in reviewing the engineering design documents or supervising the construction.

The environmental condition information will be provided to you by the owner, and you can see it on site and collect the data about the environmental condition from other sources and take into consideration all the data: weather temperature in summer and winter, morning and night, and the wind and wave effect.

If you are an owner, you must be sure that all the data are delivered to the consultant who does the inspection, as any data are valuable to him.

Any engineering work is like a computer: good in good out and rubbish in rubbish out. So you must verify all the data which you deliver to the expert who performs the building assessment.

In industrial facilities plants, the conditions and mode of operations are not clear to the consultant engineer, so you must deliver it to him; the vibration of the machines, the heat of liquid in pipelines, and so on has more impact on the concrete structure assessment result.

Take into consideration when collecting data from the process plant or factory that the mode of operations may change from time to time, so all these histories must be delivered to the engineer who does the evaluation process as he is usually a structural engineer and unfamiliar and cannot imagine these situations.

If the data of wind and wave does not exist or is not known, you must obtain it from a third party specialized in met ocean criteria. These data are very important in case of offshore structures or marine structure; their assessment depends on these data if wind and wave represent the loads most affecting the structure.

The concrete compressive strength test results are usually performed in any building by cylinder specimens (as in American code) or cube specimens (as in the British standard), but the problem is finding these data after 20 years or more, depending on the age of the building.

If these data are not available, the test must be informed by using an ultrasonic pulse velocity test, Schmidt hammer, or core test.

9.4 DETAILED INSPECTION

The purpose of the detailed inspection is to accurately determine whenever possible the degree of severity and the deterioration of the concrete.

Therefore, we need to know the number of collapses that have occurred, the cause of the deterioration in the concrete, and the amount of repair, as such quantities will be placed in the tender that will be put forward to contractors for repair.

At that stage we will need to be familiar with the reasons for the deterioration in detail and have the capability to perform a failure analysis technique.

Initially, we will use a method of visual inspection, maybe in conjunction with the use of a small hammer, going to other measurements to determine the depth of the transformation of carbon in the concrete and the degree of corrosion of steel in concrete and how much it extends in the steel bars. Moreover in this stage, it is important to define the degree of concrete electric resistivity to predict the corrosion rate; all these required measurements will be discussed in this chapter.

As you know, the weather condition that affects the building is the main factor that affects the measurement readings, and also affects the selection of the method of repair.

9.4.1 Methods of Structure Assessment

There are several methods of assessing the structure, as mentioned in terms of the degree of extension of corrosion in steel reinforcement bars and its impact on the whole structure. The first and most important method is the visual inspection, as it does not require a high cost and is an easy method to use.

This is followed by other methods, which require capabilities and are often used in cases of structures of special importance, which require high costs for repair. Therefore, the use of sophisticated technology is required to identify the degree of corrosion of the steel reinforcement in concrete with high accuracy, because that would affect the total cost of the state of structure; for example, the concrete bridges or special buildings such as parking garages, tunnels, and other special concrete structures.

Each of the measurement tools that are used has a specific accuracy, with advantages and disadvantages. In general, the process of the structure assessment due to corrosion must be conducted by a person who has an enough experience in the process of corrosion as the experience is the master key and played a big role in success assessment, but there are some devices that rely on specialized only for the use and because of their experience to know the accuracy and environmental factor that affect the equipment readings and how to overcome them.

The following methods can be used to define the condition of the structure. We are focusing here on the methods that can be used in our oil and gas plant, as other methods shall be used for bridges, such as radar, for example. The summary of these methods and their functions is presented in Table 9.2.

These measurements will determine the cause of corrosion and the degree of passive protection layer for the steel bars, as well as expect the corrosion rate in every part of the building. From this information, one can determine what kind of repair the structure is needed, its repair method statement, and calculate the quantity survey for the required repair to strength the concrete member.

Table 9.2 Practical methods for evaluating concrete structures

Methods	Inspection
Chain or hammers	Void behind cover
Concrete cover measurement	Distance between steel bars and concrete surface
Phenolphthalein	Carbonation depth
Half cell	Evaluate corrosion risk
Linear polarization	Corrosion rate

9.4.2 Concrete Test Data

The following are the tests that should be performed to obtain the data of the concrete strength if these data are not available. There are variations between these test from their accuracy, their effect on the building and their cost. Therefore, you have the responsibility to decide if any of these tests is suitable to your building that you need to evaluate.

Moreover, there are some precautions for these tests when you take the samples, as they must be based on the visual inspection results that were done before.

Moreover, when you take the samples or do the loading test, it must be accompanied by a suitable temporary support to the structure and the adjacent members. The temporary support locations should be defined by calculation considering the possibility of member failure during the test. These tests will follow ACI228-89 R1 and BS1881.

9.4.2.1 Core Test

This test is considered one of the semidestructive tests. This test is very important and popular for the study of the safety of structure as a result of changing the system of loading, a deterioration of the structure as a result of accidents such as fire or weather factors, or the need for temporary support for repair when there is no accurate data about concrete strength.

This test is not too expensive, and it is the most accurate test to measure the strength of the concrete actually carried out.

Core testing is performed by cutting cylinders from a concrete member, which could affect the integrity of structure. Therefore the samples must be taken in strict adherence to the standard. This will ensure accuracy of the results without weakening the structure.

In our case, for the deteriorated structure due to corrosion of steel bars, the structure has lost most of its strength due to reduction on the steel cross-section area, so more caution must be considered when performing this test and selecting the proper concrete member to perform this test on which will not affect the whole building's safety.

The codes and specifications provide some guidance to the number of cores to test; these values are as follows:

- Volume of concrete member $(V) \leq 150\ m^3$ take 3 cores.
- Volume of concrete member $(V) > 150\ m^3$ take $(3 + (V - 150/50))$ cores.

Before you choose the location of the sample, you must first define the location of the steel bars. To assist you, select the location of the sample away from the steel bars to avoid taking a sample containing steel reinforcement bars.

We must carefully know the integrity of structure therefore this test should be performed by an experienced engineer to conduct such an experiment should be taking precautions and to determine the responsibility of individuals and review the nondestructive testing which had been conducted in accurate manner.

Fig. 9.2 presents the process of taking the core from reinforced concrete bridge girder.

Figure 9.2 Taking a core sample.

Table 9.3 Core size with possible problem

Test	Diameter (mm)	Length (mm)	Possible Problem
One	150	150	May contain steel reinforcement
Two	150	300	May cause more cutting depth to concrete member
Third	100	100	– Not allowed if the maximum aggregate size is 25 mm – May cut with depth less than required
Last	100	200	Less accurate data

9.4.2.1.1 Core Size

Note that the permitted diameter is 100 mm in case the maximum aggregate size is 25 and 150 mm in case the maximum aggregate size does not exceed 40 mm. It is preferable to use a 150 mm diameter whenever possible, as it gives more accurate results as shown in Table 9.3, which presents the relationship between the dimensions of the sample and potential problems. This table should be referred to when choosing the reasonable core size.

Some research states that the core test can be done with a core diameter of 50 mm in the case of a maximum aggregate size not more than 20 mm overall, but it was noted that small core sizes gave different results than large sizes.

Because of the seriousness of the test and the inability to take a high number of samples, it requires good supervision when taking the sample. Moreover, the laboratory test must be certified, and the test equipment must be calibrated with a certificate of calibration from a certified company.

Sample extraction are using a cylinder which is a different cylinders equipped with ransom of special alloy mixture with diamonds powder to do the concrete cutting during the rotation of the cylinder through the concrete member. Precautions sampling method should be a match and be consistent pressure to bear on the match by appropriate and it depends on the expertise of the technical.

After that, the core will be filled by pour grouting. Another solution depends on using epoxy and injects it in the hole before inserting a concrete core of the same size to close the hole.

In any way, the filling must be done soon after making the cutting, by other meaning, the material of filing will be ready beside to the technician who do the cutting because this core is effecting on the integrity of the structure.

The lab must examine and photograph every core; noted gaps identified within the core are either small void if measured between 0.5 and 3 mm, medium void if the average measurement is between 3 and 6 mm, or a big void if measured at more than 6 mm. And also to be examined the shape, kind and color gradient of aggregates any apparent qualities of the phenomenon of sand as well.

In the laboratory, the dimensions, weight of each core, the density, steel bar diameter, and distance between the bars will be measured.

9.4.2.1.2 Sample Preparation for Test

After cutting the core from the concrete element and processing the sample testing by leveling the surface of the core, take a core which has a length not less than 95% of the diameter and not more than double the diameter.

For the leveling surface, use a chainsaw spare concrete or steel cutting disk. After that, prepare the two ends of the sample by covering them with mortar or sulfide and submerge the sample in water at a temperature of $20 \pm 2^{\circ}C$ for at least 48 h before testing the sample.

Put the sample in a machine test and apply gradually an influence load with the rate of regular and continuous range of 0.2 to 0.4 N/mm^2 until it reaches the maximum load at which the sample has been crushed.

To determine the estimated actual strength for cube by knowing the crushing stress that is obtained from the test using the following equation where, λ is the divided the core length to its diameter.

- In the case of a horizontal core, the strength calculation will be as follows:

  $$\text{Estimated actual strength for cube} = \frac{2.5}{1/\lambda} + 1.5 \times \text{core strength}$$

  where λ = core length/core diameter.

- In the case of a vertical core, the strength calculation will be as follows:

$$\text{Estimated actual strength for cube} = \frac{2.3}{1/\lambda} + 1.5 \times \text{core strength} \quad (9.1)$$

In cases of existing steel in the core perpendicular to the core axis, the previous equations will be multiplied by the following correction factor:

$$\text{Correction factor} = 1 + 1.5\ \frac{s\varphi}{LD} \tag{9.2}$$

where:

L = core length,

D = core diameter,

S = distance from steel bar to edge of core,

φ = steel bar diameter.

Cores are preferred to be free of steel, but if steel is found, you must use the correction factor, taking into account that it is taken only in the event that the value ranges from 10% to 5%. We must agree to use the core results, but if the correction factor is more than 10%, the results of the cores cannot be trusted, and you should take another core.

When examining the test results, the following points must be taken into account:

- If the test sample is submerged in water it will lead to decrease in strength to up to about 15% for the strength of dry concrete.
- The equation to calculate the expected actual concrete strength does not take into account any differences in direction between the core and the standard cube direction.
- It states that the concrete was acceptable if the average strength to the cores is at least 75% of the required strength and the calculated strength for any core is not less than 65% of the required strength.

It states that in cases of prestressed concrete, the concrete strength was acceptable if the average strength of the cores is at least 80% of the required strength and the calculated strength of any core is less than 75% of the required strength.

9.4.2.2 Rebound Hammer

This is non-destructive testing, so it is useful in determining the estimated concrete compressive strength.

This is the most common as it is easy to do and cheap compared with other tests, but it also gives a less precise outcome of data results.

This test relies on measuring the concrete strength by measuring the hardening from the surface. It will identify the concrete compressive strength of the concrete member by using calibration curves of the relationship between reading the concrete hardening and concrete compressive strength. There are different types of rebound hammers and most commonly gives impact energy of 2.2 N/mm. There is more than one way to show results based on the manufacturer; in some cases, the reading will be analog or digital, or it will be taken by connecting to a memory to record the measurement readings.

Inspect the device before using it through the calibration tools that are attached to the device when you purchase it, and the calibration should be within the allowable limit based on manufacturer recommendations.

The first and most important step in the test is that we clean and smooth the concrete surface at the sites which will be tested by homing in on an area of about 300 mm × 300 mm; preferably test on the surface that has no change after cast or the surface which has not had any smoothing done during casting process.

On the surface to be tested, draw a net of perpendicular lines in both directions 20 to 50 mm apart; the intersection points will be the points to make the test on. The test point in any case must be away from the edge by about 2 cm.

Fig. 9.3 shows that the surface must be cleaned before doing the test, and the rebound hammer is perpendicular to the surface as shown in the figure.

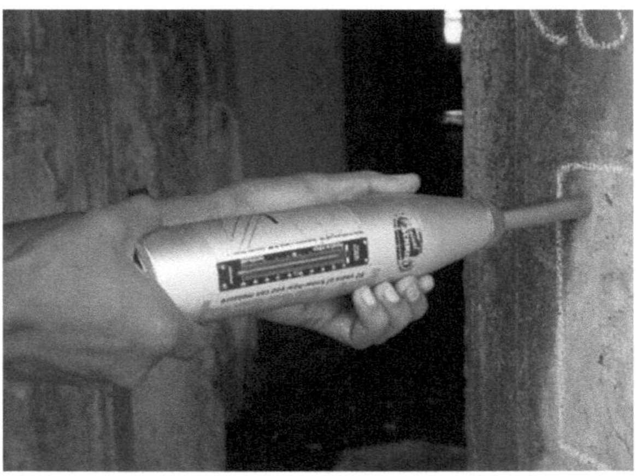

Figure 9.3 Test and read the number.

The following recommendations should be done during the test:

- The hammer must be perpendicular to the surface that will be tested in any conditions because the direction of the hammer affects the value of the rebound number as a result of the impact of hammer weight.
- The wet surface gives a reading of a rebound hammer less than reading a dry surface significantly, by up to 20%.
- The tested concrete member must be fixed and cannot vibrate.
- You must not use the curves for the relationship between concrete compressive strength and the rebound number as given from the manufacturer directly, but you must calibrate the hammer by taking the reading on concrete cubes and crushing the concrete cubes to obtain the calibration of the curves. This calibration is important to be done from time to time as the spring inside the rebound hammer will lose some of its stiffness with time.
- You must use only one hammer when making a comparison between the quality of concrete in different sites.
- The type of cement affects the readings, as in the case of concrete with high-alumina; this cement can yield higher results than concrete with ordinary Portland cement by about 100%.
- Concrete with sulfate-resistant cement can yield results of about 50% less than ordinary Portland cement.
- Higher cement content gives a lower reading than concrete with less cement content; in any case it will give a gross error of only 10%.

The number of readings must by high enough to give reasonably accurate results. The minimum number of readings is 10, but usually we take 15 readings.

The extreme values will be excluded, and we will take the average of the other remaining values. From this, the concrete compressive strength will be known.

9.4.2.3 Inherent Variations in In-Situ Strength

Much research performed in this area and based on BS6089 tests on in situ concrete indicate the following:

- In situ strength can vary within a structural member both randomly and, often, in an ordered fashion.
- The magnitude of variations of in situ strength within structural members varies from one member to another in a random fashion.

- With the height of a concrete lift, in situ strength decreases towards the top of a lift, even for slabs, and can be 25% less at the top than in the body of the concrete. Concrete of lower strength is often concentrated in the top 300 mm or 20% of the depth, whichever is the less.
- At 28 days after casting, columns can have a mean in situ strength of 65% of the mean standard cube strength, with strength in the individual columns carrying from 50% to 80% (7). The little evidence available for floor slabs suggests that mean in situ strength may only be 50% of the mean standard cube strength.
- The gain in in situ strength from 28 days onwards is not consistent. At 6 months, the increase in mean strength can vary from 0% to 25% and at 1 year from 0% to 35% of the 28-day strength.

Thus, the normal variation of an in situ strength within and between structural members has to be borne in mind when evaluating in situ results.

Examination of individual test results will identify whether variations between results are excessive. Further tests may be required to establish whether certain results are rogue values or not.

The result of the test is the estimated in situ cube strength, which is needed to compare to the design strength to evaluate if a structure is acceptable or not. The design strength is given by f_{cu}/g_m, where $\gamma_m = 1.5$ for ultimate strength.

Based on BS6089, the level of in situ cube strength that may be considered acceptable in any particular case is a matter for engineering judgment but should not normally be less than 1.2 times the design strength.

The particular strength level selected should include an allowance for possible future deterioration of the strength of concrete that may result from chemical attack, weathering, vibration, or some unforeseeable impact or other circumstances.

Apply the following equation in case of $\gamma_m = 1.5$ to accept or refuse the concrete.

Estimated in situ cube strength $\geq 1.2\dfrac{(f_{cu})}{1.5} \geq \dfrac{f_{cu}}{1.25}$.

9.4.2.4 Comparison Between Different Tests

The different methods to determine the hardened concrete strength have advantages and disadvantages, which are summarized in Table 9.4.

Moreover, these tests are different with regards to their costs, their representative to the concrete member, and the accuracy of the measured strength compared to the actual concrete strength.

Table 9.4 Comparison between different test methods

Test method	Probable damage	Precaution requirement	Representative to concrete	Accuracy	Test speed	Cost
Overload test	Possible member loss	– Member must be isolated or allowed to distribute the load to the adjacent members – Extensive safety precaution just in case collapse failure	Good	Good	Slow	High
Cores	Holes to be made good	– Limitations of core size and numbers – Safety precaution for critical member	Moderate	Good	Slow	High
Rebound hammer	None	Need a smooth surface	Surface only	Poor	Fast	Very low

Table 9.5 Variation between standard cube strength and actual strength on site

Concrete member	Actual strength/standard cube strength after 28 days	
	Average	**Range**
Column	0.65	0.55–0.75
Wall	0.65	0.45–0.95
Beam	0.75	0.6–1.0
Slab	0.5	0.4–0.6

After you obtain the result of the tests, you may find that the data of strength is lower than the concrete strength which specify in the drawings or project specifications. The value of concrete strength in the drawing is the standard cube (cylinder) compressive strength after 28 days. Noting that the cubes or the cylinder will be pour and compacting based on the standard and the curing process was for 28 days so it is different that will happen in the site. For example the curing will not be in 28 days and you cannot immerse the concrete slabs or columns in water continuous for 28 days. The temperature on site is more different and varied than that on the laboratory. In addition to that, the concrete pouring and compaction method is different for the cubes or cylinder than that in the actual member size. All these variations were taken into consideration inside the factors in the design codes, but when you make the test you measure the strength after complete hardening to the concrete. Table 9.5 shows a comparison between a standard cube test and an in situ test for different reinforced concrete members.

9.4.3 Sources of Concrete Failure

Corrosion is not the only source of failure, but there are many other sources that cause determination on the reinforced concrete structures which must be kept in mind when you do the inspection and know about it very well. These sources of failure will be as follows:

I. Unsuitable materials
- unsound aggregate,
- reactive aggregate,
- contaminated aggregate,
- the wrong type of cement,
- cement manufacturer error,
- the wrong type of admixture,
- substandard admixture,
- contaminated admixture,

- organic contaminated water,
- chemical contaminated water,
- wrong kind of reinforcement,

II. Improper workmanship
- faulty design,
- incorrect concrete mixture (low or high cement content and incorrect admixture dosage),
- unstable formwork,
- misplaced reinforcement,
- error in handling and placing concrete (segregation, bad placing, and inadequate compact),
- curing incomplete.
- size error of steel bars.

III. Environmental
- soil alkali,
- seawater or sewage,
- acid industry,
- freezing and thawing.

IV. Structural
- load exceeds design,
- accident, such as ballast load or dropped object,
- earthquake load.

9.5 TEST METHODS ON CORRODED STEEL IN CONCRETE

During the corrosion process, the volume of steel bars increases; due to the volume increase, the internal stresses format and cause the concrete cracks. These cracks may be vertical or horizontal and these cracks increase in length and width until the fall of the concrete cover.

By this concept to determine the parts of concrete element which is corroded can be define by examine the location which has a separation between the concrete cover and the concrete section.

9.5.1 Manual Method

Therefore, the deteriorated area can be defined by using hammer to hit the concrete cover and hearing the sound. If the sound indicates air behind the surface of the concrete cover, it means that the steel bars are corroded with the possibility of a concrete cover separation.

In cases of a large area of concrete surface, you can use steel chain; let it move on the surface and define in the same manner the separated cover and the damaged parts.

There are other, more complicated methods than using a hammer or steel chain, such as by using infrared or ultrasonic.

Using the above inspection methods, you can define the area that is expected to be corroded and predict concrete cover spall after a short time so it can be considered in the next repair.

Equipment used in that way is the manual hammer or steel chain on the surface deck of the bridge. This method is fast and easy to use; on the other side, it is lower-cost than the other methods such as radar, ultrasonic, or infrared ray, and this method can be used in big structures or when evaluating a huge area such as a bridge's deck.

It should be noted that the use of the hammer is often associated with the process of visual inspection where it is easily carried the hammer as it is low-cost also quick to locate defects, and dirty that make markers at the places where the sound of a vacuum behind the concrete surface is hear.

Generally, it is also noted that these methods, whether manual, radar, or infrared, need professionally trained workers with high experience and competence to read the measurement and define the defected area accurately.

There are precautions and limits on the use of these methods where, in the case of the existence of water in the cracks or that the separation is in deep inside the concrete section, as well as in the case of the steel bars embedded with depth within the concrete will be difficult to define the separate areas. Moreover, the existence of water and the depth of separation is something that affects the readings in the case of radar and infrared.

The execution of the repair process should remove area of a size more than the defected area that has been defined before by different methods because there is time between a structure's evaluation and the start of repairs due to administration work, define the funds, and obtaining approval, so this duration is sufficient to increase the volume of concrete, which enables more separation.

In brief, using infrared has main disadvantages, as it needs appropriate weather conditions to provide reasonable accuracy; for this reason, it is rarely used in industrial projects.

9.5.2 Concrete Cover Measurements

The thickness of the concrete cover measure is usually part of the concrete quality control in modern construction so as to ensure that the thickness of the concrete cover in conformity with the specifications.

The process of measuring the thickness of the concrete cover in structures was noted by the beginning of the corrosion, as the lack of concrete cover thickness increased the corrosion rate for corrosion as a result of chlorides or carbonation, where it expedited the propagation inside the concrete, causing the speed of steel corrosion, as well as the lack of cover helps the propagation of moisture and oxygen, which are form the main basis for the corrosion process.

The measurement of the concrete cover thickness explains the causes of corrosion and identifies areas that have the capability to corrode faster.

The measurement of the thickness of the concrete cover needs to be defined the *x*- and *y*-axes in order to determine the thickness of the concrete cover at every point on the structure.

The equipment which measures the thickness of the concrete cover is simple and high technology, as you can obtain measurements as numbers. Fig. 9.4 presents an electromagnetic cover meter shape and the method of reading.

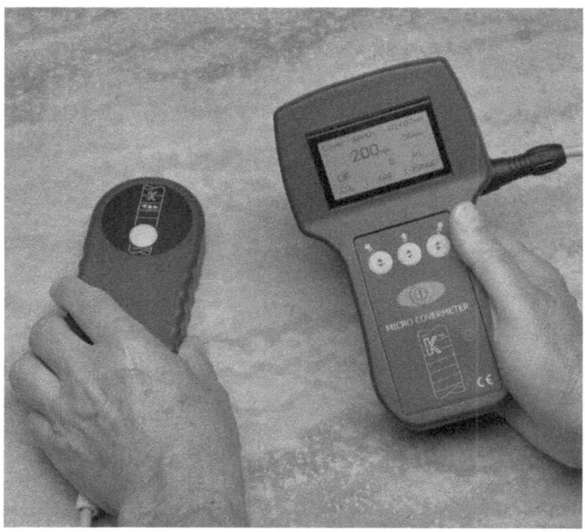

Figure 9.4 Shape of concrete cover thickness measurement machine.

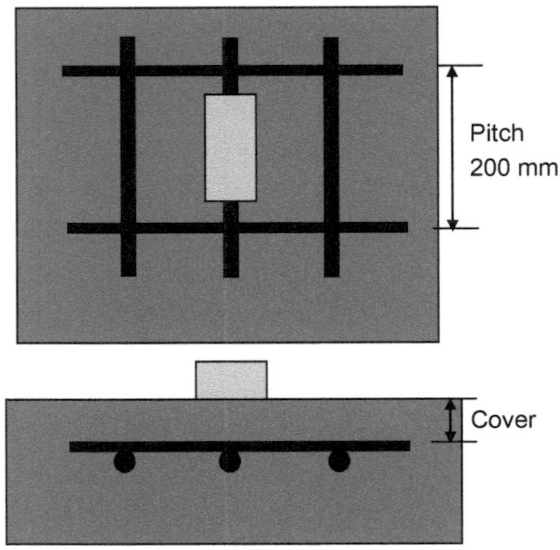

Figure 9.5 Device location on concrete surface.

Radiographs can be used on bridges but are quite expensive (Cadry Gamnon, 1992; Bungey, 1993).

The magnetic cover method is a simple method, but the accuracy is affected by the distance between steel bars, and when the thickness of the concrete cover a large influence in the readings significantly. Since this method depends on the supply of electricity through the 9-V battery, the second side of measurement for the potential voltage envelope when completed by the steel bars is buried, as shown in Fig. 9.5.

The British standard is the only standard that cares about measurement of concrete cover after construction in BS1881, Part 204. In 1993, Alldred studied the accuracy of the measurement of the cover when there are more steel bars close to each other, and he suggested using more than one head of measurement, as that will increase the accuracy of the reading; the small head has an impact on the accuracy of the equipment.

Therefore, the problem of this method is in cases of dense steel reinforcement in the concrete section that will give nonaccurate data; in this case, the equipment will be calibrated based on the existing steel bars as the reading will be effected by the type of steel. The person who works on the equipment must be competent and aware of anything that can affect the reading such as bolts, steel wires, etc.

9.5.3 Half-Cell Potential Measurements

This measurement unit is traditionally used in oil and gas plants to monitor the cathodic protection system for the pipeline and the facilities, and you will find a well-trained corrosion team can assist you in these measurements.

This system is used to determine the bars which have lost their passive protection. As explained in Chapter 2, Loads on the Industrial Structures, the bars in the concrete that have lost passive protection, have been exposed to carbonating or existing chlorides, which makes chemical react with the alkaline around the steel, which causes it to lose its passive layer.

The half cell equipment (ASTM C876 1991) is a rod of metal in a solution by the same metal ions such as a copper rod in a saturated solution of copper sulfate, and this is considered one of the most common equipments; there are some other types such as a bar of silver in silver chloride solution linked through the voltmeter and other terminal of the voltmeter is connected to the steel bars as in Fig. 9.6.

If the steel bars have a passive protection layer, the potential volt is small, with a range from 0 to 200 MV copper/copper sulfate. If there has been a breakdown of the passive protection layer and some quantity of steel has melted as a result of the movement of ions, the potential volt is about 350 MV; when the value is higher than 350 MV, it means that the steel has already started to corrode.

Figure 9.6 Half cell machine.

It is recommended to use half cell equipment consisting of silver and silver chloride solution or mercury and mercury oxide solution. The equipment consisting of copper and copper sulfate solution is used, but it does not recommend using it because it requires regular maintenance and thus readings will contain some mistakes.

To do a survey of a large surface area of concrete to determine the bars that have been corroded. Define the location of the surface that is required to measure by do signs and then do the measurement of the potential volt. From these results one draws contour lines on the surface and every line present an equal value of potential volt.

The measurement of half cell to the steel reinforcement bars is to define the probability of causing corrosion to the steel bars; the ASTM C867 specifications present a method to clarify the data which is obtained from measuring the potential volt, and these values are as shown in Table 9.6 for copper/copper sulfate equipment and silver/chloride silver equipment, and explain the condition of corrosion in each case of corresponding value.

The inaccuracy of such measurements is due to the presence of water, which increases the negative values without happened corrosion in steel as wetting concrete columns or walls give high negative values for potential volt regardless of the degree of corrosion steel and negative voltage

Table 9.6 ASTM specification for steel corrosion for different half-cells

Corrosion condition	Silver/silver chloride (mV)	Copper/copper sulfate (mV)	Standard hydrogen electrode (mV)	Calomel (mV)
Low (10% risk of corrosion)	> -106	> -200	>116	> -126
Intermediate corrosion risk	-106 to -256	-200 to -350	116 to -34	-126 to -276
High (<90% risk of corrosion)	< -256	< -350	< -34	< -276
Severe corrosion	< -406	< -500	< -184	< -426

difference over significantly under water in the offshore structures in spite of the lack of oxygen reduces the rate of corrosion.

9.5.4 Electrical Resistivity Measurement

The process of steel corrosion is the process of a combination between electrical and chemical reaction, as the electrical resistance of concrete is a major factor in the rate of steel corrosion. The corrosion rate in the steel reinforcement depends on the movement of ions from the anode to the cathode, causing corrosion.

A four probe resistivity is used to measure the soil electrical resistivity, and the same equipment with modification is used to measure the electrical resistance of concrete.

The measured resistance of concrete by the four probe bypass electric current between outside two probes and measuring the potential volt between the inner two and from this can measure the electrical resistivity of the concrete. Some equipment is needed to make a hole in the concrete to fix the probes, but the new equipment only puts it on the surface.

The electrical resistance affected so much with the moisture in the concrete internal porous and it significantly affected by the quality of the content its mean the water to cement ratio in the concrete mix, curing and using additives. The level of chlorides does not have such a great impact on the electrical resistance of concrete.

Must be taken into account during the measurement to be greater distance between the probes of the largest size of the aggregate, where we avoid the measurement of electrical resistance of a piece of aggregate only. It is important to measure away from the steel bar so must take the readings perpendicular to the steel reinforcement.

The measurement of electrical resistance of the concrete cover only through the use of two probes only and make the steel reinforcement the first probe and the second probe moving in the concrete surface.

It is shown by the results obtained on the electrical resistance of concrete and can be distinguished rate situation corrosion steel through the comparison as follows according to according to Broomfield et al. (1987, 1993).

- 120 kΩ cm low corrosion rate,
- 10–120 kΩ cm low to medium corrosion rate,
- 5–110 kΩ cm high corrosion rate,

- < 5 kΩ cm very high corrosion rate.

Many researchers studied the relation between electrical resistivity and corrosion rate by using two electron probes and concluded the following:

- 100 kΩ cm cannot differentiate between passive and active steel reinforcement.
- 50–100 kΩ cm slow corrosion rate.
- 50–10 kΩ cm from the medium to high corrosion rate with active steel to corrosion.
- < 10 kΩ cm electrical resistivity has no impact on the corrosion process.

From the previous relation, this method is considered reasonable to know the corrosion rate in steel reinforcement.

The electrical resistance is expressed in the corrosion rate of steel reinforcement, and is also a good representative of the concrete quality control and the density of the concrete, which is depend on the compaction method on site.

The major problem for the four poles to measure the electrical resistance is the effect of steel on reading with that according to the study from the University of Liverpool that if angle of measurement perpendicular to the steel the reading error can be reduced to less extent possible. It also found when the electrical resistance is more than 10 kΩ cm, it cannot distinguish between reinforcing steel active or passive protection by and if *I* electrical resistance of 10 kΩ cm, the resistance is not effective in the process of corrosion.

9.5.5 Measurement of Carbonation Depth

Carbonation depth is formulated in the propagation of carbon dioxide inside the cover of the concrete, which reduces the concrete alkalinity value. The concrete alkalinity value is expressed by pH values between 12 and 13, with carbon dioxide propagation causing a reaction which reduces the alkalinity; this will affect the steel surface, damage the passive protection layer around the steel bars, and then start the corrosion process.

Therefore, it is very important to define the depth of the concrete that has transformed due to carbonation and how far away or close it is to the steel bars.

This test is performed by spraying the surface of broken concrete or letting it break with special tools to obtain the carbonation depth by Phenolphthalein dissolved in alcohol. This solution color becomes pink when touching the surface of concrete with an alkalinity value of pH 12 to 13.5; and the color turns gray or blue if the concrete has lost alkalinity, and its pH value is less than 9. In this case, steel bars have lost their passive protection layer.

The test must make sure that the part that will test newly broken and must be done during the test to test whether beams, columns, or slabs. After measuring the carbonation depth that has occurred to the concrete cover and the distance between carbonation depth and the steel bars, it is easy to evaluate the corrosion risk to the steel reinforcement.

The best solution is to use Phenolphthalein with alcohol and water to be 1 mg phenolphthalein with 100 mm alcohol/water (50:50 mixing ratio) or more alcohol than water (Building Research Center, 1981; and Parrott, 1987) if the concrete is completely dry, humidifying the surface by water.

If the thickness of carbonation measurement reveals that the carbonation reaches until around 5 mm from the steel bars which is known by change of the concrete surface color, then this means that the break of passive layer is under progress.

Some the aggregate or concrete mixtures have a dark color, which makes reading Phenolphthalein difficult; therefore, we must ensure that the surface is clean when testing.

This test must be done in the areas accessible to break the required concrete cover, as well as facilitate the work of repair easily for the part, which was broken.

9.5.6 Chloride Test

A chloride test (AASHTO T260-84 1984; ASTM D1411-82 1982) relies on an analysis of the samples of the concrete powder to determine the quantity of chloride. This test is done by making a hole with a drilling machine to rotary dig inside the concrete and extract concrete powder out of the hole or through the broken part of the concrete.

Several separate samples must be taken at different depths. The depth of the hole will be varied to increase the accuracy and always taken from 2 to 5 mm. First 5 mm from the surface of concrete often the chlorides concentration is very high, particularly in structures exposed to seawater and chlorides.

The holes are made by special devices to collect concrete powder created as a product of the process of drilling the hole. The concrete powder, which will be taken at every drilling depth, will be added to a solution of acid, which is determined by the amount of chloride concentration. There are two principal ways of measuring in situ (Quanta Strips) and the method to determine the electron-ion and the last experience more accurate, but we need high-cost equipment.

In chloride concentration experiments, professionally trained workers are needed.

When the readings of chloride concentration at various depths have been assembled, draw the shape to chloride concentrations with depth from the surface of concrete into the concrete.

From the general figure could determine whether chlorides are inside the concrete or from the impact of air and environmental factors affecting the concrete from outside. In case of chloride attack from outside, it is found that the chloride concentration on the surface is high and then gradually decrease with the depth until reach constant value.

On the other hand, if the chloride exists inside the concrete if the concrete mix used had salt water or aggregate had high chloride, in this case near the surface the chloride concentration is at a minimum and increases with depth until reaching constant values with the depth.

The chloride test method is done through melted in acid and calibration work based on specifications (BS1881, Part 124). In addition to the method of melting in the melting acid can work in the water and follow these tests are follow ASTM D1411, 1982, and AASHTOT260, 1984.

The effect of chloride concentration on the steel reinforcement will be defined as if the chloride concentration exceeds 0.6 from hydroxyl concentration, that will cause failure in the passive layer around the steel reinforcement. This ratio is approximately 0.2% to 0.4% chlorides by weight of cement or 1 lb/yard cubic concrete or 0.5% chloride by weight of concrete.

Through the distribution of chloride concentration with depth, we can determine the degree of impact of chlorides in the steel reinforcement and the impact of chlorides; specifically, we can determine whether it is from inside the concrete or from the outside.

9.6 STRUCTURE EVALUATION TECHNIQUE

9.6.1 Case Study 1: For Structure Evaluation

As per visual inspection, a longitudinal crack is found in the pedestal concrete that carries a steel structure, as shown in Fig. 9.7.

Figure 9.7 Machine that measures the shape of concrete cover thickness.

By performing a close visual inspection, it is found that the crack in the fireproofing which have a light steel reinforcement and the cracks is not in the concrete foundation. The second point is to inspect the connection of the steel bracing, which will be critical.

9.6.2 Case Study 2: For Structural Assessment

There are cracks on the ring beam around the steel tanks. This tank is carrying oil mixed with a high percentage of water. The location of the tanks is not in a bad environmental condition. It was constructed in 1983.

When making an observation, a leak is found in the tanks which contain hydrocarbon and water that affects the reinforced concrete beam.

As shown in Fig. 9.8, the corrosion is parallel to the stirrups, as it is every 200 mm and on the corner steel bars.

The reason and the shape of the cracks reveal that there is corrosion on the steel bars. Some areas have a spalling of concrete cover, and the reduction on the steel bar cross-sectional area is more than 20% as shown in Fig. 9.9, so it needs complete repair.

Figure 9.8 Cracks in the ring beam.

Figure 9.9 Reduction in steel cross-sectional area.

From a structural point of view, the steel bars have a tension force, as the ring beam is always under tension.

9.6.3 Case Study 3: For Structure Assessment

As we mentioned, the first step is to collect the data. This site is located in the desert with no high effect on temperature. A foundation under a reciprocating gas lift compressor was installed, and after 1 year they installed another one with a similar load and design dimensions. After installing the second machine, the site engineers observed a high vibration in the administration building.

The administration building about 100 m away from the machine. The cracks were presenting between the reinforced concrete columns and the bricks; this building was constructed in 1975.

After collecting data, the reason is due to resonance and the vibration wave transfer from the machine to the building as may be a rock underneath the foundations. In any way all jump to the civil and structure solution by putting a rubber around the machine foundation or by any idea cutting the wave propagation to the building. Second step is to perform repair to these cracks.

After cooperation with the mechanical engineers, we can change the machine speed by reducing the revolutions per minute (RPM) for one of the machines to avoid the resonance; this solution is the right one as the vibration was down, and the repair for the cracks was done successfully. Therefore, it is worth to mention that, do not study the solution form the structural engineering point of view only specific in oil and gas industry as the solution may be cheap if done by another discipline.

9.6.4 Case Study 4: For Structure Assessment

As we will discuss in this chapter regarding the repair and maintenance plan for the structure, Fig. 9.10 presents a crack on the concrete

Figure 9.10 Concrete foundation cracks.

foundation at the edge in gas plant. Before jumping to a solution, the first requirement is to collect data. This plant is near the Mediterranean Sea. After a review of the soil data and based on our experience, this area has a bad soil type; the soil report mentioned clearly that the predicted settlement is about 70 mm in 20 years. Moreover, the plant is near about 100 m from the sea, which means the relative humidity is very high.

So the probability of settlement and corrosion problem is very high. The foundation shown in the photo is one of three foundations carrying a steel I-beam section to carry a manifold. The problem is a settlement in this foundation without a tie beam. The clear decision is to increase the foundation size, connecting them by a beam if the steel is in good condition; if it is not in good condition we may need to construct another foundation. Let us stop and ask if these supports are really important, as it seem they are not in the original drawings or in the as built. After performing the piping stress analysis one can find that these foundation are not required but need just a minor change in piping configuration. Again, do not run quickly toward a structure solution.

9.7 STRUCTURE ASSESSMENT

The tests that have been clarified in the preceding sections are only tools to assist in the process of evaluating the structure. The main factor in this process is the experience of the engineer who is examining the structure.

The evaluation of the building is to determine whether the structure is or is not subject to repair, or if the cost of repair is close to the cost of the demolition and reconstruction of the building.

Therefore, we note that the evaluation of the building is vital to the decision process as the decision-making will be based on an assessment of the building in terms of both safety and the repair cost.

Therefore, the repair scope of work is based on the accuracy of the test results and the previous experience of the skilled labor who conduct these tests.

When determining the method of repair and the determination of the necessary cost value of the repair process, do not forget to keep in mind the methods required to protect the structure from corrosion after repair.

9.7.1 Maintenance Plan

In any organization, the maintenance team is responsible for many buildings. The responsibility of this team is to maintain the reliability of the structures. The responsibility of this team in the ministry of transportation is to do the bridges maintenance. In the United States, for instance, the famous golden gate bridge has a dedicated team who paint the bridge nonstop.

When defining the time of performing the maintenance, we should consider the structural probability of failure. From a practical point of view it is complicated to calculate the accurate structure probability of failure as it is a research matter rather than practical issue. The calculation of the structure's probability of failure and the consequence of obtaining the structure's risk as called the qualitative risk assessment. The other popular method is the quantitative risk assessment, as it is easy and can be handled through the maintenance team without outside resources.

9.7.2 Assessment Process

The assessment of the concrete structure is depending on its structure type, location, and existing load and operation requirements. On the other hand, for studying the risk assessment and the different critical items, the economic effects must be considered. The general definition of risk is summarized in the following equation:

$$\text{Risk} = \text{probability of failure} \times \text{consequences}.$$

So to define the risk assessment of any concrete structure, you must take into consideration the factors that may affect the business economics itself. The structural risk assessment is represented by the probability of failure and the consequences of the failure happening, noting that the failure may be the whole building, part of it, or only a concrete member.

The first step to defining and calculating the quantitative risk assessment is to hold a meeting with the maintenance team, possibly with a consultant engineering company in case of huge structures.

This meeting will take several hours, and contain discussions to define the factors that affect the structure's probability of failure, such as, corrosion on the steel reinforcement bars, over load than the design, new data revealed low concrete strength than the design concrete strength in the drawings and project specifications and the redundancy of the structure it self.

The required data about the corrosion of steel bars, concrete compressive strength, and other data are discussed before, and it can be obtained by collecting the data and performing the visual inspection and performing the detailed inspection by measurement equipment such as the Ultrasonic pulse velocity technique, rebound hammer, or other available techniques. The factor that has the most influence on estimating the structure's probability of failure is the structure redundancy, which can be explained by Fig. 9.11.

If you need to design the beam and you receive the following two solutions, the first is a structure system to be fixed in the two ends and the second structure system is hinged in the two ends. Which system will you choose? Give yourself 5 min to think.

There are many factors that control your decision in selecting the suitable system. The following are the advantages and disadvantages for these two systems.

Structural system 1:

- The beam cross section will be smaller.
- The connections will be big and complicated as it has a shearing force and moment on it.
- It is reasonable from architectural point of view.
- The construction is complicated in connection.

Structural system 2:

- The beam cross section will be big.
- The connection will be small as it designs to shear force only.
- It is easy to construct it as the connection is simple.

So by discussing the above two systems, if you are afraid of construction onsite, or maybe you will carry perform the construction with inhouse engineers and workers to avoid any problems, you will go toward the simple beam option. This type of thinking happens as we always forget the maintenance point of view. Fig. 9.12 shows the steps of collapse failure.

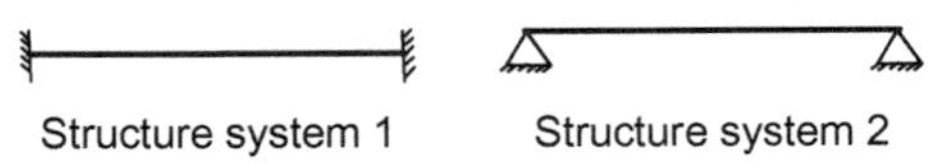

Figure 9.11 Two static structure systems.

Figure 9.12 Comparison between failure steps for the two different structure system.

The structure system 2 as in Fig. 9.11 which is the simple beam assume you increase the load gradually the beam will carry the load until form a plastic hinge in the middle of the beam at the point of maximum bending moment and then collapse will occur.

On the other side, for the structure system 1, which is fixed in the two ends, when increasing the load gradually the weak of left or right connection will fail first as shown in Fig. 9.12; the plastic hinge will form on the left connection and then after increasing the load, the other connection will form a plastic hinge, so it is now working as simple beam then by increasing the load the plastic hinge 3 will be formulated in the middle of the beam and then collapse failure will occur.

From this discussion, one can find that structural system 1 will take more time to collapse (as it fails after three stages), as opposed to structural system 2, which fails at the first stage; so structure system 1 is more redundant than system 2.

Moreover, when you make a comparison between different reinforced concrete members such as slabs, beams, columns, and cantilevers, one can obtain that some members are more critical than others. As cantilever is the most critical structure member because any weak in its capacity will lead to a high deflection and then failure so it is critical, in addition to that the column when column is fail the load distributed to other columns until failure of the whole building so the column is low redundancy and very critical as any failure on it will fail the whole building however the cantilever failure it will be a member failure only so we can go through the consequences from this approach.

In case of beam and slab, when you design a slab, you assume a simple beam and calculate the maximum moment in the middle; then you design the concrete slab by choosing the slab thickness and the steel reinforcement, and the selected steel reinforcement will be distributed along the whole span. So theoretically by increasing the load, the failure will be in one point; but actually, the surrounding area will carry part of this load, so the redundancy of the slab is very high.

Some research mentions that the reinforced concrete slab can accommodate a load twice the design load.

The pushover analysis is used to obtain the redundancy of the structure; this analysis is nonlinear analysis and exists for any structure analysis software in the market. From this analysis, one can obtain how much more load the structure can carry than the design load before failure; in addition to that, we can obtain the location that the first plastic hinge will occur and from this we can know the critical member in our structure.

Therefore, to start doing the maintenance plan for a specific concrete structure, the first step is to discuss all the structural part that need to be inspect and repair will be set in a table and for this table the team will put all the factor that effect the structure probability of failure.

Assign a value to each factor; for example, in the case of a redundancy factor, assign a value from 1 to 10, where a column will be 10 and a slab will be 2. For example the entire team must agree on these numbers based on their experience and repeat the same values or higher for other factors.

The other factors are the age, the engineering office which performed the design, the contractor who constructed it, and the code that is used in design. Some countries have used seawater in mixing water, and some codes in the past have agreed to use 6-mm-diameter steel in stirrups; so if there is corrosion, you will not see the stirrups, as they will be completely corroded.

The following Table 9.7 contain the structure's probability of failure factors, but when you use it, make it match with your structure requirement. You must know that it is not limited to only the factors in the following table.

This table is a simple example of calculating the quantity of the probability of failure, but when you hold a meeting you must put a criteria on the number. For example if the engineering office who makes the design is professional and famous in the market for not having any problems, and you have a good experience with it, so give it the lower values; but if it is not competent and you have an example of a problem in the same

Table 9.7 Structure likelihood of failure

Structure	Redundancy 1-10	Age 1-10	Designer 1-10	Contractor 1-10r	Code 1-10	Last inspection 1-10	Total score
Structure (A)	8	10	6	6	9	10	49
Structure (B)	5	2	7	3	3	10	30
Structure (C)	3	5	8	10	5	2	33
Structure (D)	7	4	2	5	5	1	24
Structure (E)	1	3	4	2	5	10	25
Structure (F)	3	1	4	5	6	7	26

type of structure you put it the higher value. This procedure will also apply to the contractor.

The values in the table depend on the experience of the team, so may be another source can cause a probability of failure is out of your experience. Therefore, in some complicated structures or huge facilities it is worth contacting a competent engineering office for a maintenance plan.

Define the record of the last inspection, in case you do not do any inspection it is mean the building is a black box for you so you give it a highest value. As the thing you know it bad is less critical than you know nothing about it.

Then it is necessary to go through the consequences by the same approach, put the economic impact. Economic impact is the answer of what will be the dangerous effect in economic point of view if the failure of the whole structure or part on it.

There are many hazard factors which will guide structure risk assessment. The first factor is the location itself, the structure expenses, and whether or not it carries process facilities for industry so it must be taken into consideration as the failure may cause hazard or stop the production as in case of foundation under machines as any failure on it will influence the performance of the machine and will be the main cause of the process plant shutdown and stop production.

The same as the probability of failure we calculate the consequences quantity by the same procedure as presented in Table 9.8.

Now, we have the quantity of the probability of failure and the consequences, so from the following Table 9.9 we calculate the quantity risk assessment.

Table 9.8 Impact due to structure failure

Structure	Impact on person (on a scale of 1–10)	Impact on cost (on a scale of 1–10)	Impact on environment (on a scale of 1–10)	Impact on repetition (on a scale of 1–10)	Total score
Structure (A)	8	10	6	6	30
Structure (B)	5	2	7	3	17
Structure (C)	3	5	8	10	26
Structure (D)	7	4	2	5	18
Structure (E)	1	3	4	2	10
Structure (F)	4	5	3	5	17

Table 9.9 Risk for structure failure

Structure	Probability of failure	Consequences	Risk
Structure (A)	49	30	1470
Structure (B)	30	17	510
Structure (C)	33	26	858
Structure (D)	24	18	432
Structure (E)	25	10	250
Structure (F)	26	17	442

Table 9.10 The structures priority maintenance plan

Structure priority	Color code	Risk values
Structure (A)	Red	1470
Structure (C)	Red	858
Structure (B)	Yellow	510
Structure (F)	Yellow	442
Structure (D)	Green	432
Structure (E)	Green	250

9.7.3 Risk-Based Inspection Maintenance Plan

After we calculate the quantity risk assessment, as shown in the previous table, we classify the maintenance plan. The top third of the risk structure will be red, the second third will be yellow, and the remaining third numbers of structures will be green; the propriety list will be as in Table 9.10.

From this table, Structures (A) and (C) are considered the critical structures which you will inspect first. Structures (B) and (F) will be secondary priority, so they need to be inspected after the first two structures. The last structures—Structures (D) and (F)—will have less risk, so they do not require an inspection in this time.

As shown by this table, this is a simplified method that can be used to plan your maintenance for inspection and repair. Take into account that the budget is an important factor in your maintenance plan; in the previous example, the budget may be enough this year to inspect Structure (A) only, so you must plan for subsequent years (up to five).

Nowadays, most international oil and gas companies go toward applying the integrity management system from around 1990; they found that applying this system is very important to maintain the structure along its lifetime, taking into consideration all the historical data.

The closed loop between structure risk assessment, an inspection program, a repair program, and reassessment is very important in order to maintain the structure in good condition and reliably meet the operations requirements.

The total quality control management system is very important to match with the operation requirements when reviewing design, construction, and maintenance. When proposing development, making an extension to the building, or increasing the load, considerations must be made by a system of management of change procedure in order to upgrade the previous risk-ranking table.

REFERENCES

AASHTO T260-84, 1984. Standard Method of Sampling and Testing for Total Chloride Ion in Concrete Ratio Materials. American Association of State Highway Transportation Officers, Washington, DC.

ACI 318-97, Standard Building Code Requirements for Reinforced Concrete, Detroit, 1997.

ACI 228-89-R1 In-place methods for determination of strength of concrete.

Alldred, J.C., 1993. Quantifying The Losses In Cover-Meter Accuracy Due To Congestion of Reinforcement, Proceedings of the Fifth International Conference on structural Faults and Repair, Vol. 2. Engineering Techniques Press, Edinburgh, pp. 125–130.

American Concrete Institute, 1990. Corrosion of Metals in Concrete, Report by ACI Committee 222. ACI222R-89. American Concrete Institute, Detroit, MI.

ASTM C876, 1991. Standard Test Method for Half-Cell Potentials of Uncoated Reinforcing Steel in Concrete. American Society for Testing and Materials, Philadelphia, PA.

ASTM D1411-82, 1411. Standard Test Methods for Water Soluble Chlorides Present as Admixes in Graded Aggregate Road Mixes. American Society for Testing and Materials, Philadelphia, PA.

ASTM C 42-90m Standard test method for obtaining strength and testing drilled cores and sawed beams of concrete.

Broomfield, J.P., Langford, P.E. and McAnoy, R., Cathodic protection for reinforced concrete: its application to buildings and marine structures, in Corrosion of Metals in Concrete, Proceedings of Corrosion/87 Symposium, Paper 142, NACE, Houston, TX, pp. 222–325, 1987.

Broomfield, J.P., Rodriguez, J., Ortega, L.M., Garcia, A.M., 1993. Corrosion rate measurement and life prediction for reinforced concrete structures, Proceeding of Structural Faults and Repair – 93, Vol. 2. Engineering Technical Press, University of Edinburgh, pp. 155–164.

British Standard Institution-BS 1881-1983-Part 120. Method for Determination of the compressive strength of concrete cores.

BS:1881:Part6-1971, Methods of testing Concrete, Analysis of Hardened Concrete.

BS8110 Structural user of Concrete-Part 1:1985 Code of Practice for Design and Construction, London.

BS 1881 Testing Concrete: Part 5:Methods for testing hardened concrete for other than strength.

Bungey, J.H. ed., Non-Destructive Testing in Civil Engineering, International Conference by The British Institute of Non-Destructive Testing, Liverpool University, 1993.

Cady, P.D., Gannon, E.J., 1992. Condition evaluation of concrete bridges relative to reinforcement corrosion. Vol. 1: State of the art of existing methods. National Research Council, Washington, D.C, SHRP-S-330.

Concrete Society, Repair of concrete Damaged by Reinforcement Corrosion, Technical Report No. 26., 1984.

Concrete society Technical report No.1 including addendum (1987). Concrete core testing for strength cement and concrete association, Wexham spring, sloogh SL3 6PL.

El-Reedy, M.A., 2009. Advanced materials and techniques for reinforced concrete structure. CRC press.

Parrott, L.J., 1987. A Review of Carbonation in Reinforced Concrete, a review carried out by C&CA under a BRE contract. British Cement Association, Slough, UK.

The Concrete Society, 2000. Diagnosis of deterioration in concrete structures. Technical Report 54. The Concrete Society, Crowthorne.

INDEX

Note: Page numbers followed by "*f*" and "*t*" refer to figures and tables, respectively.

C

D

FSC
www.fsc.org
MIX
Papier | Fördert gute Waldnutzung
FSC® C083411

Kolibri 360 GmbH
Ferdinand-Jühlke-Straße 7
99095 Erfurt, Deutschland
produktsicherheit@kolibri360.de